Wireless Data Services

Technologies, Business Models and Global Markets

By the time this book is published, there are likely to be over 1.3 billion mobile subscribers around the world. Despite this phenomenal global growth, wireless technologies have progressed in very different ways in the key territories of Asia, Europe, and North America. Technologies such as i-mode in Japan, SMS in Europe, PDAs and BlackBerry in North America point us to the fact that wireless applications and services are often unique to the culture and business models of a region. This book takes a deeper look into why certain technologies, business models, and adoption strategies succeed while others fail, and how all these elements will impact on the future of wireless communications. With the help of examples, case studies and interviews with industry luminaries, the authors identify the key factors behind the success or failure of different strategies and provide insights into how to match wireless technology and services to global markets.

Chetan Sharma is a recognized industry expert in the strategy and implementation of wireless data and pervasive computing ideas and solutions. Mr. Sharma was formerly Leader of the Emerging Solutions practice at Luminant Worldwide. Prior to this, he served as founder and principal of Luminant's wireless practice. In this role, he oversaw global client engagements, provided vision and strategic direction, conducted wireless research and development, and established partnerships with leading industry players.

Yasuhisa Nakamura is Senior Vice President at NTT DoCoMo USA. He is a widely recognized specialist of wireless telecommunications industry with over 20 years of experience. Dr. Nakamura holds more than 50 patents (awarded and pending) and is a frequent speaker at international conferences.

Wireless Data Services

Technologies, Business Models and Global Markets

Chetan Sharma

Yasuhisa Nakamura

CAMBRIDGE
UNIVERSITY PRESS

PUBLISHED BY THE PRESS SYNDICATE OF THE UNIVERSITY OF CAMBRIDGE
The Pitt Building, Trumpington Street, Cambridge, United Kingdom

CAMBRIDGE UNIVERSITY PRESS
The Edinburgh Building, Cambridge CB2 2RU, UK
40 West 20th Street, New York, NY 10011–4211, USA
477 Williamstown Road, Port Melbourne, VIC 3207, Australia
Ruiz de Alarcón 13, 28014 Madrid, Spain
Dock House, The Waterfront, Cape Town 8001, South Africa

http://www.cambridge.org

First published 2003

Printed in the United Kingdom at the University Press, Cambridge

Typeface Minion 10.5/13.5 pt.   *System* LATEX 2_ε [TB]

A catalogue record for this book is available from the British Library

ISBN 0 521 82843 0 hardback

To my lifelong sweetheart, my staunch supporter, my angel, my Sarla
Chetan

To my parents (Toshikazu, Mariko) family (Izumi, Asami, Ryo, Yu) and
all who have enlightened my life
Yasu

Contents

Figures

About the authors

Chetan Sharma is a recognized industry expert in the strategy and implementation of wireless Internet and pervasive computing ideas and solutions. He has a strong background in developing distributed network-based technologies for the wireless industry, including extensive experience in managing and delivering all phases of the system development cycle. His particular fields of interest are in wireless Internet applications, multi-modal access, pervasive computing, and Internet security. Mr. Sharma has assisted some of the most prominent telecom operators in North America, Europe, and Asia.

Most recently, he was Leader of Emerging Solutions practice at Luminant Worldwide. Prior to his leadership of the R&D group, he served as founder and principal of Luminant's wireless practice. In this role, he oversaw global client engagements, provided vision and strategic direction, conducted wireless research and development, and established partnerships with leading industry players. From 1994 to 1999, Mr. Sharma was a systems engineer and product manager, focusing on mobile communications, at Cellular Technical Services, a Seattle firm, then at Philips Consumer Communications.

Mr. Sharma is the author of two other books on wireless and voice communications. He is author of the best-seller *Wireless Internet Enterprise Applications* (John Wiley & Sons, 2000), and co-author of *VoiceXML: Strategies and Techniques for Effective Voice Application Development* (John Wiley & Sons, 2002), Mr. Sharma has patents in wireless communications, is regularly invited to speak at conferences worldwide, has appeared as a technology expert witness in intellectual property litigation cases, and is an active member in industry bodies and committees. Mr. Sharma has an MSEE from Kansas State University and a BE from University of Roorkee.

Yasuhisa Nakamura is a widely recognized specialist of the wireless telecommunications industry with over 20 years of experience. He graduated from University of Tokyo in 1980 and joined NTT Electrical Communication labs doing research in digital radio transmission systems. In 1986 and 1987, he was invited to join France Telecom tabs (CNET) as a researcher. He obtained his Ph.D. from the University of

Tokyo in 1991. From 1995 to 1998, he served NTT Central Personal Communications, Inc., as a senior manager and was engaged in the development and overseas deployment of the 1.9 GHz micro-cellular system – PHS. During this period, he was invited as a lecturer to the Malaysian Institute of Technology, Malaysia.

Dr. Nakamura joined NTT DoCoMo in December 1998 and served as executive senior manager of the PHS business planning division. In March 1999, he went to Rio de Janeiro, Brazil, as a technical director of NTT DoCoMo Telecomunicacoes do Brasil, 100% subsidiary of NTT DoCoMo. His assignment was to work with local authorities and companies to facilitate the deployment of DCM technology into Latin America. He and his staff developed and launched a mobile portal service called WMAP for the Brazilian market. He was a lead negotiator with the Brazilian Government in introducing the GSM system into the Latin America region.

In June 2001, he was nominated as a senior vice president of NTT DoCoMo USA. Dr. Nakamura has more than 50 patents (awarded and pending) and papers on wireless technologies and is a frequent speaker at international conferences. He was a member of TG8/1, ITU-R on FPLMTS (Future Public Land Mobile Telecommunication Systems; former name of IMT-2000) and Asia Pacific delegate for the radio committee of the IEEE. He is a member of the IEEE and IETE, Japan. He is considered an expert in dealing with cross-cultural business and technology challenges in the wireless industry as he has the unique experience of working on four different continents.

Foreword

Rapid and widespread growth of wireless technology in the 1990s shaped one of the largest technology markets after the PC revolution in the 1980s. Untethered connectivity, any time anywhere, fueled a major market and technology disruption, which permeated almost every consumer market worldwide. The domino effect of the success of wireless technology resulted in a unique opportunity for innovation and creativity in technology, marketing, and business strategy.

Unceasing innovation in technologies ranging from the semiconductor industry to network design set the stage for the immense success of wireless technology such that, within a few years of its inception, the wireless phone transcended from a luxury gadget or business item to a necessary tool in everyday life. The personalized aspects of mobile phones, along with ease of use in voice-centric applications, helped to make the mobile phone an indispensable part of our life beyond age, gender, or even social class.

The next commonsense step was perceived to be further wireless services in addition to voice. Yet wireless, like any other major technology and market breakthrough, is no exception to the cyclic nature of the high tech economy. Rapid market growth driven by innovation and competition diminished the profit margins of the terminal market and network access rate to a level that compelled the technologists and marketers to search for new applications and markets. Sluggish, or in some cases negative, growth of ARPU for voice services and attractive promises of 3G systems motivated many wireless operators to explore new opportunities of the wireless data market and embrace the tremendous spectrum and infrastructure costs of next-generation networks.

Offering services beyond voice-centric applications (such as wireless data, wireless Internet, and video streaming) has been the main focus of new systems such as 3G and wireless local area networks. However, data services require a new paradigm shift on both the technology and business fronts. Technical challenges of data services such as the bursty nature of data traffic, high data rate, variable quality of service measures, complex application software requirements, and network design/optimization demand new standards and creative solutions. Equally important business issues such as the complex wireless value chain for data services, pricing policies, and the critical role of content and applications have sent business strategists back to the drawing boards.

I marvel at the breathtaking rate of growth in wireless technology from the bulky brick-sized phone with basic network features of a few years ago to the pocket-sized advanced communicator and complex networks of today. Yet I believe wireless technology will bear many future wondrous achievements beyond our simplistic predictions. Ubiquitous wireless connectivity will go beyond today's applications and devices to the extent that upcoming breakthroughs will dwarf the achievements we have seen so far, thanks to the multiple effects that I briefly outline below.

- Astonishing advances in the semiconductor industry prompted the diminishing cost of silicon real-estate in complex systems on chip solutions without compromising power consumption. Therefore, wireless connectivity will not constitute a prohibitive cost factor of future devices, beyond the wireless phone. Also, complex terminals including multimode and adaptable devices capable of supporting multiple applications and networks are more practical.

- A major upward shift of complexity to application processors from modem processors is eminent. Software complexity continues to tilt the complexity balance. Emergence of advanced real-time operating systems with the goal of joint optimization of hardware and software will facilitate support of many new application, some of which we may not have envisaged yet, in future wireless devices.

- Heterogonous networks offer an adaptive, scalable, and cost-effective backbone for new and otherwise overwhelming wireless services. Seamless support of a variety of services with different quality of service requirements and multiple air interfaces demands an intelligent network. Advances in IP backbone can facilitate the fastest/shortest access link with a combination of different wireless and wired links to provide the most cost-effective and reliable link for respective services. Integration of cellular, wireless LAN and other *ad hoc* networks can expedite the transition. Also, security, as one of the key concerns of pervasive wireless network, can more efficiently be addressed in an all-IP network.

- The wireless value chain has been evolving from a basic voice-centric model to a more comprehensive structure to support new services and technologies. Content providers, virtual service providers, and service aggregators are among the new elements of the new pyramid, requiring a different business model and interaction.

A global perspective of the wireless industry gives us a more comprehensive view of the direction that the overall industry is taking. ITU efforts during the past few years have proved that the roadmap toward a single worldwide solution is more challenging than many anticipated as cultural differences, intellectual properties, economical conditions, and many other factors can impact on the direction of this market.

In this book, the authors have with great care analyzed and articulated the business and technology aspects of wireless data. The unique global perspective of wireless market and technology presented in this work offers a comprehensive view of the challenges and future promises of one the most dynamic fields of modern technology.

Dr. A. Bahai

Acknowledgements

My life's journey has been blessed with the touch of so many individuals that it is hard to capture them all or do justice to their influence on my thinking and my work. I am forever indebted to a number of important contributors listed below.

This project was inspired by the vision of Dr. Nakamura. Working with him on this book was truly a rewarding experience. My thanks for the sojourn and for our precious friendship.

Many thanks to Eric Willner and his team at Cambridge University Press for their continuous enthusiasm, support, and feedback that made this project a reality. I would also like to express our gratitude to Beverley Lawrence for her painstaking copyediting work and attention to detail.

I thank Ed Reilly, Michel Gaultier, Neeraj Chawla, Steve Elston, Miten Mehta, Connie Wong, Scott Weller, Dave Keller, Charlie Matheson, Naveen Jain, Latif Nathani and Sunil Jain for their assistance with the project. My thanks also to my friends who have supported me all along – Mani, Sumeet, John, Verma, Anil, Rajeev, Bill, Rajesh, Bruce, AJ, Ramani, Ashu, Nitish, Rana, Amit, Sunny, Tom, Nitin, Sudhir, Jayant, Prashant, and countless others.

I also thank our interviewees (and their organizations) – Steve Wood (Wireless Services Corporation), Mark Tapling (Everypath), Mark Anderson (Strategic News Service), Jon Prial (IBM), Frank Yester (Motorola), Scott McBurney (BusinessLink), and Satoshi Nakajima (UI Evolution) – for kindly taking some time from their busy schedules to share their vision and perspectives.

I thank my mentor Dmitry Kaplan and good friend Joe Herzog who helped on this project as if it were their own. Their thorough review, insights, comments, corrections, and suggestions throughout the course of the project were invaluable. Their help is very much appreciated.

I am also thankful for the blessings of my parents Dr. C.L. Sharma and Prem Lata Sharma. I wouldn't be here without them. I would also like to thank Rahul, my brother, and Dropadi, Deepti, Aditya, and Rakesh Sharma for their support and encouragement.

There is one person in my life who urged me on by way of her untiring support and seemingly unlimited belief in me; compared with that person, all else pales. This project wouldn't have even begun if it weren't for the sacrifice and patience of my wife Sarla, who is my biggest cheerleader. I owe her more than I can ever put in words. My thanks to the Almighty for our union and my deepest gratitude to Sarla for her presence, support, and everlasting smile.

Chetan Sharma

This book project is a result of joint collaboration with my co-author Mr. Chetan Sharma, who I met by chance after we moved from "Always Sunny" Rio de Janeiro to "Always Rainy" Seattle in late 2001. I had been meaning to work on a book project for several years. The gloomy weather provided the perfect diversion and opportunity to work on this project.

First, I thank Dr. Ahmad Bahai, CTO National Semiconductor and Professor at Stanford, for kindly taking the time to pen the foreword for this book.

I thank all of my ex-bosses and mentors, especially Dr. Kenji Kohiyama, Dr. Takeshi Hattori, Dr. Takehiro Murase, Dr. Syuzo Komaki, Dr. Osamu Kurita, and Mr. Tadao Kobayashi who have taught me all about work ethics.

I have a deep respect for the NTT DoCoMo management, especially, Dr. Keiji Tachikawa, Mr. Shiro Tuda, Dr. Kota Kinoshita, Mr. Nobuharu Ono, Mr. Keichi Enoki, and Mr. Kiyoyuki Tsujimura, who are leading a great global company. It is more than an honor to work as a wireless engineer for such an awesome company.

I am also thankful to my parents Toshikazu and Mariko Nakamura and to my sister Kazuko. I hope for their eternal health.

Lastly, I am eternally grateful to my wife and my family, Izumi, Asami, Ryo and Yu for their patience to help maintain our cross-continental lifestyles. Thank you for all your support and encouragement. I do believe your experience in different cultures will be very useful in the future.

Dr. Yasuhisa Nakamura

Abbreviations

2G	Second-Generation Mobile Network
2G+ or 2.5G	Second-Generation Enhanced
3GPP	Third-Generation Partnership Project
3WC	Three Way calling
4GL	Fourth-Generation Language
802.x	Series of LAN standards developed by IEEE
911	Emergency telephone number in North America
AAA	Authentication, Authorization and Accounting (as specified by the IETF)
ABR	Available Bit Rate
AC	Access Control, or Alternating Current, or Authentication Center (also AUC)
ACD	Automatic Call Distributor
ACE	Authentication Encryption
ACEK	Authentication Encryption Key
ACK	Acknowledgement
ACP	Access Control Point
ACRE	Authorization and Call Routing Equipment
ADPCM	Adaptive Differential Pulse Code Modulation
ADSL	Asymmetric Digital Subscriber Line
A-GPS	Assisted GPS
AGRAS	Air–Ground Radiotelephone Automated Service
AIDC	Automatic Identification and Data Capture
AIN	Advanced Intelligent Network
A-Key	Authentication Key
ALI	Automatic Location Information
AM	Amplitude Modulation
AMPS	Advanced Mobile Phone System
ANI	Automatic Number Identification
ANSI	American National Standards Institute

ANSI-41	ANSI standard for mobile management (ANSI/TIA/EIA-41)
AOA	Angle of Arrival
API	Application Programming Interface
ARDIS	Advanced Radio Data Information Service
ARIB	Association of Radio Industries and Businesses (Japan)
ARP	Address Resolution Protocol
ARPA	Advanced Research Projects Agency
ARPANET	ARPA Network
ARPU	Average Revenue Per User
ARQ	Automatic Repeat Request
ARS	Automatic Route Selection
ASCII	American Standard Code for Information Interchange
ASIC	Application-Specific Integrated Circuit
ASN.1	Abstract Syntax Notation 1 (ISO)
ASP	Application Service Provider
ASR	Automatic Send Receive, Automatic Speech Recognition
ATIS	Alliance for Telecommunications Industry Solutions (formerly ECSA)
ATM	Asynchronous Transfer Mode, Automatic Teller Machine
ATVEF	Advanced Television Enhancement Format
AuC	Authetication Center
B2B	Business-to-Business
B2C	Business-to-Consumer
BER	Bit Error Rate
BIOS	Basic Input–Output System
B-ISDN	Broadband ISDN
BOC	Bell Operating Company (USA)
BONDING	Bandwidth On Demand Interoperability Group
bps	bits per second
BRI	Basic Rate Interface (ISDN)
BS	Base Station
BSC	Base Station Controller
BSS	Basic Service Set
BSS	Business Support Systems
BTS	Base Transceiver Station
CAD	Computer-Aided Design
CAM	Computer-Aided Manufacturing
CAMEL	Customized Applications for Mobile-network Enhanced Logic (GSM/ETSI)
CAN	Campus Area Network
CAVE	Cellular Authentication and Voice Encryption

CBR	Constant Bit Rate
CCC	Clear Channel Capability
CCCH	Common Control Channel
CCD	Charge Coupled Device
CDMA	Code Division Multiple Access
CDPD	Cellular Data Packet Data
CDR	Call Detail Record
c-HTML	compact-HTML
CK	Ciphering Key
CLEC	Competitive Local Exchange Carrier
CLI	Calling Line Identification
CMIP	Common Management Information Protocol (ISO)
CNET	Centre National d'Études en Télécommunications (France Telecom)
CNI	Calling Number Identification
CO	Central Office
CODEC	Compression/Decompression
CORBA	Common Object Request Broker Architecture (OMG)
CPIM	Common Presence and Instant Messaging
CPP	Calling Party Pay
CPS	Cycles per second (hertz)
CPU	Central Processing Unit
CSR	Customer Service Representative
CSS	Cascading Style Language
CSU	Channel Service Unit
CT	Cordless Telecommunications
CTI	Computer Telephony Integration
CTIA	Cellular Telecommunications Industry Association
CTIA	Cellular Telecommunications Industry Association
DAB	Digital Audio Broadcasting
D-AMPS	Digital Advanced Mobile Phone Service
DARPA	Defense Advanced Research Projects Agency
DAT	Digital Audio Tape
dB	deciBel
DBMS	Database Management System
DBS	Direct Broadcast Satellite
DCE	Data Communications Equipment, or Distributed Computing Environment
DDD	Direct Distance Dialing
DDS	Dataphone Digital Service
DECT	Digital Enhanced (formerly, European) Cordless Telecommunication

DES	Data Encryption Standard
DHCP	Dynamic Host Configuration Protocol
DLCS	Digital Loop Carrier System
DLL	Dynamic Link Library
DNA	Digital Network Architecture
DNS	Domain Name Server
DPCM	Differential Pulse Code Modulator
DRM	Digital Rights Management
DS	Digital Service, Digital Signal
DS-0	Digital Signal, Level 0 (64 kbps)
DS-1	Digital Signal, Level 1 (1.544 Mbps)
DS-1C	Digital Signal, Level 1C (3.152 Mbps)
DS-2	Digital Signal, Level 2 (6.312 Mbps)
DS-3	Digital Signal, Level 3 (44.736 Mbps)
DS-4	Digital Signal, Level 4 (274.176 Mbps)
DSI	Digital Speech Interpolation
DSL	Digital Subscriber Line
DSP	Digital Signal Processor
DSS	Decision Support System
DSU	Digital Service Unit
DTAM	Document Transfer and Manipulation (ISO)
DTE	Data Terminal Equipment
DTMF	Dual-Tone MultiFrequency
DTU	Data Transfer Unit
DTV	Digital Television
DV	Digital Video
DVB	Digital Video Broadcasting
DWDM	Dense Wavelength Division Multiplexing
E2E	End-to-End
E911	Enhanced 911
EAI	Enterprise Application Integration
EBCDIC	Extended Binary Coded Decimal Interchange Code
EBPP	Electronic Bill Presentation and Payment
EBU	European Broadcasting Union
EbXML	e-business XML
EC	European Community
ECMA	European Community Manufacturers Association
EDACS	Enhanced Digital Access Communication Systems
EDGE	Enhanced Data Rates for Global Evolution
EDI	Electronic Data Interchange
EFT	Electronic Funds Transfer

EHF	Extremely High Frequency (more than 30 GHz)
EIA	Electronic Industries Association
EIR	Equipment Identity Register
EISA	Extended Industry Standard Architecture
EJB	Enterprise Java Beans
EMS	Enhanced Messaging Services
ENS	Enhanced Network Services
EOT	End of Transmission
E-OTD	Enhanced Observed Time Difference
EPI	External Provisioning Interface
ERP	Enterprise Resource Management
ESMR	Enhanced Specialized Mobile Radio
E-TDMA	Expanded Time Division Multiple Access
ETSI	European Telecommunications Standard Institute
EU	European Union
FCC	Federal Communications Commission (USA)
FDD	Frequency Division Duplex
FDDI	Fiber Distributed Data Interface
FDMMA	Frequency Division Multiplexing Multiple Access
FHSS	Frequency Hopping Spread Spectrum
FIFO	First-In First-Out
FM	Frequency Modulation
FMS	Fraud Management System
FoIP	Fax over IP
FPLMTS	Future Public Land Mobile Telecommunications System
FR	Frame Relay
FRS	Family Radio Service
FSK	Frequency-Shift Keying
FSS	Fixed Satellite System
FTP	File Transfer Protocol (TCP/IP)
FWA	Fixed Wireless Access
GAN	Global Area Network
Gb	gigabit
GB	gigabyte
GEO	Geostationary-Earth-Orbit
GGSN	Gateway GPRS Support Node
GHz	gigahertz (billions of cycles per second)
GIF	Graphics Interchange Format
GIS	Geographic Information Systems
GMLC	Gateway Mobile Location Centre
GMRS	General Mobile Radio Service

GMSC	Gateway Mobile Switching Centre
GOSIP	Government Open Systems Interconnection Profile (USA)
GPRS	General Packet Radio Service
GPS	Global Positioning System
GRX	GPRS Roaming Exchange
GSA	Global mobile Suppliers Association
GSM	Global System for Mobile telecommunications, or Groupe Spécial Mobile
GUI	Graphical User Interface
GVPN	Global Virtual Private Network
H323	ANSI standard for format over IP
HDLC	High-level Data Link Control (ISO)
HDML	Handheld Device Markup Language
HDSL	High-bit-rate Digital Subscriber Line
HDTV	High-Definition Television
HF	High Frequency (3 MHz to 30 MHz)
HFC	Hybrid Fiber/Coax
HiFi	High Fidelity
HIPERLAN	High-Performance Radio Local Area Network
HIPERLAN/2	High-Performance Radio LAN Type 2
HLR	Home Location Register
HMI	Human–Machine Interface
HomePNA	Home Phoneline Networking Alliance
HPPI	High-Performance Parallel Interface
HSCSD	High Speed Circuit-switched Data
HSDPA	High Speed Download Packet Access
HTML	Hyper Text Markup Language
HTTP	Hyper Text Transfer Protocol
HTTP-NG	HTTP Next Generation
HTTPS	HTTP Secure
HVAC	Heating, Ventilation, and Air Conditioning
Hz	hertz (cycles per second)
IAB	Internet Architecture Board
IANA	Internet Assigned Numbers Authority
ICANN	Internet Corporation for Assigned Names and Numbers
ICI	Interexchange Carrier Interface
ICMP	Internet Control Message Protocol
ICR	Intelligent Call Routing
ICS	Intelligent Calling System ID Identification
IDDD	International Direct Dialing Designator
iDEN	integrated Dispatch Enhanced Network

IDPR	Interdomain Policy Routing
IECTC-100	Multimedia Systems and Equipments Standardization Committee
IEC	International Electrotechnical Commission, or InterExchange Carrier
IEEE	Institute of Electrical and Electronics Engineers
IESG	Internet Engineering Steering Group
IETF	Internet Engineering Task Force
IIN	Issuer Identification Number
ILEC	Incumbent Local Exchange Carrier
IM	Instant Messaging
IMAP	Internet Mail Access Protocol (IETF)
IMEI	International Mobile Equipment Identity
IMN	Intelligent Mobile Network
IMPS	Instant Messaging and Presence Services
IMSI	International Mobile Subscriber Identity
IMT	International Mobile Telecommunications
IMT-2000	International Mobile Telecommunications 2000 (ITU)
IN	Intelligent Network
IN7	Compaq SS7 stack (previously DECss7)
INAP	Intelligent Network Application Part (IN)
INMS	Integrated Network Management System
INTELSAT	International Telecommunications Satellite Organization
IOC	InterOffice Channel
IOPS	Internet Operators' Providers Services
IOTP	Internet Open Trading Protocol
IP	Internet Protocol (IETF), or Intelligent Peripheral (IN)
IPDR	Internet Protocol Detail Record Organization
IPN	Intelligent Peripheral Node
IPR	Intellectual Property Rights
IPSec	IP Security
IPv4	Internet Protocol version 4
IPv6	Internet Protocol version 6
IPX	Internetwork Packet Exchange
IRC	International Record Carrier
IrDA	Infrared Data Association
IrLAN	Infrared LAN
IrLAP	Infrared Link Access Protocol
IRQ	Interrupt Request
IRTF	Internet Research Task Force
IS	Information System, or Interim Standard (TIA/EIA)
IS-41	TIA/EIA Interim Standard for Mobile management

ISA	Industry Standard Architecture
ISC	International Switching Center
ISDN	Integrated Services Digital Network
ISM	Industrial, Scientific, and Medical (frequency bands)
ISO	International Standards Organization
ISOC	Internet Society
ISP	Internet Service Provider
ISSS	Information Society Standardization System
ISUP	ISDN Service User Part
IT	Information Technology
ITR	Intelligent Text Retrieval
ITU	International Telecommunications Union
IVR	Interactive Voice Response
IXC	IntereXchange Carrier (also IEC)
J2ME	Java 2 Micro Edition
JDC	Japan Digital Cellular
JIT	Just in Time
JPEG	Joint Photographic Experts Group
kHz	kilohertz (thousands of cycles per second)
LAN	Local Area Network
LAP	Link Access Protocol (or Procedure)
LAT	Local Area Transport (Digital Equipment Corporation)
LATA	Local Access and Transport Area (USA)
LBA	Location Based Advertising
LBS	Location Based Services
LCD	Liquid Crystal Display
LCN	Local Channel Number
LCR	Least Cost Routing
LCS	LoCation-based Services
LD	Laser Diode
LDC	Long Distance Carrier
LEC	Local Exchange Carrier
LED	Light-Emitting Diode
LEO	Low-Earth-Orbit
LF	Low Frequency (30 kHz to 300 kHz)
LIR	Local Internet Registry
LLC	Logical Link Control
LMDS	Local Multipoint Distribution Services
LMS	Location and Monitoring Service
LMU	Location Measurement Unit
LRC	Longitudinal Redundancy Check

LSI	Large-Scale Integration
LU	Logical Unit (IBM)
MAC	Media Access Control
MAGIC	Mobile Multimedia; Anytime, Anywhere, Anyone; Global Mobility Support; Integrated Wireless Solution; and Customized Personal Service
MAN	Metropolitan Area Network
MAP	Manufacturers Application Protocol (ISO, General Motors), or Mobile Applications Port
MAPI	Messaging Applications Programming Interface (Microsoft)
MAPS	Maps Application Provisioning System
MAU	Multistation Access Unit
Mb	Mega bits
MB	Mega bytes
MBS	Mobile Broadband Systems
MC	Management Center or Message Center
MCA	Micro Channel Architecture (IBM)
MCC	Mobile Country Code
MCU	Multipoint Control Unit
MD	Mediation Device
MDF	Main Distribution Frame
MDN	Mobile Directory Number
MEO	Middle-Earth-Orbit
MES	Master Earth Station
MF	Mediation Function, or Medium Frequency (300 kHz to 3 MHz)
MHS	Message Handling Source MHz megahertz (millions of cycles per second)
MIB	Management Information Base
MIC	Management Integration Consortium
MIME	Multipurpose Internet Mail Extensions
MIN	Mobile Identity Number
MIPS	Millions of Instructions per second
MIS	Management Information Services, or Marketing Intelligence System
MJPEG	Moving JPEG
MLS	Mobile Location Services
MMAC	Multimedia Mobile Access Communication systems
MMDS	Multichannel, Multipoint Distribution Service
MMF	Mobile Manufacturers Forum
MMI	Man–Machine Interface
MMS	Multimedia Message Service

MNC	Mobile Network Code
MO	Magneto-Optical Modem Modulator/demodulator
MP3	Music Player
MPEG	Motion Picture Experts Group
MPLS	Multi Protocol Label Switching (IETF)
MPPP	Multilink Point-to-Point Protocol
MRI	Magnetic Resonance Imaging
ms	millisecond (thousandth of a second)
MS	Mobile Station
MSC	Mobile Switching Center
MSIN	Mobile Station Identification Number
MSISDN	Mobile Subscriber Integrated Services Digital Network
MSN	Microsoft Network
MSRN	Mobile Station Roaming Number
MSS	Mobile Satellite Service
MTBF	Mean Time Between Failure
MTS	Message Telecommunications Service
MTSO	Mobile Telephone Switching Office
MTU	Maximum Transmission Unit
MVNO	Mobile Virtual Network Operator
MVPRP	Multivendor Problem Resolution Process
MWIF	Mobile Wireless Internet Forum
NAI	Network Access Identifier
N-AMPS	Narrowband Advanced Mobile Phone Service
NAP	Network Access Point
NAS	Network Attached Storage
NATA	North American Telecommunications Allocation
NBS	National Bureau of Standards
NCC	Network Control Center
NCP	Network Control Program (IBM SNA), or Network Control Point
NE	Network Element
NEBS	New Equipment Building Specifications
NECA	National Exchange Carrier Association
NEF	Network Element Function
NetBIOS	Network Basic Input–Output System (Microsoft)
NFS	Network File System (or Server)
NIC	Network Interface Card, or Network Information Center (Internet Registry)
NiCad	Nickel Cadmium
NiMH	Nickel-Metal Hydride
NIST	National Institute of Standards and Technology

NIU	Network Interface Unit
NLM	NetWare Loadable Module (Novell)
NLP	Natural Language Processing
nm	nanometer (10^{-9} meter)
NM	Network Manager
NMC	Network Management Center
NML/NMS	Network Management Layer/Network Management Service
NMS	NetWare Management System (Novell)
NMSI	National Mobile Station Identifier
NMT	Nordic Mobile Telephone
NO	Network Operator
NOC	Network Operations Center
NOS	Network Operating System
NRZ	Non-Return to Zero
NTE	Network Terminal Equipment
NTSA	Networking Technical Support Alliance
NTSC	National Television Standards Committee
NTU	Network Termination Unit
OA&M	Operations, Administration and Maintenance
OAM&P	Operations, Administration, Maintenance and Provisioning
OC	Optical Carrier
OC-1	Optical Carrier Signal, Level 1 (51.84 Mbps)
OC-3	Optical Carrier Signal, Level 3 (155.52 Mbps)
OC-9	Optical Carrier Signal, Level 9 (466.56 Mbps)
OC-12	Optical Carrier Signal, Level 12 (622.08 Mbps)
OC-18	Optical Carrier Signal, Level 18 (933.12 Mbps)
OC-24	Optical Carrier Signal, Level 24 (1.244 Gbps)
OC-36	Optical Carrier Signal, Level 36 (1.866 Gbps)
OC-48	Optical Carrier Signal, Level 48 (2.488 Gbps)
OC-96	Optical Carrier Signal, Level 96 (4.976 Gbps)
OC-192	Optical Carrier Signal, Level 192 (9.952 Gbps)
OC-256	Optical Carrier Signal, Level 256 (13.271 Gbps)
OCR	Optical Character Recognition
ODBC	Open Database Connectivity (Microsoft)
ODS	Operational Data Store
OEM	Original Equipment Manufacturer
OLAP	Online Analytical Processing
OLE	Object Linking and Embedding (Microsoft)
OMA	Object Management Architecture
OMAP	Open Multimedia Applications Platform
OMC	Operations and Maintenance Center

OMF	Object Management Framework
OMG	Object Management Group
OOP	Object-Oriented Programming
ORB	Object Request Broker
OS	Operating System
OS/2	Operating System/2 (IBM)
OSA	Open Services Architecture
OSF	Open Software Foundation
OSI	Open Systems Interconnection (ISO), or Objective Systems Integrator
OSI	Open System Interconnection
OSP	Online Service Provider
OSP	OutSide Plant
OSPF	Open Shortest Path First
OSS	Operation Support System
OTA	Over The Air Activation
OTDOA	Observed Time Difference of Arrival
P2P	Person-to-Person
PABX	Private Automatic Branch eXchange
PACS	Personal Access Communications System
PAD	Packet Assembler–Disassemble
PAL	Phase Alternating by Line
PAM	Pulse Amplitude Modulation
PAMR	Public Access Mobile Radio
PAN	Personal Area Network
PAP	Password Authentication Protocol
PBX	Private Branch eXchange
PC	Personal Computer
PCB	Printed Circuit Board
PCI	Peripheral Component Interconnect
PCIA	Personal Communications Industry Association
PCM	Pulse Code Modulation
PCMCIA	Personal Computer Memory Card Industry Association
PCN	Personal Communications Networks
PCS	Personal Communications Services
PDA	Personal Digital Assistant
PDC	Personal Digital Cellular (Japan)
PDE	Positioning Determination Entity
PDN	Packet Data Network or Public Data Network
PDU	Protocol Data Unit
PGP	Pretty Good Privacy

PHY	Physical Layer
PIM	Personal Information Management
PIN	Personal Identification Number
PKI	Public Key Infrastructure
PLMN	Public Land Mobile Network
PM	Phase Modulation
PMD	Physical Media-Dependent
PMR	Private Mobile Radio
PNG	Portable Network Graphics
PnP	Plug-and-Play
POI	Point of Interest
POP	Point Of Presence
POS	Point Of Sale
POTS	Plain Old Telephone Service
PPC	Pay-Per-Call
PPP	Point-to-Point Protocol
PQA	Palm Query Applications
PRI	Primary Rate Interface (ISDN)
PROM	Programmable Read-Only Memory
PSK	Phase Shift Keying
PSN	Packet-Switched Network
PSO	Protocol Supporting Organization
PSTN	Public Switched Telephone Network
PT	Payload Type
PVN	Private Virtual Network
QoS	Quality of Service
RACE	Research for Advanced Communications in Europe
RAID	Redundant Array of Inexpensive Disks
RAM	Random Access Memory
RAN	Radio Access Network
RASP	Remote Access Security Program
RBES	Rule-Based Expert Systems
RBOC	Regional Bell Operating Company (USA)
RCU	Remote Control Unit
RDBMS	Relational Database Management System
RF	Radio Frequency
RFC	Request For Comment (IETF)
RIP	Routing Information Protocol
RIR	Regional Internet Registry
RISC	Reduced Instruction Set Computing
RJE	Remote Job Entry

RMON	Remote Monitoring
ROI	Return On Investment
ROM	Read-Only Memory
RPC	Remote Procedure Call
RSS	Radio Sub-System
RSVP	Resource ReSerVation Protocol (IETF)
RT	Remote Terminal
RX	Receive
SDSL	Symmetrical Digital Subscriber Line
SGML	Standard Generalized Mark-up Language
SGSN	Serving GPRS Support Node
S-HTTP	Service Hypertext Transport Protocol
SIM	Subscriber Identification Module
SIMPLE	SIP for Instant Messaging and Presence Leveraging
SIP	Session Initiation Protocol (IETF)
SKU	Stock-Keeping Unit
SLA	Service Level Agreement
SLIC	Subscriber Loop Interface Circuit
SLIP	Serial Line Internet Protocol (IETF)
SLP	Service Location Protocol
SMDS	Switched Multimegabit Data Services
SMIL	Synchronized Multimedia Markup Language
SML/SMS	Service Management Layer/Service Management Service
SMR	Specialized Mobile Radio
SMS	Short Message Service
SMSC	Short Message Service Center
SMT	Station Management
SMTP	Simple Mail Transfer Protocol (IETF)
SNMP	Simple Network Management Protocol
SNR	Signal to Noise Ratio
SONET	Synchronous Optical Network
SQL	Structured Query Language
SS#7	Signaling System # 7 (ITU-T)
SS7	Signaling System 7 (ANSI)
SSL	Secure Sockets Layer (IETF)
SSP	Service Switching Point (IN)
STDM	Statistical Time Division Multiplexing
STM	Synchronous Transfer Mode
STP	Shielded Twisted-Pair, or Signal Transfer Point (IN)
SWAP	Shared Wireless Access Protocol
SyncML	Synchronization Markup Language

T1	Transmission service at the DS1 rate of 1.544 Mbps
T3	Transmission service at the DS3 rate of 44.736 Mbps
TACS	Total Access Communications System
TAPI	Telephony Application Programming Interface (Microsoft)
TB	terabyte (trillion bytes)
Tbps	terabit per second
TCAP	Transaction Capabilities Application Part (IN)
T-Carrier	Trunk Carrier
TCP/IP	Transmission Control Protocol/Internet Protocol (IETF)
TDD	Time Division Duplex
TDMA	Time Division Multiple Access
TDOA	Time Difference of Arrival
TETRA	Terrestial Trunked Radio
TIA	Telecommunications Industry Association
TIFF	Tag Image File Format
TLA	Top Level Aggregate
TLD	Top Level Domain
TLS	Transport Layer Security (IETF)
TMF	Telecommunications Management Forum
TMN	Telecommunications Management Network
TOA	Time of Arrival
TQM	Total Quality Management
TRS	Telecommunications Relay Services
UDP	User Datagram Protocol (IETF)
UHF	Ultra High Frequency (300 MHz to 3 GHz)
UIML	User Interface Markup Language
UMTS	Universal Mobile Telecommunications System (ETSI)
UMTS	Universal Mobile Telecommunications System
URL	Universal Resource Locator (WWW)
USIM	Universal Subscriber Identity Module
UTRA	UMTS Terrestrial Radio Access
UTRAN	UMTS Terrestrial Radio Access Network
VHE	Virtual Home Environment
VHF	Very High Frequency (30 MHz to 300 MHz)
VLF	Very Low Frequency (less than 30 kHz)
VLR	Visitor Location Register
VLSI	Very Large-Scale Integration
VMSC	Visited Mobile Switching Centre
VoD	Video on Demand
VoIP	Voice over IP
VP3	High-quality variable bit rate video

VPN	Virtual Private Network
VXML	Voice eXtensible Markup Language
W3C	World Wide Web Consortium (also WWW3)
WACS	Wireless Access Communications System
WAE	Wireless Application Environment
WAN	Wide Area Network
WAP	Wireless Application Protocol
WASP	Wireless Application Service Provider
WATM	Wireless Asynchronous Transfer Mode
WATS	Wide Area Telecommunications Service
WCDMA	Wideband – Code Division Multiple Access
WCS	Wireless Communications Service
WDM	Wavelength Division Multiplexing
WDP	Wireless Datagram Protocol
WEP	Wired Equivalent Privacy
WG	Working Group
WIM	Wireless Identification Module
WLAN	Wireless Local Area Network
WLL	Wireless Local Loop
WLS	Wireless Location Services
WML	Wireless Markup Language
WPKI	Wireless Public Key Infrastructure
WSF	Workstation Function
WSP	Wireless Session Protocol
WTA	Wireless Telephony Application
WTLS	Wireless Transport Security Layer
WTLS	Wireless Transport Layer Security
WTP	Wireless Transaction Protocol
WWAN	Wireless Wide Area Network
WWW	World Wide Web
WWW3	World Wide Web Consortium (also W3C)
WYSIWYG	What You See Is What You Get
xDSL	"generic" Digital Subscriber Line
XDSL	Digital Subscriber Line
XHTML	Extensible Hypertext Markup Language
XML	Extensible Markup Language
XMT	eXtensible MPEG-4 Textual Format
XNS	Extensible Name Service
XSL	Extensible Stylesheets Language

1 Introduction

If we look at the progress made by *Homo sapiens* over the past 1000 years, the past 100 years have had more impact on us than the rest of the centuries combined. The breathtaking pace of advances in various fields of technology has transformed the human landscape. As we move into the twentyfirst century, the impact of *wireless technologies and globalization* will have a profound effect on the way we interact culturally, socially and intellectually across geopolitical boundaries.

The new millennium will continue the feverish pace of globalization, and bring its denizens closer to each other than ever before. One of the critical factors at the heart of this revolution is mobility; both wireless voice and data technologies will continue to enhance our daily lives and help transform the business and consumer market place over the course of next several decades. The global community continues to embrace wireless applications and services and globalization, irrespective of region, gender, culture, or age. This trend is not limited to human beings; even machines are exchanging information wirelessly.

The recent explosive development of wireless technology has contributed not only to the acceleration of globalization of the world economy, but has also changed our lifestyles. Conversely, the rapid globalization of the world has also made a great impact on the wireless industry. For wireless engineers like Chetan Sharma and Yasuhisa Nakamura, the present authors, physical borders mean nothing. Everyday, a new wireless device is being introduced and a new service starts somewhere in the world. This is especially true with the introduction of global third-generation (3G) standards serving as a great catalyst in establishing borderless global markets and greater service commonality of wireless services across regions.

When Yasuhisa Nakamura was a 10-year-old elementary school student, he was taken to a movie theater in Yokohama, Japan, by his father to watch the movie *2001: A Space Odyssey* by Stanley Kubrick. At that time, he understood almost nothing about the essential message of the movie. As he grew up, he watched the movie repeatedly and discovered many of its messages along the way. In this movie, we can find a full line-up of advanced wireless technologies such as smart antenna, voice recognition, video phone and data communications, etc. Most of the products and technology

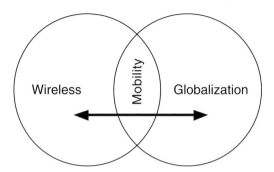

Figure 1.1. Relationship of the wireless industry with globalization.

shown in the movie are already available (3G, voice recognition, broadband access and biometrics). In most cases, we have actually done better. This shows how the progress in wireless technology keeps on exceeding the imagination of human beings.

Nobody can ignore the globalization of the world. Economical activities and lifestyles are becoming borderless in spite of cultural differences and physical boundaries. It often happens that a Japanese singer becomes a superstar in Hong Kong after a single performance. This is true for economical activity as well. We can not forget the tragic events of 11th September 2001 in New York, Washington, DC, and Pennsylvania, which had a serious impact on the global economy. A situation in one part of the world can have a tremendous effect thousands of miles away. Industries such as banking, tourism and manufacturing are closely knit across continents. The world is becoming a smaller place every day.

As briefly mentioned before, one of the fundamental threads that tie wireless and globalization together is mobility. Wireless access enables the mobility of the end-users and services. Globalization can not be achieved without mobility of information and human activities. Mobility is a unique function of the wireless system that no wired system can enjoy. Wireless systems accelerate the mobility, and mobility in turns accelerates globalization. Mobility bridges wireless and globalization, which makes it a win–win relationship (Figure 1.1).

As we take a deeper look at the wireless industry, the growth in usage and popularity of mobile phones has just been phenomenal. By the time this book is published it is estimated that there will be over 1.3 billion mobile subscribers (2 billion by 2006) around the world. That is just astonishing. Another fact that is often overlooked is that wireless technologies have progressed differently in Asia, Europe and North America. Technologies such as i-mode in Japan, SMS in Europe, PDAs (personal digital assistants) and Blackberry in North America point us to the fact that wireless applications and services are unique to the culture and business models of a region. When thinking about transferring successes and learning lessons, one must consider the effects of globalization on wireless communications and vice versa. This is one of the focus areas for this book. We will look at various wireless technologies impacting on the businesses and consumers alike, various business models of different players in the value chain, and the effect of globalization on wireless and vice versa. This book takes a deeper look into why certain technologies, business models and adoption strategies succeed whereas others fail, and how all these elements will impact on the future of wireless communications. With the help of examples, case studies and interviews with industry luminaries, we will learn some lessons and derive some conclusions and recommendations. Books often concentrate only on technology, but technology is only part of the picture. By looking at different business models and cultural nuances around various geographies, we hope to provide a complete picture that will benefit the planners and implementers around the world.

1.1   Progress in wireless technology

Since the early 1980s, when brick-size mobile phones were introduced on analog networks, we have come a long way. Today an average cell-phone weighs less than 100 oz., has a rich graphical user interface, the processing power of a desktop computer, and is loaded with features that a few years ago seemed possible only in fiction movies. With the advent of 3G services like FOMA (Freedom of Mobile multimedia Access) in Japan and around the world, we are looking at broadband networks able to support video streaming, voice recognition, international roaming, high-resolution GUI, and much more. We have not only made significant strides in wireless WAN (wide area network) technologies, but also in wireless LAN (local area network) with IEEE 802.11 or Wi-Fi and wireless PAN (personal area network) with Bluetooth, UltraWideBand, HomeRF and RFID (radio frequency identification) tag technologies.

An almost unprecedented adoption of wireless devices has led to the growth and popularity of wireless applications and services over the past decade. Just as the Internet enabled legacy systems to be linked to and accessed from the Web and led the e-business revolution; access to information anytime, anywhere using wireless devices such as mobile phones, PDAs, autoPCs, Kiosks, etc., is enabling the next phase in pervasive computing – the mCommerce, or mobile commerce, revolution. The wireless applications and services enable consumers and enterprises to access information via their hand-held devices such as palm or pocket PCs or WAP phones, thus empowering the individual.

Advances in technologies in wireless and computing are energizing the vision of making Internet device independence a reality. No longer must we be shackled to our desktops to access information. It is now possible to access that same information from an endless number of devices – PDAs, palmtops, smart phones, landline phones, TVs, elevator kiosks, airline entertainment seats, pagers, autoPCs, gameboys, refrigerators, exercise machine screens and video phones.

Also, the convergence of the computing and communications industries has rapidly blurred the lines between the devices produced by each industry. Phones and PDAs now possess the power and sophistication of desktops. This, coupled with an astonishing surge in mobile subscribers worldwide, is fueling the tremendous momentum of a world without wires. Players in all industries seek ways to lever the wireless Internet phenomenon to reinvent themselves and extend their reach to customers, partners and opportunities. Those who do not will risk being obliterated by competitors.

Advances in wireless technologies are redefining the existing Internet model and its services. Technologies such as WAP (wireless application protocol) are bringing Internet e-commerce closer to the mobile world. A bank that extended its operations arm to provide services over the Internet now provides access to user data and services over the air. The bank uses WAP-based applications, essentially putting its services at consumers' fingertips. As e-commerce vendors connect their warehouses to these wireless devices, they provide continual, anywhere service.

Leaders in all major industry segments are participating in trials and launching experimental services to learn, and adapt to, customer behavior and worldwide market opportunities. Sample applications include trading stocks; checking weather, news, and traffic conditions; buying books, running auctions, providing location-based customized information and playing games – all while being mobile.

1.2   The business model and global wireless competition

Like other global industries such as finance, manufacturing and information technology (IT), the wireless telecommunication industry is also facing intensive global competition. Cellular terminal vendors such as Nokia, Motorola and Panasonic have global strategies to increase their market share. International wireless operators such as Vodafone and NTT DoCoMo are trying to get as many customers as possible, through acquisitions and strategic alliances, because they know that the key issue to win the game is the scale of merit and service commonality in the global market.

Presently, there seem to be two fundamental business strategies in this endless war. One is to increase profit per end-user (consumer and corporate) by adding new values to the existing services. A typical example of such a strategy is to integrate wireless and Internet to introduce new revenue streams by adding new applications and services. The unprecedented success of i-mode by DoCoMo is a typical example of such wireless Internet integration. Before i-mode, there was no real business model for wireless Internet; however, after i-mode DoCoMo enjoyed an increase in ARPU (average revenue per user) in spite of the extreme disappointment of WAP in Europe and North America.

Another strategy is to enjoy the scale of merit, since the greater the market becomes, the less is the cost for operations and the price of product. Vodafone and Nokia are typical promoters of this strategy. By the end of 2001, the Vodafone group had extended its reach to 28 countries and by mid-2002 the total number of subscribers to this group exceeded 91 million, which is twice as many as DoCoMo. Based on this huge number of subscribers and its footprint in the world, Vodafone is increasing its sales every year.

Unlike the PC software or aircraft manufacturing markets, where Microsoft and Boeing pretty much own the market, there are no dominant winners in global cellular competition at present. Neither Vodafone nor DoCoMo have established a stable position and there are many challengers every day. This represents a very attractive opportunity for ambitious entrepreneurs and also for existing wireless players in the global arena. Telecommunications players and Internet business players are trying to catch this huge potential market.

In this book, we will carefully study the various business models (both successes and failures) and give the reader some insights. We will also take an in-depth look into the success of i-mode in Japan and see what lessons can be drawn from the experience.

1.3   Cross-cultural challenges

We will always face cross-cultural challenges in the process of globalization of any kind of business. The development of wireless business is no exception. These challenges often come from the difference in business customs, languages, cultural behavior and also life-style.

In the USA, almost 60–70% of the total voice traffic of the cellular service is placed or received in a vehicle, which is very different from the cell-phone usage trends in Asia. In Japan and in several European countries, the train is the most convenient form of transport, so while commuting the consumers use that time to place or receive calls or to play games or check news on their cell-phones. The language poses significant localization and transcoding challenges for service providers. One Chinese or Japanese-Kanji character has multiple meanings. A sentence of 15 Kanji characters may have the same meaning as a sentence of 50 Roman alphabet characters. This affects the design of display of cell-phone terminals as well as the applications and services (see Figure 1.2).

The high penetration rates of Internet-enabled desktop PCs into residential homes in the USA exceed the rest of the world by several orders of magnitude. For most consumers in other countries, their first (and may be only) interaction with the Internet is via their wireless devices.

あなたが私にプレゼントをくれなかったので、
私はあなたにビールを買わないつもりです。

לא נתת לי מתנה ולכן לא אקנה לך בירה

*You didn't give me a present, therefore I am not going to buy
you a beer.*

版權所有 不得翻印 如有雷同 實屬巧合

זכויות יוצרים. אין להעתיק או לשכפל ללא אישור.
כל דמיון בין אנשים או ארועים אמיתיים, הינו מקרי לחלוטין

*Copyright material. No unauthorised reproduction. Any
resemblence to real-life events or people is incidental.*

Figure 1.2. Internationalization – intelligent translation is more than word-for-word (top);
layout and real-estate can be markedly different (bottom) (source: *BT Technology Journal*, 2002).

These are just a few examples. To promote a wireless project across the continents,
careful investigation is needed into what the respective market wants. This approach
is similar to that of PC software products. Newly developed PC software is carefully
localized, including the human interface aspects, for each respective country in order
to satisfy the demand of the market.

In spite of cultural differences across the world, the explosion in the increase of
the number of cell-phones is a common phenomenon. In 2001, China became the
country with the largest number of cellular subscribers. A similar explosion is happening in other Asian and South American countries. Many young people walk along
Ipanema beach in Rio de Janeiro, Brazil, talking with cell-phones. This scene is as
common as it is in New York, Beijing or Paris.

1.4   What makes this book unique?

This book aims to take a wider view of evolution of wireless[1] services around the
world. We look at its impact on society from three critical angles: *technology innovation, business models, and cultural nuances that characterize global markets.* We believe
that only by taking these three aspects into consideration can any business strategy,
application or service become successful and truly global. We study the impact of
globalization on the wireless industry and how it keeps shaping the industry's future

---

[1] In this book, we are going to be mostly talking about wireless data related technologies and business models.
Unless otherwise stated, wireless refers to wireless data (and wireless Internet).

by analyzing the effect of global markets on key business sectors such as operators, equipment manufacturers, and the computing industry. To follow up on our global view theme, we take a deeper look at the trends across various important geographical areas such as the USA, South America, Japan, South Korea, China and Europe. We discuss the various computing and communications technologies that are of great interest to the industry, and also those that are going to pave the way for new exciting applications and services. A large portion of the book is dedicated to understanding and discussing the business side of running a wireless business, whether you are dedicated 100% to it or just 10%. We do this by analyzing the wireless value chain in detail and discussing the business models adopted by players in different segments of the chain, and by looking at what has worked and what has not. No wireless Internet discussion is complete without the inclusion of the growth and impact of i-mode on the industry. In various sections throughout the book we draw upon the lessons learned and dispel some i-mode myths on the way.

Another important aspect of the book is the case studies and executive interviews. The examples and views presented and argued represent a wide spectrum of thoughts and ideas as to how to empower the end-user with more information and functionality. The executives represent some of the most accomplished players in various segments of the value chain and should provide the reader with a unique collection of perspectives. They share their knowledge and perspective on the future of the wireless industry, applications, wireless devices, 3G, the international landscape, regulatory issues and much more.

Finally, we look at the future of wireless technologies, applications and services, and draw upon the lessons learned in taking an educated guess about what is to come in the future. Our aim has been to provide a perspective of the industry from all angles to help the reader understand the complexities, vagaries and opportunities that wireless applications and services present.

We believe that by taking the broad view of the wireless industry, we offer valuable insights to a wide spectrum of readers around the world.

1.5   How is the book organized?

Wireless Data Services is organized into 14 chapters, including this Introduction, as well as suggestions for further reading. As with any technology, to build wireless applications and services successfully, one needs to understand the market trends, history, supporting technologies and the industry in general. We have tried to cover all of these areas to give the reader a good understanding of the wireless space in the context of technology, business models and global markets. The chapters of the book are organized as follows.

Chapter 2 The impact of globalization
   This chapter looks at the impact of globalization on the wireless industry and vice versa.
Chapter 3 Adoption trends and analysis by region
   In this chapter, we will look at the adoption trends across various regions around the world. We will take a look at the USA, South America, Japan, South Korea, China, Hong Kong and Europe. In addition we will analyze the affect of wireless data on different business sectors such as the telecommunications industry, equipment manufacturers, and the computing industry.
Chapter 4 Subscriber needs and expectations
   We will discuss the needs and expectations of consumers when it comes to using wireless data on a daily basis – whether it is for personal or business applications.
Chapter 5 The wireless value chain
   In this chapter we discuss the wireless value chain in some detail.
Chapter 6: Global wireless technologies: systems and architecture
   The wireless technology ecosystem will be discussed in more detail. As will become obvious from the discussion, with the convergence of the communications and computing industries, the wireless industry is becoming more dynamic and the value chain is becoming more extended and vibrant. We have divided the discussion into systems and architectures where we talk about various network architectures (2G–3G) and device technology. We will also look at various WAN, LAN, PAN, IP and related technologies.
Chapter 7 Global wireless technologies: network, access, and software
   In this chapter we look at how network, access, and software technologies are changing the face of the industry.
Chapter 8 Business models and strategies
   In this chapter we will discuss in more detail the various business models being adopted. We will also take a deeper look at i-mode, the most successful wireless Internet service to date.
Chapter 9 Business issues and challenges
   This chapter will focus on business issues. Although some industries have some common idiosyncrasies, the wireless industry is unique and sometimes issues and problems are specific to a region or country. For instance, spectrum (which we will discuss later) is squarely an issue for the USA, while their counterparts in Japan do not encounter such problems, and 3G spectrum auctions are driving many carriers out of business in Europe. While Europe and Japan are approaching saturation in the consumer market, the US market is largely untapped.
Chapter 10 Technology issues and and challenges
   This chapter will address some of the challenges in the technology arena.
Chapter 11 Case studies
   So far we have talked about the wireless industry – its peculiarities, the challenges and issues with business and technologies, case studies and where

the industry is heading. In this chapter we will look at some case studies that show how companies have adopted wireless to solve problems, enhance their business and become more efficient.

Chapter 12 Perspectives

For this chapter, we sat down with several industry leaders belonging to different parts of the value chain, operating in various parts of the world, and with people who have been around the industry for a long time and witnessed the trials and tribulations of the wireless sector over the years:

1. Steve Wood, CEO, Wireless Services Corporation
2. Mark Tapling, CEO, Everypath
3. Mark Anderson, President and Editor, SNS
4. Jon Prial, VP Pervasive Computing, IBM
5. Frank Yester, VP Motorola Labs, Motorola.

This chapter presents our one-on-one discussions with them on the case studies, various aspects of the industry, their own organizations, and where they see wireless industry in the future.

Chapter 13 Future of wireless technologies, applications and services

This chapter aims to focus on the discussion of future of wireless technologies, applications and services in the twentyfirst century. In Chapter 10, we discussed some of the technology issues and challenges for the industry. In this chapter we will continue the discussion on evolution and challenges of wireless technology in the context of the next 5–10 years.

Chapter 14 Conclusions and recommendations

Finally, we leave the reader with our thoughts on how the industry might evolve under various scenarios and what the future might be like a decade from now.

References and further readings

For those of you interested in further exploring some of the topics discussed in this book, we have included a detailed list of references and web links at the end of the book.

1.6   Who should read this book?

Anyone who is interested in learning about the wireless market and its future will find this book helpful. This book is aimed at the following readership.

(a) Corporate technical managers around the world who are responsible for understanding and implementing wireless data solutions. These people are from different segments of the wireless value chain, from handset manufacturers to carriers to content providers.

(b) Marketing, sales and other non-technical staff members. Successful technologies, applications or services generally do not translate well across geographies unless attention is paid to business models and cultural nuances. Marketing and sales executives need to understand this the most.

There are many books focusing on wireless technology and how it works and how one should be using it, but rarely do the discussions on *why* go into any detail. If one can't answer the why (should it be used?) of technology, something is wrong with the picture. Technology for technology's sake does not work very well, we have all witnessed that. It is the "value proposition", it is the "what's in it for me" question, that makes or breaks the solution offering. Although we delve into technical issues and discussion, this is mostly a business book about wireless technology, applications and services. We strive to answer the question: *why should you care about wireless data technologies?*

The salient features of this book are as follows.

- A review of various wireless WAN, LAN, PAN related and upcoming technologies.
- A discussion of globalization and its impact on the wireless industry.
- The strategy behind i-mode's success in Japan and lessons drawn.
- A look at various business models for players across the value chain.
- The transfer of technology issues and business models across borders.
- An analysis of successes and failures in the wireless industry.
- The convergence of the computing and communications industries.
- A discussion of the wireless Internet value chain.
- A preview of next-generation (3G and beyond) wireless technologies.
- The strategies for 2.5/3G and beyond.
- A detailed discussion of issues and challenges of the wireless data industry.
- Detailed case studies: consumer and enterprise.
- Interviews with industry executives and experts from IBM, Motorola, and others.
- Insights from professionals who have built systems and implemented technologies around the world.

1.7   Summary

Former Canadian English-literature scholar Marshall Mcluhan once said "the world will become a Global Village" and predicted the arrival of new social systems. He also said "Electronic media will make it happen". The development of electronic media, especially wireless technology, has helped support the new social systems. The broadband wireless networks around us will soon become an "air-like" infrastructure, which we will just use subconsciously. We will not be conscious of the charge, location, time or sometimes even the device, but we will enjoy high-speed integrated broadband services under ubiquitous networks. As time goes by, more and more people will become acclimatized to the omnipresent wireless infrastructure and environment, and the physical boundaries will be less clear. Some may become confused and get lost in the situation, but we believe the advantages will overcome the disadvantages for most of us.

In the global village of the twentyfirst century, there will be no boundaries. People will obtain full mobility in their daily lives. Any physical barriers such as location, language and age are going to become irrelevant. However, there will be significant challenges in this pursuit. Some challenges will require a pure technical approach and thinking. Others are more driven by regulation, sociological values, and business models that will shape the markets.

We invite the reader to delve into the discussions that we present in this book. We hope he or she will benefit from the insights and conversations that are drawn from our experiences as well as lessons from success and failure stories around the world, from the executives and field staff we talked to, and from our colleagues in the industry. We start by talking about globalization and its impact on the wireless industry in Chapter 2.

It is our hope that the information, discussions and ideas presented in this book will help the reader's understanding of the wireless data industry, inspire new products and ideas, and become a useful text for consultation from time to time. We, the authors, would be very happy to hear from you, the reader. Your feedback will help us with future work. We may be contacted at chetan@ieee.org and nakamuyasu@aol.com. Also, a companion website for this book is maintained at http://www.wireless-data.net.

2 The impact of globalization

With each passing year, geographical borders between nations appear to be shrinking. The masses might be separated by culture heritage, religious beliefs, or ethnicity, but almost all of the human race has been transformed and touched by the global phenomenon called globalization. In general, globalization refers to the adaptation of business practices and processes to take a business, product, application or service to global markets. It refers to the internationalization[1] and localization[2] of these products and applications, so that they are ready for the global market. As economies continue to become interconnected and interdependent, an event in one corner of the globe can have an effect of much larger magnitude than ever before. One does not have to look far to fathom the ripple effect of world events. The collapse of the Asian financial markets in the mid-1990s or the horrible events of 9/11 in the USA touched the lives of people living far beyond. With trade barriers lifted (most recently in China), products from around the world compete on quality and price in any local market. With the advent of the first wireless technology and then the Internet, physical boundaries have become almost meaningless as far as trade and business are concerned. The European adoption of a common GSM standard for wireless communication fostered the growth in the industry that is unparalleled in recent times. As communications and the computing industry continue to merge, the boundaries continue to blur to the point where it is meaningless to talk about globalization with mentioning wireless. Globalization and wireless are tightly interlinked. IDC estimates that, by 2005, more than 70% of humans will speak a language other than English, and so as the non-English speaking population grows, the need to design localized applications and services will increase as well. And it is not just about applications and services; business practices and processes also need to adapt to the

---

[1] Internationalization is the process of designing a software or Web application to handle different linguistic and cultural conventions without additional engineering.
[2] Localization is the process of adapting a product to the requirements of a target locale. This involves the translation of the user interface (UI) – including text messages, icons, buttons, etc. – of the online help, and of any documentation and packaging, and the addition of cultural data and language-dependent components, such as spell-checkers, input methods and so forth.

locale. Let us look at some of the challenges of internationalization:

- ensuring that all localizable elements are extracted from the source code (e.g. by collecting all localizable items in external resource files);
- ensuring that the design of the user interface (UI) is flexible and neutral;
- ensuring that the relevant character set is supported;
- ensuring that regional standards are supported;
- for applications and services, ensuring that text embedded in graphics is easily localizable.

Internationalization issues may arise at any time from political and economic changes. For instance, the adoption of the common Euro currency in 2002 by 11 members of the European Union involved major changes in business practices and financial systems, and consequently on related software and Web applications.

In this chapter, several examples of globalization are discussed, from the viewpoint of business activity in the wireless industry. Having a global perspective while expanding and promoting its business is one of the key elements of success for growing companies. This is not only true for IT and the telecommunication industry but also for all kinds of business areas.

In 1995, Starbucks, founded in 1971 in Seattle, opened its first coffee house outside of the USA in Tokyo, where average retail rent cost is two to three times more expensive than in the northwestern USA. To date, the success of the business in Japan is phenomenal and Starbucks is recognized as one of the most preferred restaurant chains in Japan. It may be true that the fact that the Japanese young generation love American brands greatly helped the business activity, but we can not forget that Starbucks management did their best to open the Japanese market.

When a person or a company goes across borders, they sometimes have to fight challenges such as language, culture and social environments because of diversity. But diversity may provide us with a great potential chance for success.

Recently, owing to the standardization of technology, especially in the area of telecommunications, the technology development of 3G and wireless Internet has become mainstream. However, global business activity needs to be carefully promoted considering the diversity mentioned above and localization.

Also, with their home markets saturated, some of the major OIC carriers are looking to global markets to expand their reach. Vodafone, NTT DoCoMo, and Deutsche Telekom have been aggressively making investments in foreign markets to drive revenue and consumer acceptance.

2.1   The impact of globalization and wireless technology on education

Education forms the fundamental pillar of our existence. Irrespective of culture, country, class or creed, the majority of parents hope for a good education for their children. Some parents are so driven by this goal that they devote their entire lives and resources to their children's education for they feel it will add character and knowledge and help them to reach their goals and dreams in life. Education is important not only for children, but continuous education is indispensable for adults as well. In order to survive in the global business scene, individuals may need to acquire expertise of businesses, a second language, and marketing and technical areas through education (after graduation).

As businesses are becoming more global and dynamic day by day, we need to enhance these skills through taking advanced education such as Masters degree courses or attending seminars and conferences that are customized to our needs and goals. This is not only true for advanced countries like the USA, because the yearning for education and learning remains true throughout the world.

The impact of globalization can be seen on the campus of any major university. During the summer of 2001, an Executive Program for Growing Companies at Stanford University, a two-week short program designed for the CEO, CTO, CFO level of growing companies in the world had 80 participants. Only 33 students were from the USA. The rest came from Switzerland, Japan, Argentina, Saudi Arabia, Mexico, France, Canada, Taiwan, Italy, Brazil, Australia, the UK, Spain, Netherlands, Korea, Sri Lanka, Fiji Islands, China, Chile, Nepal, Hong Kong, India, Venezuela and New Zealand. This list clearly shows how businesses, especially growing businesses such as IT, Bio, financing, marketing, telecommunications, etc., are becoming worldwide and borderless. It is not a fallacy that we learn from diversity in culture and background, and hence the education experience becomes richer. Education systems worldwide are clearly the beneficiaries of the ever-proliferating globalization.

According to Stanford University, around 5% of the total undergraduate students come from outside the USA and the number of counties reaches 55, of which 48% are from Asia, 19% from Europe and 22% from the Americas. Among graduate students, 32% of the total number are international students from 86 countries. The trend of global participation is only going to increase.

2.1.1   The Internet at school

In many countries, the Internet is becoming one of the most essential tools for education. The Internet provides an excellent forum for students to research and exchange ideas. In highly educated communities such as universities, the Internet is the most common way to communicate, study, search for information and send/receive reports. Also, the Internet is a great tool for students to learn to think globally. As a result, there is a tremendous increase in penetration of Internet usage at school and this has accelerated the global way of thinking among students.

However, there are some critics who do not like to mix the Internet and schools. Firstly, they argue that the distribution of fixed computers to every school/student requires a lot of space and budget for the school. Secondly, teachers are neither always well trained nor skilled to teach students about computers and the Internet. Thirdly, as students are usually moving around at school, they are reluctant to sit for a long time in computer rooms. As mobility is an indispensable characteristic of students, they do not want to lose it because of computers. For any student, mobility is a symbol of freedom!

With the advent of wireless Internet solutions, some of the challenges that schools face can be resolved. Firstly, cell-phones or a wireless PDA are much cheaper than a desktop PC; they are "affordable." Secondly, teachers do not need to learn how to use them, because students can already easily handle these devices. Thirdly, they give the students full mobility inside and outside the school campus. Wireless can be a great tool to enhance the productivity of any education system. In Japan, an English teaching school gives personal English lesson using FOMA video phones. With high-speed data transmission, an English-speaking teacher can easily give a lesson with voice and images.

High-speed wireless LANs can help schools set up an Internet infrastructure very quickly at a fraction of the usual cost. As the coverage of a WLAN access point (100 m) is on relatively the same scale as a school facility, an access point can make good coverage with relatively low cost.

2.2   The impact of 9/11 on the world

The horrible events of 11th September 2001 shook the world like no other event in recent memory. The scale and precision of destruction was just unfathomable. While we continue to express shock and dismay over this human tragedy, several key debates have surfaced that continue to shape the economies, perspectives, and implementation of solutions to better our lives. If there were any doubts about globalization becoming an important part of our lives, the events of 9/11 left nothing to doubt. The events had a ripple effect that was felt across the world. Many things that we used to take for granted were no longer there. It is now more obvious than ever before that isolation of an economy or country is impossible, and the world's future will be defined by global relationships and cooperation. Let us briefly touch upon some of the key impacts of 9/11, not only on the USA but also on the world at large.

2.2.1   The impact on subscriber adoption and how wireless phones emerge as a necessity

In the Thursday edition of *USA Today* following the 9/11 attack, journalist Olivia Barker wrote: "In the wake of Tuesday's attack, cell phones have shifted from social nuisance to necessity in pockets of our consciousness. Across the USA, those who ever dismissed the phones as status symbols are recognizing them as a lifeline." The stories of passengers on the ill-fated flights having last-minute phone conversations with their loved ones were both poignant and telling. As Olivia Barker pointed out, wireless phones have since then become more than an accessory. They have become indispensable devices that we can not live without. It is not only the wireless voice services that become important, but also the data services. During emergencies like earthquakes, accidents, volcano eruptions, etc., wireless voice networks very quickly become saturated but the data networks can handle more traffic. Within minutes of the tragedy on 9/11, thousands of SMS messages and emails were being sent out and received by loved ones around the world. The value of knowing that your son or brother or your friend is safe amidst the initial chaos is just invaluable. A wireless device can also serve as a very important search and rescue tool. By using position location technologies, the whereabouts of victims can be pinpointed, and this can be the difference between life and death (more on E911 in Section 2.2.3). It is because of these reasons that all the markets in the USA have seen a rise in subscribers, and the effect has also been seen in other parts of the world.

2.2.2   Globalization

The so-called "war against terrorism" has brought home one point very crisply and strongly: that we share a common planet, and that we are dependent on each other in more ways than we care to admit, and hence we need to work together as nations and citizens of this planet. The events of 11th September 2001 have affected the dialogue of globalization in more ways than one. The live non-stop television coverage brought the events closer to home for everybody. There is a sense of more cooperation between governments and citizens of countries bound by geographical and cultural borders. The free market enterprise is one where everyone can compete based on merit and quality. Even the Western nations are willing to let go of their perceived double standards in trying to keep foreign goods that undersold their home products out of the market and yet push developing nations to open their markets. Leaders and heads of states are willing to talk about issues and problems that confront us all, like the spread of diseases, global warming, growth in terrorism, economical stability, disparity between rich and poor, and the Internet and communications infrastructure. These are no longer regional issues, they are global issues.

In the wireless industry, players like Vodafone, NTT DoCoMo, Nokia, Ericsson, Sony and others are more concerned with building their brands worldwide than just focusing their efforts in their home country. We will continue to discuss the impact of globalization on the industry throughout this book.

2.2.3   E911

Recognizing the need for better position-location technology, the FCC mandated US carriers to implement enhanced 911 (E911) emergency infrastructure by 1st October

2001. Unsure about how to fund the costs, and owing to a lack of FCC strictness, carriers continued to delay the deployments. As a result, the deadline of 1st October came and went, and no US carriers were ready with their position-location technology. The wireless industry has taken its share of criticism for not meeting E911 mandates and providing the ability to pinpoint more accurately the location of wireless 911 callers. The industry has faced increased scrutiny on E911 and other public policy issues, such as priority network access in emergencies, since recent terrorism attacks. The attacks of 11th September 2001 on the World Trade Center and the Pentagon make public safety technologies, such as an enhanced 911 service for mobile phones, even more important to everyone.

Despite that sense of urgency, on 5th October 2001 the FCC granted carriers dispensation from meeting the E911 mandates, saying in separate orders for each carrier that the new compliance plans were specific, focused and limited enough in scope to meet the FCC's waiver test. The new interim deadlines for carriers vary, but the full compliance date of 2005 applies to all. The FCC issued the following orders.

- Nextel Communications Inc., Sprint PCS and Verizon Wireless get relief for some of the initial 2001 and 2002 deployment milestones because of showings related to equipment availability.
- AT&T Wireless and Cingular Wireless get similar relief for the GSM portions of their networks, subject to enforcement of the new schedules.
- Carriers must adhere to modified plans or they face enforcement action.
- E911 deployment must be complete by 2005.
- Carriers were required to supply quarterly reports on their implementation of phase I and phase II mandates of E911 as from 1st February 2002.
- The FCC will conduct an ongoing inquiry into E911 technical issues, including evaluating vendor and manufacturer claims.
- Small and rural carriers that cannot comply with the commission's E911 deployment rules had until 30th November 2001 to file for a waiver.

A recent survey commissioned by LetsTalk.com found that 59% of those questioned said the ability for rescue workers to locate them in an emergency was the most important feature for a mobile phone to have. The second most important feature was email, at 23%.

By April 2002, 65% of Verizon Wireless' network was capable of meeting E911 requirements. In keeping with that, the no. 1 US wireless carrier has introduced the E911-capable Samsung SCH-N300 phone. When local and state public safety call centers upgrade their systems, they will be able to pinpoint the location of emergency callers using the SCH-N300. Coast-to-coast network upgrades were completed by the second quarter of 2003, the company says. Verizon Wireless, along with most other carriers, was granted a waiver by the FCC from meeting the 1st October 2001 E911 deadline. The operators argued that technical hurdles and the lack of adequate numbers of handsets made it impossible to meet the target date.

Although things in the E911 space are still moving more slowly than most consumers want, there is a renewed urgency amongst carriers and legislators to get something, albeit in phases, in place very soon. Carriers who can roll out their E911 infrastructure (network or handsets or both) can easily attract customers from their rivals, touting security. Similarly, carriers in Europe and Asia are either launching new technologies or improving the accuracy of their existing infrastructure to pinpoint wireless devices, especially in cases of emergency and life-and-death situations.

2.2.4   Cell towers

Emergency crews struggled to put out fires, clear streets and find survivors following the two crashes at the World Trade Center, a third at the Pentagon and a fourth in Pennsylvania. Even before the dust settled, wireless carriers responded with temporary cell sites, reconfigured network operations and antenna sites. The 9/11 terrorist attack against the World Trade Center towers in New York City may have wreaked havoc on the wireless communications infrastructure equipment, cell sites and cell site towers in the surrounding area, but it certainly did not destroy the goodwill of the tower industry. Instead, the tragedy has pulled together players from disparate groups in this sector, many of which traditionally are at odds when it comes to tower construction and securing rooftop rights. In addition, many real-estate landlords, once opposed to putting antennas on their rooftops, are now reconsidering. People put off by leasing antenna space are more likely to listen to a deal now.

2.2.5   Wireless data and messaging

As discussed above, the events of 11th September 2001 brought home the point that, in addition to wireless voice communications, there is also a role for wireless data and messaging solutions. Email, SMS, and instant messaging (IM) accounted for a surge in data traffic of over 50% on the data networks of various carriers like Cingular and Motient. Further, the value was in the quick delivery and reliability of the networks to deliver the message. Wireless networks proved their resilience for data transport, and messaging services proved their usefulness.

Wireless data networks also proved very important and crucial to several of the enterprises whose corporate networks were blown away and destroyed. The high concentration of the popular Blackberry devices in the financial services industry in the New York City area drew attention to the benefits of packaged corporate email solutions. Also, some corporations used IEEE 802.11 networks to get their businesses up and running in a very short time. As a result, a wide range of industries in all vertical sectors are seriously looking to evaluate and implement wireless data solutions.

During 2001, US carriers started to launch SMS solutions and made their packaged data offerings uncomplicated and more useful to the consumers. A majority of the handsets sold today are data and messaging capable. However, the lack of interoperability amongst US carriers has been hindering the sky-rocketing consumer adoption that we have witnessed in Europe. The lack of interoperability across carriers and networks hinders SMS's utility during disasters and crisis situations when time is of the essence and consumers are not going to be on the same networks. The real-time, immediacy, and presence-feature advantages, as compared with other non-real-time messaging solutions, have drawn attention to instant messaging. Today, both consumer and enterprises are adopting instant messaging as a way of communication in their daily lives. However, the industry needs to work on "seamless" integration of network capability to deliver messages irrespective of carrier network technology.

2.2.6   Carrier marketing

As we discussed earlier, coverage and reliability of a carrier network has become increasingly important to the consumer at large. In a survey done by the Yankee group right after the 11th September tragedy, consumers voice "coverage and reliability" as their top two concerns when making purchasing decisions. When asked, "What is the main reason wireless service was purchased for children under 18?", parents overwhelmingly (over 54%) said that security and getting in touch with their children during an emergency was the prime reason. The Yankee group concluded that carriers who provide more reliable service and robust coverage, and can establish and communicate those attributes to potential subscribers, will gain an edge in a crowded and homogeneous marketplace. Carriers are also being forced to work with each other to provide network and in particular messaging interoperability. Also, carriers who efficiently implement E911 solutions will have an edge in taking their story to the consumer market. Position location solutions will prove beneficial to the enterprise market.

2.2.7   Privacy vs security

Appalled by the 2001 hijackings, many Americans have declared themselves willing to give up civil liberties in the name of security. With the renewed focus on security, the questions of privacy that plagued biometrics and the security industry took a back seat. Every major news organization has been talking about the balance between privacy and security, discussing whether various increases of security are worth the losses of privacy and civil liberty. While the pros and cons of the two sides are debated, there has been an acceleration of biometrics woven into the fabric of e-commerce and security in the USA and elsewhere. Biometric solutions and identification, tracking, and profiling technologies are rapidly being prototyped, tested, and put into use.

The debate is going to affect the wireless industry deeply as well. With the introduction of position-location and presence-based solutions and technologies, carriers can theoretically know more about you than you are willing to let them know. Public-interest groups have been fighting to legislate the privacy issue for the wireless industry so that the subscriber information can not be used without his or her knowledge. In addition, with the advent of wireless messaging, consumers are concerned about spam and advertisers sending those messages that they do not want and might even have to pay for.

The debate of liberty vs security will be a fierce one. Public perception of privacy vs security is likely to change and there will be more willingness to embrace security solutions like biometrics and convenience solutions such as position-location for both commercial and non-commercial applications. We will further discuss this issue in Chapter 9.

2.3   Globalization: culture and lifestyle

A detailed discussion of the pros and cons of globalization is beyond the scope of this book. However, it is difficult to ignore the impact of globalization on our daily lives, or the continuous coverage of the globalization debate. We can not forget the WTO riots in Seattle in 2000 and the protests that have become part and parcel of these global meetings. Some of the confusion and fear stems from the fact that the economics of globalization affect the existing framework of some businesses and industries like agriculture, trading, and manufacturing. Whichever side of the debate one might be on, it is difficult not to be impacted by globalization.

Any major metropolitan city boasts of its international culinary delights, whether it is the streets of New York, Paris, Tokyo, London, or Rio. Hollywood movies and music by US artists are among the strongest export products of the USA. The fast-food chain giant MacDonald's is one of the most internationally recognized brands in the world with its presence in every major country in the world. The Japanese "Pokemon" cartoon characters are a favorite amongst children and adults alike, even in a far off place like South America which is diametrically opposite to Japan. Tiger Woods is one of the best-known sports superstars in Asia and there is hardly anyone who is not familiar with him, even though they might not know the name of their own prime minister. The caliber and strength of the French national soccer team, for example, receives admiration of school boys playing soccer all around the world. Hardly a day goes by without events that touch our lives, providing proof that we are all part of a global community, and filling us with a wonderful joy.

Every country has its own unique culture and heritage. In some countries like India, individual states are like a different country when it comes to culture and differences. The differences might be in the form of language, history, food, music, literature or anything else that helps to characterize a culture. Also, the culture is closely related to the way people think, live and act. It is a great joy to experience different cultures and their heritage in our daily surroundings; that great joy is the primary motivating force for us to travel or read. In this context, the value of culture lies in its differences. So, how do globalization and differences in cultures mingle with each other? The answers are quite complex; however, let us get a perspective from a wireless viewpoint.

2.3.1   Nuances in culture make popular content

The differentiation factors in the global standard era are service and content, not technology itself. Knowledge about culture can be used to provide the most popular and prominent content for global wireless business sectors because each end-user belongs to their respective culture(s).

2.3.2   Wireless accelerates globalization

It is fully recognized that the (wired) Internet has contributed greatly to the globalization of the world. The Internet allows us the freedom of access to information easily from anywhere and at anytime. Nobody can stop the operation of the Internet. It is the infrastructure that helps the circulation of digital content around the world. If this wireless-based access is coupled to information, the value mobility provides to the users not only enhances the experience but provides the freedom and delight of being connected to your world and information from anywhere. As such, wireless Internet is having a profound impact on the globalization of information and the acceleration of Internet adoption around the world.

2.3.3   Wireless lifestyles

Mobile life represents a lifestyle that uses wireless services as an essential lifeline for businesses and consumers. In countries where wireless technology has been embraced as part and parcel of daily life, such as Japan, Scandinavian countries like Finland, and HongKong, the actual penetration rates are reaching over 90%. People carry their phones just as they would carry their wallets and driver's license. Teenagers addicted to i-mode in Japan send and receive several e-messages a day to and from their friends. Intranet access capability with i-mode Java phones makes them an indispensable part of a sales person's equipment for customer care. Niche businesses such as language teaching schools are going to start using advanced FOMA video phones to provide training with state of the art tools; this represents the embrace of wireless technology in our daily lives.

The essential difference between a fixed (phone/Internet) device and a wireless device is the length of the time in a day for which the end-user is touching the device. An end-user of a wireless device is usually in very close proximity to his or her handset for almost 15 hours a day on average (except sleeping time). However, in the case of a fixed device, it is at most 8–10 hours a day at the office or at home. This indicates how we spend longer with portable wireless devices than with fixed ones, and as such are willing to interact with the device much more than we would with a wired phone. This analogy is similar to the case study of wrist watches against wall mounted clocks. Just like wrist watches, mobile devices have become a part of our daily lives – or we could say that our lives have morphed into mobile or M-life.

2.4   Effect on business sectors

We have reviewed the wireless growth trends across different continents, and we have had a glimpse of how various players are interacting with each other to increase their value in the wireless value chain (which is covered in greater detail in Chapter 5). We will now further discuss the interaction between various wireless players in the industry and globalization, based on several case studies. As we shall see in later chapters, most of the key promising wireless technologies will be implemented based on global standards such as WCDMA (ITU and 3G-PP), WAP2.0 (OMA), UltraWideBand (UWB) and WLAN (IEEE 802). Working on a global standard like WCDMA is no easy task. It requires dedicated work and negotiations among players from around the globe. Sometimes it takes years before a common global standard can be agreed upon by all its members.

However, we can safely say that there are more benefits to a global standard than there are disadvantages. A case in point is the GSM standard from Europe that allows seamless roaming amongst GSM carriers worldwide, from all continents. A global standard system provides the following benefits.

To operators/carriers:
(1) traffic increase by seamless service capability beyond the border,
(2) traffic increase by global roaming (Terminal, SIM card, voice, data),
(3) number portability,
(4) reduce the equipment procurement cost owing to mass production,
(5) increase the chance to select best vendors around the world, and
(6) shorten the IOT (interoperability testing).

To equipment vendors:
(1) increase the chance to sell their product worldwide,
(2) open and standardized specification is more conducive for R&D,
(3) having the chance of international R&D collaboration,
(4) reduce device procurement cost owing to mass production, and
(5) shorten the IOT (interoperability testing).

To end-users:
(1) seamless service capability beyond the border,
(2) global roaming (voice, data),
(3) number portability,
(4) low-price equipment owing to mass production,
(5) worldwide contents access.

2.5   Operator perspective

During the 1G (analog) and 2G (GSM, TDMA, CDMA, PDC, DECT, PHS, etc.) technology time-period, operators could (and still can) establish their own strategy for
service discrimination with competitors. The reason was that they were free to select
their favorite systems and technologies. In many cases, the competition was limited
to national levels and no global competition was considered. So, the performance of
the respective technologies itself becomes a factor for service discrimination between
operators.

Where multiple standards are allowed for the 2G system (such as in the USA,
Hong Kong and Japan), some operators opt for TDMA (AT&T Wireless), GSM
(VoiceStream), CDMA (Sprint PCS) or PDC/PHS (NTT DoCoMo). A TDMA operator offers a line of TDMA terminals and so does a CDMA operator. So the technology
in a sense differentiates the operators in a given market and hence it is relatively easy
for end-customers to select their carrier. Often, this could be based just on handset
and coverage appeal. That situation will change in the era of global standard systems
like 3G. Europe has already been in similar situation, since GSM is a unique system
in Europe. In this situation, the types of terminal that multiple operators provide for
end-users are very similar or sometimes exactly the same except for the logo with the
operator's name and brand.

So, what will be the major factor of service differentiation between the operators?
To some extent, it will be pricing. Operators will try to increase the number of subscribers by reducing the tariff plan and terminal price. It may be effective during
the initial phase of competition, but we can recall a similar story of competition in
the airline industry in the 1980s. In 1981, American Airlines introduced the first
frequent-flier program in the world and then almost immediately all airline companies introduced their own programs. What happened finally? The rate for miles
increased twice and has even tripled. Finally, most airlines encountered disastrous
financial situations. A price war with the same service quality always results in this
negative feedback from the market.

So, what is the most important factor for operators to compete with? After the
initial phase, it clearly will not be the differentiation based on price; attractiveness of a
carrier will be based on services, content and applications that can be provided using
a global standard platform. Compelling content and applications is what is most
desirable by end-users, in both the consumer and enterprise markets. Quality and
easy-to-use content will clearly be a differentiating factor for the carriers. In the race
to 3G, BWCS (a technology consulting company) warns wireless operators not to
lose sight of the most important factor in all this – the consumer. Using the example
of WAP, BWCS says that consumers are not interested in the underlying technologies
but simply want attractive, useful services at the right price. The UK-based firm

strongly encourages operators not to focus on 3G data speeds and technology at the expense of the development of 3G content and applications, if they want to make a profit.

Now let us discuss the above points using some multinational big wireless carriers like Vodafone (UK), NTT DoCoMo (Japan), and Telefonica (Spain). These companies have similar business scales but their approach to becoming global is very different. The similarity in their business strategy is that these carriers pursue global markets to grow and expand. The differentiation is apparent in their investment strategy.

2.5.1   Vodafone

The UK-based Vodafone AirTouch Group (VF) is the largest carrier in the world. In *The Guinness Book of Records 2001*, the company was introduced as "the biggest cell phone company created on Feb.5, 2000 by a $ 171.1 Billion merger between British Vodafone and Germany Mannesmann."

The strategy of VF's global development and expansion is based on M&A. In 1999, VF made an acquisition of AirTouch Communications of the USA, which instantly doubled the market share of VF. Then, VF acquired Mannesmann of Germany, which has again doubled its global market share. In 2001, J-phone (one of the three 3G licensed operators in Japan) also came under the control of VF. J-phone is well known for its aggressive marketing and its number of subscribers reaches 12 million in Japan. The purpose of this acquisition of J-phone is to get the know-how of advanced mobile Internet experience and 3G services in Japan (see Section 3.3.3).

The number of companies (except VF in the UK) in which VF has invested more than 50% is 13, and the number of companies in which VF has invested less than 50% is 13. As a result, the total number of VF group subscribers had reached 112.5 million by the end of 2002. This figure is counted based on investment ratio. The key of VF's global strategy is the pursuit of scale of merit by majority investment.

2.5.2   NTT DoCoMo

NTT DoCoMo (DoCoMo means Do Communications by Mobile) was established in 1992 as the spin-off mobile business unit of giant NTT. At the time of the spin-off, the total number of subscribers to DoCoMo (also known as DCM) was only 80 000 and its population penetration rate was less than 1%. Eleven years later, at the end of February 2003, the total number of subscribers had reached 44.6 million in Japan alone. DoCoMo made the biggest sales numbers amongst carriers worldwide, coming in the vicinity of 4686 billion Yen, and their profit reached a massive 77 billion Yen (note that US $1 is takes to be the equivalent of 120 Yen).

The strength of DoCoMo is its powerful R&D arm. DoCoMo had inherited most of NTT's R&D power in wireless technology at the time of spin-off in 1992 and the

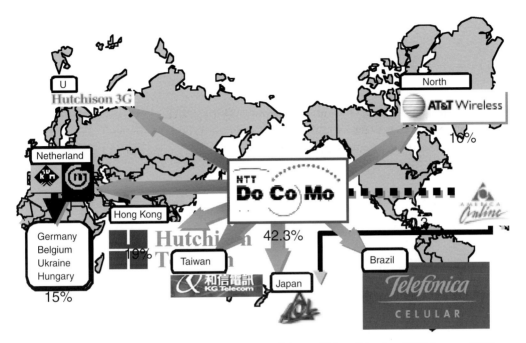

Figure 2.1. NTT DoCoMo's investment around the world (as of October 2002) (source: NTT DoCoMo).

strength of their R&D continues to increased as the sales increase post-1992. The budget of DoCoMo's R&D was 100 300 million yen in 2002 (1 US \$ = 120 Yen) and it includes more than 1000 pure research workers. DoCoMo owns three R&D centers, in YRR (Yokosuka Research Park) in Japan, Silicon Valley in the USA, and in Munich, Germany (for more details, please see www.docomousalab.com).

The strategy of DoCoMo's global development is based on minority investment (less than 20% of total stock) supported by powerful R&D and the technology coming out of their laboratories. DoCoMo has already selected and invested in some of the most prominent operators on each continent (Figure 2.1). They are AT&T Wireless (North America), Telefonica Cellular (South America), KPN Mobile (Europe), KG Telecom (Taiwan), Hutchison Telecom Cellular (Hong Kong) and Hutchison 3G UK (UK). KPN Mobile started its i-mode service in March 2002. It is the first operator outside Japan to introduce i-mode.

Six weeks after KPN Mobile launched i-mode in the Netherlands, 85% of all users said they were satisfied with the new mobile data service. A profile has emerged of the typical i-mode customer at the present time. The majority of i-mode users are aged between 20 and 35. The current i-mode customer base consists of 82% private users and 18% business users. The data bundles of 200 kB and 1 MB are the starting point for 72% of i-mode customers while 28% have opted for larger bundles in excess of 5 MB. Many i-mode customers currently subscribe to an average of three services.

Since the i-mode launch, 20 new content partners have come forward. The number of i-mode services is expected to increase rapidly. Content providers setting up their own distribution channels include the Albert Heijn supermarket chain (via XL store and a website) and Radio 538 (website).

KG Telecom (Taiwan) had followed KPN Mobile into i-mode by the middle of 2002. The key for DoCoMo's global strategy is the pursuit of an increase in corporate value by the transfer of advanced services and technology (i-mode, W-CDMA) by minority investment.

2.5.3   Telefonica Cellular

The Telefonica Cellular (TC) group is a Spanish telecommunications giant and it is well known as a strong promoter/supporter of GSM service. TC's investment strategy is unique compared with VF and DoCoMo. VF and DoCoMo's business strategy is to focus on opportunities across all continents and markets equally, but TC's main interest seems to be focused on Spanish-speaking Central/Latin American countries. In Central/South America, TF has invested in Argentina, Brazil, Chile, Peru, Guatemala, El Salvador and Puerto Rico. It is great expansion strategy since commonality in culture and language lowers the barrier between cross-cultural organizations.

In Europe, TC operates GSM/GPRS systems very successfully. As a result of this success, the total stock price of TC in Spain once reached almost one fifth of the total stock value index of Spain. However, TC has been obliged to operate non-GSM systems such as CDMA and TDMA in Central/Latin America countries. This is because the authorities in these countries did not allow GSM systems to be operated for a long time. This created headaches for TC's management in Madrid (where it has its headquarters) as they could not transfer their GSM expertise to Central and Latin American countries and were forced to operate alien systems. The good news for TC is that Brazil decided to open its market for GSM in 2001.

To expand their partnerships, in June 2002, Telefonica Cellular and NTT DoCoMo announced a strategic alliance to promote i-mode services. The decision will make TF reconsider its network architecture strategy in South America. Nevertheless, the key for TC's global strategy is investment focusing on common culture/language areas.

2.6   Equipment manufacturer perspective

As we have discussed before, globalization of the economy and technology has had a great impact on the wireless industry and that includes manufacturing. In this section, we will touch on some events and developments that have affected the wireless manufacturing industry.

Table 2.1. Suppliers of key cell phone components

Product	Key suppliers
Reference design	Texas Instruments, Intel, Motorola, Qualcomm, Ericsson, Microsoft
Operating software	Motorola, Nokia, Ericsson, Microsoft, Symbian
Flash memory	Intel, STMicroelectronics
Baseband chip	Texas Instruments, Intel, Qualcomm
Radio chip	Texas Instruments, Motorola, Qualcomm
Complete handset	Nokia, Motorola, Sony Ericsson, Samsung, NEC, Panasonic

2.6.1   Electronics manufacturing services

In the 1990s, outsourcing of the PC components such as DRAM was extremely critical for PC makers to cut down costs and remain competitive in the market. For some manufacturers, it was essential for their survival. In the late 1990s, the same situation was happening to the cellular terminal manufacturers. This tendency came from the fact that most cellular terminals can be made of a combination of standard chipsets and devices. Because of this development, it became very easy for cell-phone manufacturers to promote outsourcing of cellular-terminal manufacturing. Also, the above, coupled with recent depression of the IT industry and the cellular market, strongly forced the market to reduce prices and this has accelerated the introduction of electronics manufacturing services (EMS).

As an example, Ericsson and Motorola have selected a Singaporean company called Flextronics International as their EMS partner. It is clear that the amount of manufacturing outsourcing will continue to increase every year in the wireless industry, just as it increased in the PC industry. The resources of cellular-terminal manufacturers will be more focused on design and planning of their business strategy, marketing and product design. In early 2002, Microsoft and Intel, in an effort to replicate their success in the PC market, announced that they were working together to design a "template" for a high-end phone that they will license to cell-phone makers. The template includes software from Microsoft and chips from Intel. As indicated in Table 2.1, the handset manufacturers are increasingly buying portions of technology from other companies and even competitors.

This represents a significant shift in the cell-phone manufacturing industry as the domain once considered to be dominated by the leading manufacturers like Nokia and Motorola is being invaded by computer technology players like Microsoft, Intel and SUN.

2.6.2   Acceleration of joint development beyond the border

In order to establish a stranglehold in the 3G race, major players in wireless industries are trying to establish partnerships along various business lines that go beyond their own geographical borders. The areas of cooperation include standardization, joint research and development of software, content and terminals. The major benefit of the joint development is to make R&D more efficient and speedy and at the same time less risky. Now let us consider a couple of examples.

Case 1: software technology

NTT DoCoMo and Nokia announced the start of collaborative work on unification of communication middleware technology for 3G. This project aims to establish common specifications focused on three essential technical areas: browsing, messaging and the application operating platform for 3G services.

Also, Nokia initiated the Open Mobile Architecture initiative with six wireless carriers and twelve vendors. Qualcomm have proposed a software platform called BREW (Binary Runtime Environment for Wireless), which aims at building interactive applications and services.

Case 2: terminal development

The terminal business is one of the most competitive segments of the market because the life cycle of a cellular terminal is so short (6–9 months) that terminal vendors have to shorten the development cycle and at the same time reduce cost. Sony and Ericsson have established a joint company called Sony Ericsson Mobile Communications to build terminals jointly. Also, Panasonic and NEC have announced the joint development of software technology for 3G phones. Fujitsu and France-based Sagem have also announced the joint development of GSM/GPRS and UMTS dual-mode phones. There are a lot of similar examples.

2.6.3 Software crisis

In an era of global standard systems like 3G, it is difficult for wireless carriers to differentiate themselves from other carriers just by technology. So, in order to show the difference of applications and services to the market, wireless carriers need to develop their own services. The success of i-mode has proved the importance of interactive services and applications rather than data speed itself and this result indicates that software will be the key factor in the next 3G war.

In the voice-centric market, wireless carriers have long been obliged to compete only by tariffs, but in the mixed market of voice and data (2.5/3G), service, content and applications are the key to revenue generation and profitability. This trend will create a great demand for software development, not only for terminal vendors, but also for contents developers, wireless carriers and application developers.

Another major difference between 3G and 2G terminals is that the 3G phone must be equipped with two processor configurations whereas voice-centric 2G phones need only one processor. 3G phones need one processor for communication and the other non-real-time processor for applications. As a result, the total volume of software and the test items to check the overall software function for 3G becomes several times larger than 2G. In order to develop the software for 3G phones in a timely manner and with fewer bugs, the software for 3G must be modularized further.

2.7   Computer industry perspective

One of the fundamental transitions that started during the past decade is that of convergence of computing and the communications industry, not only in concept, but in reality. From hardware to terminal devices, from software to support back-end to wireless enablement of PDAs and wireless phones, this phenomenon lies at the center of the pervasive computing era. The convergence of the computing and communications industries promises exciting applications. Vendors from these growing sectors are working together on standards and technologies that blur the lines between desktop PCs and mobile communication devices.

The costs of memory, power and processing are falling rapidly. This cost reduction allowed yesterday's mobile devices to become powerful and useful extensions of the computing world. Processing power can be measured in millions of instructions per second (MIPS). The higher the number of MIPS, the faster the processor. The lower the MIPS-per-dollar cost, the more MIPS a vendor can provide. More MIPS means more complex applications can be made available. In addition, if we can take advantage of new technologies that allow better usage of MIPS, there will be an improvement in MIPS per dollar over time. The number of MIPS per dollar is expected to double in the very near future. With dropping costs and the miniaturization of electronics, it is becoming possible to provide enough processing power to handhelds and phones to allow them to perform like desktop computers.

Memory bandwidth is the key to complex applications and services. With more memory available for a dollar (a trend expected to continue into the next decade), handhelds and phones are able to handle multiple applications at the same time. For example, you can simultaneously have a three-way conference call and transfer email headers. Also, battery resources are scarce for mobile devices. Unlike desktops, mobile units have limited power resources. Complex applications drain battery resources quickly, much to the users' dismay. Improvements in power budgets will foster the development of exciting new applications.

As is evident from Figure 2.2, computing and communications devices are converging into devices that not only communicate but that can also be used for many complex computing applications such as address book, email, calendar, browsing, games, and so on. Devices such as Nokia's Communicator or Pocket PC phones are popular examples of such convergence. Each of these devices is a PDA and phone in

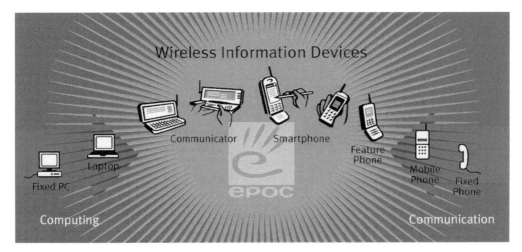

Figure 2.2. Convergence of computing and communications platforms (source: Symbian).

one unit. With time, their costs will come down and the form factor will improve to make them acceptable to mass markets.

The convergence of the two major industries has allowed the computing players to enter the communications industry and vice versa. Microsoft's stated vision for the future emphasizes wireless and other means of information access with equal importance. AOL has acquired a number of small wireless companies to prepare itself for the wireless applications battle. Standard application packages such as Oracle, MS SQL, Sybase, DB2 and Java are all available for a variety of handsets and phones. Similarly, carriers have been playing in the applications field either by partnering online content companies or by developing content in-house. This change is making the industries more competitive, forcing the rate of innovation to go up a notch.

In the chaos discussed above, there is no dominant player such as Microsoft (in desktop software) and Boeing (airplanes) to date. In other words, competition and industry is wide open.

In this chapter, we discussed the evolution in wireless Internet technologies, applications, and services from the point of view of operators, equipment manufacturers, and the computing industry. We saw how the convergence is disrupting the value chain and established paradigms.

In this chapter, we also looked at the impact of globalization on the wireless industry and vice versa. In later chapters, this will become even clearer as we talk about business models and the wireless value chain (Chapter 5), and discuss case studies and present perspectives from various executives representing different segments of the value chain (Chapters 11 and 12). Some of the issues become more transparent when we come to discuss the issues and challenges facing the wireless industry in

Chapters 9 and 10. In Chapter 3 we will look at the adoption trends across various geographies around the world. We will take a look at the USA, South America, Japan, South Korea, China, Hong Kong and Europe. In addition we will analyze the effect of wireless data on different business sectors such as the telecommunications industry, equipment manufacturers and the computing industry.

3 Adoption trends and analysis by region

As wireless technology and services became more pervasive around the world over the course of the 1990s, one thing immediately became clear: the uniqueness of the major markets around the world. SMS in Europe, i-mode in Japan and Blackberry in North America all point to the diversity of consumer acceptance. Likewise the problems facing these markets are unique as well. While Scandinavian and Japanese carriers worry about market saturation, the USA wrestles with its spectrum allocation confusion and intense competition. While European carriers worry about getting out of their enormous debts due to auctions, carriers in South America worry about the technology evolution of their TDMA networks. In this chapter we will cover the salient features of the major global wireless markets – the USA, China, Japan, South Korea, South America and Europe.

Figure 3.1 and Table 3.1 show the penetration of wireless telephones and wireless data, respectively, across different regions of the world.

3.1 USA

During the past few years, the US wireless market has continued its substantial growth: from about 86 million subscribers at the end of 1999, the total number of subscribers had increased to 137 million by February 2003. Carriers are all upgrading their networks to gain more capacity and efficiency for their voice networks as well as to introduce high-bandwidth data capabilities. The year 2001 saw experimentation with wireless applications and the development of various relationships amongst the players in the wireless value chain. It is becoming more and more evident that carriers are finally accepting the need to partner with application developers with unique value propositions and offerings.

3.1.1 Market analysis

Presently, the USA is the most competitive and lowest margin market in the world with six national wireless players. Wireless penetration rate is relatively low to date

Table 3.1. Penetration of wireless data users by region

	2000	2003	2005
Western Europe	23%	72%	91%
Japan	21%	62%	90%
United States	7%	44%	83%
Asia-Pacific	1%	4%	8%
Rest of the world	0%	4%	10%
Total	3%	12%	20%

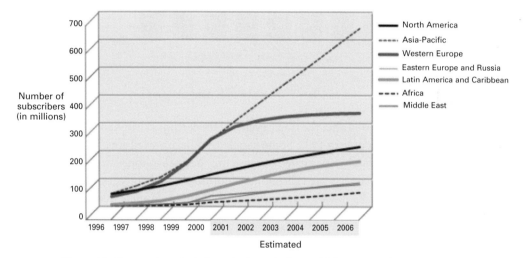

Figure 3.1. Penetration of wireless telephone in the world (source: Yankee Group).

compared with several European and Asian countries. This makes the US wireless market a very promising one. Below we discuss various issues and challenges of the US market.

3.1.2   Potential market

The USA has relatively low penetration rates in voice service and very low penetration in wireless data service (Figure 3.2). This shows great potential for the US market, especially when compared with saturated markets like Northern Europe and Japan.

3.1.3   The need for wireless devices to improve security

The tragic events of 9/11 in the USA reminded everyone of the importance of wireless devices to contact their loved ones or emergency personnel in case of emergency. This is especially true for children and teenagers. It is to be noted that the cellular

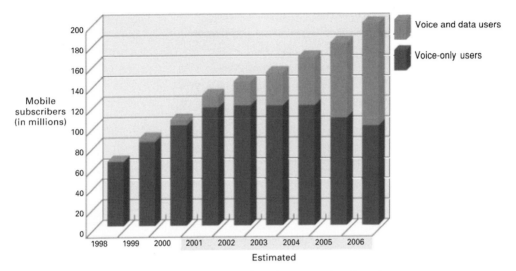

Figure 3.2. Forecast of US mobile subscribers (source: Yankee Group).

boom in Japan started after the tragic events of the earthquake in Kansai, which killed more than 5000 people in 1998. Some people who were buried alive under shattered houses and buildings were rescued because they had cell-phones with them. People in Japan recognized that the cell-phone is not only a tool for communication but also an indispensable lifeline for survival in calamities. Similar thinking has taken place in the USA after 9/11.

3.1.4   Integration of Internet and wireless

The USA is the highest penetrated country in terms of the Internet. The consumers are mature and are "Internet-savvy". With the integration of Internet and wireless communications, consumers who are used to email, address books, IM, and other content applications can easily grasp similar concepts in a wireless context. The USA is a car-centric society, so even on the road, people want access to their email or situation-specific content. This phenomenon is akin to the situation in Japan where people use their web-phones while waiting for subway trains and banking machines. So, people stuck in traffic in the USA look to access their email while they are in the car via wireless and voice means.

3.1.5   Technology roadmap

Some US operators such as AT&T Wireless have already started or clearly announced the introduction of a global standard system such as GSM/GPRS/EDGE and 3G (WCDMA). This is a great decision for consumers around the world, especially in the USA. Owing to this strategy, most of the end-users will be able to enjoy bilateral roaming services and price reduction thanks to mass production of equipment in the near future.

3.1.6   New applications are emerging

Strong US industry sectors of finance, IT, computers, contents and car electronics can create new niche services by combination with wireless infrastructure. Telematics is one of the most promising areas.

3.1.7   Access to capital remains difficult

In order to build next-generation networks, the six wireless carriers need to spend billions of dollars to upgrade and maintain their networks. As such, they require significant amounts of capital to fund their growth and new networks, but market conditions remain very challenging for both debt and equity issues.

3.1.8   Intense competition of standards and carriers

US operators currently use four non-compatible 2G technologies – CDMA, TDMA, GSM and iDEN. This creates a lot of disadvantages and service interoperability problems relative to European operators who have GSM as their primary 2G technology. The situation will improve as networks converge with 3G when only two major technologies will prevail – WCDMA and CDMA2000.

3.1.9   Limited spectrum

The current spectrum-cap rules in the USA have a negative impact on the sector. They do not allow consolidation to happen and make it more difficult for operators to deploy advanced data services. However, the FCC is considering relaxing some of these rules.

3.1.10   NextWave debacle

In addition to market dynamics and consolidation pressures, the USA is embroiled in the spectrum debate as well. The complex but important issue goes back to the wireless spectrum licenses held by the FCC (Federal Communications Commission) and auctioned from 1998 to 2001. In 1998, NextWave Telecom placed the winning bid ($5 billion) for a license to develop portions of the wireless spectrum that had been unavailable to commercial entities.

When NextWave declared bankruptcy in 1998, Verizon Wireless and several other carriers stepped in late in 2001 with bids of about $16 billion for the licenses NextWave had defaulted on its required down payments. The FCC then collected billions of dollars in down payments from the new auction winners. With the FCC and NextWave embroiled in litigation, Verizon asked the FCC to return its down payment, since the spectrum space had not been released to the winning bidders. The FCC returned 85% of the down payments ($2.8 billion) but did not agree to release Verizon from its obligations under the contract signed by the winning bidders. After the second round of auctions, the operators hoped to use the spectrum to rollout 3G voice and data services in markets like New York, Los Angeles, and San Francisco. However, the US Court of Appeals for the District of Columbia ruled in June 2002 that revoking the licenses violated bankruptcy law, and NextWave got them back. Later, the FCC asked the Supreme Court to hear an appeal against that ruling.

In January 2003, the Supreme Court ruled 8–1 that NextWave Telecom Inc. could keep the licenses it bought at auction from the FCC. NextWave is aggressively looking to sell the bandwidth to larger carriers in need of spectrum for third-generation voice and data services.

The decision draws to a close the sorry seven-year saga of the NextWave 3G spectrum licenses.

The latest Supreme Court ruling means that NextWave will soon be able to sell these licenses back to the major carriers that bid for them in the second round of auctions. However, before it can do that, the company will probably have to build out a little network first. This is because the FCC has rules against any company buying up and selling spectrum for pure speculative gain.

Plans for this are already in hand: Lucent Technologies Inc. is slated to build a high-wireless data network for NextWave. The undertaking is expected to cost $400 million. Some commentators say that once NextWave and Lucent have built the network, that should be enough to satisfy the FCC rules, leaving the company in the clear to resell its spectrum licenses to other carriers.

3.1.11 Enterprises in the USA are more amenable to wireless data applications

Unlike those in Japan or Europe, enterprises in the USA have been willing to experiment with wireless data and adopt the technology in comparison with the consumer market. While being focused on ROI, enterprises are willing to invest in projects that can improve their bottom line, increase efficiency of the work force, spread brand awareness and give customer loyalty a boost.

US corporations quite routinely allocate budgets for wireless data projects. PDA-based applications and services will continue to be more prevalent in the USA than in any other wireless market.

Additionally, corporate IT industry spending on wireless applications are expected to have reached an average of $680 000 in 2002, a dramatic 94% rise over $360 000 in 2001, according to a report by the World Information Technology and Services Alliance (WITSA) and the Wireless IT Research Group (WIRG) released at the 2002 World Congress on IT in Adelaide, Australia. The *Global Wireless IT Benchmark Report* – 2002 says that 51% of firms report increasing their wireless budgets in 2002, while only 8% will decrease their budgets, and 41% of wireless budgets will remain unchanged.

3.1.12   The USA is moving ahead of Europe in wireless

For most of the 1990s, Europe had a commanding lead over the USA in the wireless industry. But because of the 3G spectrum auctions in which the European carriers invested billions of dollars on licenses, they are now finding it difficult to spend resources on new applications and services. Additionally, the launch of WAP in Europe has not met expecatations. In the meantime, the lagging US industry has been infiltrating the heart of the wireless market. The latest attempt to turn mobile phones into pocket computers that can download stock prices, music and video is based for the most part on US technology. In a *Wall Street Journal* analysis piece titled "Europe Had Decisive Wireless Lead, But Lost It to U.S. With Poor Moves", David Pringle argued that the US wireless industry is primed to take the lead from Europe. This is supported by the concentration of wireless investments in US vs Europe. According to Rutberg & Co., a San Francisco research firm, during the period 2001–2002, venture capitalists have pumped $12.1 billion (13.8 billion euros) into North American wireless startups and only $4 billion into their European counterparts.

3.2   South America

3.2.1   Market analysis

Although the South American market is huge (Figure 3.3), here we will focus on Brazil as a case study, since Brazil is the biggest existing and potential telecommunications market in South America. Also, the policy and technology selection of the country will greatly impact on other South American countries. Like the situation in other countries in the world, the wireless telecom market of Brazil was controlled and operated for a long time by the government (called TELEBRAS) until mid-1998. In 1998, ANATEL (Brazilian MPT) decided to open its telecom market and introduce a competition policy, which produced eight fixed operators, two long distance operators (Embratel, Intelig) and 22 cellular operators (20 CDMA and TDMA operators and two GSM operators). Using this opportunity, major global wireless players such as Telefonica, TIM, Portugal Telecom(PT), BellSouth and NTT DoCoMo rushed into the market.

In 2000, ANATEL held a public open hearing and made an important decision to accept GSM/GPRS technology. As is the case in other South American countries,

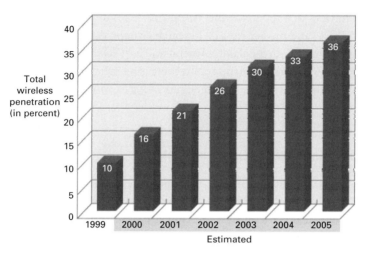

Figure 3.3. Latin American subscriber forecast (source: Yankee Group).

Brazil has been operating US-based technologies: AMPS for 1G and TDMA (IS-136: Digital AMPS) and CDMA (IS-95) for 2G, since most of the technical standards of the country have been based on US standards for years. For this reason, in the two major cities of São Paulo and Rio de Janeiro, European Giant Telefonica and PT were obliged to operate CDMA networks in spite of their expertise in GSM operation in Europe. The recent decision by ANATEL allowed the country to be a member of the GSM association. This may accelerate the introduction of 3G (WCDMA) in the near future.

3.2.2   Potential market

Brazil is an economic giant and accounts for over 50.6% of the total population in South America (2001). In the 1990s, Brazil had steady economic growth with an average annual GDP increase of 7%. As a result of the economic development and liberalization policy on telecoms, the wireless service is exploding, especially in big cities like São Paulo and Rio de Janeiro. In these two major cities, the wireless penetration was 25.2% and 26.9%, respectively, by the end of 2001. The total number of cellular subscribers reached 27.9 million and the subscriber increase rate in the year 2001 was second in the world after China.

3.2.3   Expansion of the pre-paid service

The pre-paid service is a good entrance for cellular beginners, especially the young generation, since it requires no authentication and no monthly payment. In Brazil, the market share of the pre-paid service reached 70% in 2001. However, the introduction of the pre-paid service has its pros and cons. It is estimated that the ARPU of a pre-paid user is $\frac{1}{3}$ to $\frac{1}{4}$ compared with that of a post-paid user. As the network resources needed for the pre-paid service are almost equal to or greater than those of the post-paid service, too much dependence on the pre-paid service makes the profitability of the operating company degrade. Some operators intentionally refrain from expanding the pre-paid service to keep the ARPU high in Asia and the USA. On the other hand, the pre-paid service is very popular in Europe, just like in Brazil.

3.2.4 Wireless mobile banking

Brazil is one of the most advanced countries in the world in the area of Internet banking. The reason for this advance is the continuous inflation in the 1970s and 1980s in the country. Brazilian banks have to up-date exchange and account information constantly owing to inflation. This environment makes Brazilian banks very competitive with the help of IT. Banco Bradesco is the largest Brazilian bank with 20 million customers, and is one of the most advanced banks in the world in introducing new IT technologies in its operation. The book by Bill Gates titled *Business @ the Speed of Thought* explains in detail the intense effort of Banco Bradesco (www.brandesco.com.br). Because of this, the Mobile banking service of Banco Bradesco has been very popular since 1999.

3.3 Japan

3.3.1 Market analysis

In Japan, the boom of the cellular service has come relatively late compared with other high-penetration region's such as North Europe and Hong Kong. At the end of 1994, the total number of cellular subscribers was only 4.3 million, which is penetration rate of less than 4%. Eight years later, in 2002, the total number of cellular subscribers had approached 70 million, which is a penetration rate of almost 70% (Figure 3.4). This tremendous success of the cellular business has impacted in various ways on the business and social scenes in Japan. Paging services reached their peak at 1995 with 11 million users and then declined to almost zero in 2002. The largest Internet service provider had long been "Nifty" until the service launch of i-mode by DoCoMo, and now DoCoMo is an undisputed leader in both cellular and Internet access services. The NTT fixed telephone service reached its peak at 1996 with 61 million users and since then the number is decreasing by more than one million users every year.

The impact on social life seems to be larger than the above statistics. First, a subway train carriage has become more calm and silent compared with five years ago since people (young and old) tend to watch their cell-phone display rather than talk on a

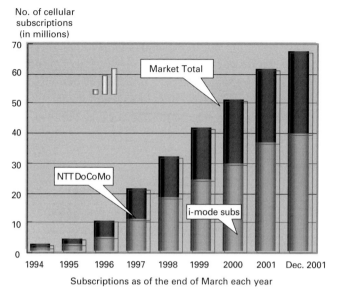

Figure 3.4. Japanese wireless market growth (source: NTT DoCoMo).

phone. They are emailing, checking websites and maybe reading a downloaded novel on the display.

It is to be noted that the first Internet phone in Japan was not i-mode, but PHS. PHS can transmit circuit-switched data with 32/64 kbps speed and it launched its Internet access service one year prior to the launch of i-mode with 9.6 kbps. However, owing to its limited coverage area, limited content and applications, and also because of its circuit-switch-based tariff system, the Internet access service operated by PHS could not create enough enthusiasm in spite of the 32/64 kbps data speed. This result gives us a very important insight. The data speed itself does not attract the market; the key factor is the total attractiveness of the service including coverage area, "cool" devices, economical tariff and of course charming contents, along with business models.

3.3.2   Device development

Japan has a long tradition of miniaturization. For more than 1000 years, noble families have enjoyed writing a Japanese character on the surface of a grain of rice. Walkman by Sony was a great success in the 1980s all over the world. Also in the 1980s, the Japanese compact car opened up the US market to Japanese car makers, who are now selling many large-scale cars like Acura and Lexus. The latest high-tech example is a Digi Q by the toy maker Takara. Digi Q is a strategic product of Takara with a volume of $50 \times 27 \times 32$ mm^3. Within this ultra-compact body, Digi Q is equipped with a receiving sensor of infrared light called MICRO IR, an 8-bit microcomputer and a super small motor. These are just some examples.

We can also find this kind of enthusiasm for miniaturization in wireless devices like i-mode or the FOMA (DoCoMo WCDMA) phone in Japan. The latest FOMA phone (P2101V) has the following technical features:

- phone numbers; max. 500,
- email address numbers; max. 1500,
- screen memos; 24,
- color LCD; 260 000 colors,
- video storage; ASF format, 800 kB capacity,
- still storage; JPEG format, 300 kB capacity,
- ringing melodies; 16 chords, and
- built-in digital camera; 110 000 pixels and 200% zoom, etc.

The specification may soon be upgraded by a next-version terminal. The rapid progress of wireless terminals is one key factor to attract the market, especially among the younger generation, because "cool" terminals are visible and are a very physical thing.

3.3.3   Wireless kingdom

TCA (Telecommunication Carriers Association of Japan) announced in February 2002 that the total number of Cellular/PHS subscribers in Japan had reached 68 million. These subscribers were distributed by carriers as follows: NTT DoCoMo 59% (40 million), KDDI 23% (16 million) and J-Phone 18% (12 million). If we look at subscribers with Internet-ready phones, i-mode users with NTT DoCoMo account for 31 million, EZ-web users at KDDI are around 9.3 million, and J-Sky users at J-Phone are around 9.7 million, taking the total aggregate to over 50 million subscribers.

The above figures show that 74% of all cellular/PHS subscribers have Internet-ready phones. Given this trend and the popularity of Internet-enabled phones, all phones from Japanese operators are Internet-enabled without exception, thus making Japan the top market for wireless data and wireless Internet applications and services. Figure 3.5 shows the growing importance of wireless data in Japan.

With regards to the 3G service and applications, Japan remains the frontrunner. The number of subscribers to FOMA (DoCoMo WCDMA), which started as a commercial service in June 2001, reached 114 500 in February 2002. This adoption trend is promising as FOMA was initially rolled out with limited coverage in the Tokyo metropolitan areas. DoCoMo extended the 3G service areas to major cities by April 2002 and intended this to go national by 2003. KDDI started a 3G service (CDMA2000) in April 2002 with maximum data-rates of 144 kbps and had 1.15 million subscribers by June 2002. KDDI announced five new cdma-2000 terminals, which included a GPS embedded-camera built-in terminal. Not to be left behind, J-phone started a 3G service (WCDMA) in mid-2002. The main reason for KDDI to be able to have strong growth with their 3G launch was that CDMA networks are

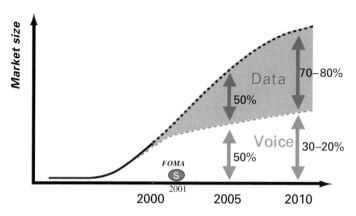

Figure 3.5. The growing importance of wireless data in Japan (source: NTT DoCoMo).

easy and less expensive to upgrade as compared with WCDMA and as such many more networks can be upgraded in parallel.

As seen above, wireless business competition in Japan has entered the second stage. In the first stage, in the 1990s, the competition was determined by new service, new technology, and high network quality. But in the second stage, from 2001 onwards, the competition will be determined by branding and marketing strategy since the traditional mobile market is almost approaching saturation. This is especially true in regards to marketing to create new user segments such as seniors (60 years old and more), young teenagers (below 15 years old) and housewives. Considering these trends, Japanese cellular operators are shifting their management resources into the marketing side to win the *endless war*. These trends will be followed by many other countries in the near future.

3.4   South Korea

3.4.1   Market analysis

South Korea has a national policy to develop and promote CDMA technology (by both operators and manufacturers). In the late 1990s, the Korean Ministry of Information and Communication decided to give three cellular and three PCS nationwide licenses when the basic study of 3G technology was just starting. Since that decision, South Korean manufactures such as Samsung and LG have put in tremendous efforts to develop the US-based CDMA technologies and finally South Korea has become a frontrunner in deploying CDMA technology.

As a result of these efforts, South Korea had over 30 million cellular subscribers (which was a penetration rate of over 63%) by mid-2002 (source: Korea MC). Dataquest reports that in the global CDMA cell-phone market of 0.17 billion devices in 2001, the total market share of South Korean manufactures reached 57%. This result implies that the national policy of focusing on CDMA technology has been successful. The mobile Internet access service in South Korea is based on WAP technology and the number of WAP terminals in South Korea has reached approximately 9.2 million, which is a penetration rate of 20%. However, the number of applications and services is still limited and the mobile Internet service in South Korea is still immature.

The World Cup in 2002 (soccer) was a great motivation for South Korea, as well as for Japan, to show their advanced telecom and IT infrastructure to the world. South Korean cellular operators are aggressive in providing global roaming services with other countries. Recently, GSMA announced the GSM/CDMA roaming service plan between French GSM operator Bouygues Telecom and KTF (Korean Telecom Freetel) using a SIM device.

3.4.2   3G service deployment

In spite of the large scale success of 2/2.5G CDMA technology, the future of 3G in South Korea is not clear. The main reason for the confusion comes from the licensing process of 3G by the South Korean MPT. In the MPT's 3G guideline in July 2000, the number of open seats for 3G operators was three: two for WCDMA and one for CDMA2000. After all three potential consortiums announced their intention to employ WCDMA, the negotiation initiated by MPT has started to change one consortium from WCDMA to CDMA2000. The main reason for this movement comes from the worry about the decline of overseas exports of South Korean-made CDMA products. In October 2000, it was announced that the 3G licenses would be given, one for WCDMA, one for CDMA2000 and one for any technology. As discussed here, the situation on 3G in South Korea is not stable presently, but it is believed that South Korea will become one of the most advanced countries in the 3G technology development race.

3.5   China, Taiwan and Hong Kong

3.5.1   Market analysis

Chinese people love to talk and they love to talk on cell-phones. For a long time, a cell-phone was not only a tool for communication, but also a status symbol.

But, the situation has changed dramatically. Cell-phone penetration in Taiwan and Hong Kong is among the highest ranking in the world and a cell-phone is no longer merely a tool to show status. It is a lifeline for them. Also, the increase of cellular subscribers in the People's Republic of China is overwhelming

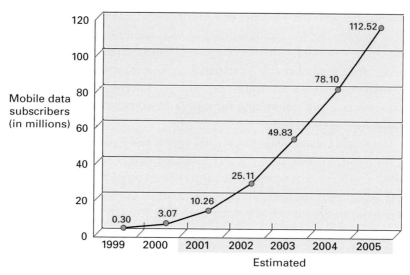

Figure 3.6. Subscriber forecast for China (source: Yankee Group).

(Figure 3.6). The increase in 2000 was over 10 million and total subscriber numbers reached 83 million at the end of 2000. However, it is only a 6% penetration rate. We can imagine how big the potential market in China is! By the end of 2002, China had over 144 million subscribers, thus moving it to the number one spot in the world in terms of number of subscribers. One of the factors of this rapid cellular increase is the delay in the deployment of a fixed network infrastructure. People in China prefer to use a cell-phone rather than waiting a long time for a fixed phone. GSM is the main technology in China, but CDMA is also in operation. With China's entry into the WTO in 2000, the presence of the global wireless industry in China is imminent. Some wireless manufacturing giants such as Motorola, Nokia and Ericsson have already made huge investments in China. With regards to 3G, China has three technical standards: WCDMA, CDMA2000 and Chinese-standard TD-SCDMA.

3.5.2   Showcase of wireless service: Hong Kong

Hong Kong has been a very important city for more than two centuries as it is located at a strategic position in Asia. After being returned to the People's Republic of China in 1997, its strategic position remains unchanged, especially in the telecommunications business, in spite of its small area (about 600 km^2). Telecom operators and manufactures consider the market as a gateway to mainland China. In Hong Kong, which is one of the most competitive markets in the world, there are multiple systems in operation for 2G (GSM, cdmaONE) and 3G licenses have been given to 3G operators.

3.6   Europe

Europe is the world's biggest wireless market. Some in the industry might even argue that GSM served as a unifying factor for the wireless industry in Europe. While the carriers in the USA were drawing battle-lines with TDMA vs CDMA rivalry during most of the 1990s, Europe was busy rolling out GSM infrastructure with excellent coverage and quality of voice service. As such, Europe has always been considered a breeding ground for the future of wireless thinking, technology, applications and services. Most of the major players in the wireless industry have a presence in Europe. The world's largest carrier – Vodafone – is based in Europe, and so are some of the major players in the value chain: Ericsson, Nokia, Logica, Seville, and many others. GSM's success has been so prominent that it is the most widely spread wireless technology standard in the world; from the Americas to Asia, from Africa to Europe, it dominates the rest of the technology standards deployed in the world today. One of the key elements of its success has also been the popularity of SMS (short message service). SMS is one application that is universal in Europe, irrespective of the carriers. Essentially a two-way paging system, it has been built in to almost all GSM phones since 1995, but its use had increased ten-fold by 2000. Some of the largest carriers are handling up to 8000 messages per second. The SMS explosion is largely the result of European operators working together to link their networks. As of 2001, there were over 200 billion SMS messages being sent every month. That is just phenomenal. Subscriber penetration in some markets, particularly Scandinavian, stands close to 70%.

Also, although WAP has largely been a failure, owing to misguided marketing and exaggeration of wireless browsing, the first commercial WAP services started in Europe. In the deployment of next-generation networks (2.5/3G), Europe is again leading the deployments in terms of consistent common standards for widespread interoperability and convenience to consumers.

3.6.1   CPP and Pre-pay

Calling Party Pays (CPP) is how European operators have handled their billing (unlike in the USA where you as a consumer pay, no matter what)[1]. Because of this, consumers leave their phones on all the time. This is in contrast to US mobile customers, where more than 30% of people do not leave their phones on, according to the Yankee Group. CPP has allowed European operators to profit at the expense of their wireline rivals, which are charged excessive interconnect fees for relaying calls to mobile phones.

---

[1] The major exception to CPP is calls made to mobile customers who are roaming abroad, in which case the recipient has to pay for the international portion of all calls.

Because most mobile users ignore the costs other people incur to call them, dialing a mobile from a landline can cost five times as such as the other way around. Similar to voice, the sender pays for the data transmission of SMS messages.

Europe also led the world in experimenting with pre-pay, a system whereby users avoid signing a service contract by buying airtime vouchers in advance. Both payment schemes have helped to drive the European wireless market. Most SMS users do not actually receive a phone bill; instead, they use vouchers to pay for their calls and messages in advance. According to the Strategies Group, pre-pay has accounted for over 75% of new mobile phone subscriptions in Europe as a whole, and over 90% in Portugal, where the system first started. However, as we have discussed in relation to South America, too much dependence on a pre-pay service may threaten the business model of cellular operators in Europe.

3.6.2   Wireless data market

According to a recent study by JP Morgan and Andersen, the European market for wireless data will be worth over $82 billion by 2010. Helped by a common standard and a desire to be the leader in wireless technology, European policy makers and value chain players are working hard to keep the GSM momentum and seize the opportunity once again. European carriers continue to acquire more assets outside Europe; for example, Vodafone's alliance in the USA, and Japan and Deutsche Telekom's purchase of VoiceStream in the USA.

Table 3.2 shows the estimated penetration figures for Europe and Figure 3.7 shows the forecast of revenues for all applications.

Table 3.2. Estimated penetration figures (in percentages)

Country	2000	2002	2005
Denmark	43	71	80
Finland	72	79	83
France	40	59	76
Germany	38	59	76
Italy	60	77	84
Netherlands	52	73	83
Norway	64	73	80
Spain	48	67	81
Sweden	62	74	82
UK	50	68	81
Scandanavia	64	74	81
Portugal	52	67	79
Belgium	38	63	79
Greece	44	67	81

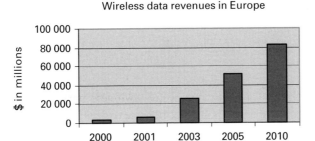

Wireless data revenues in Europe

Figure 3.7. Revenues for all applications in Europe (source: Andersen and JP Morgan).

3.6.3 Nordic markets lead the charge

The most advanced markets in Europe for wireless data services will continue to be the Nordic markets, reflecting the advanced levels of wireless penetration, Internet penetration, and the income distribution within these markets. The UK market also offers substantial upside potential, with business data applications generating a relatively high proportion of total revenues. The UK currently has the highest proportion of B2B-enabled firms in Europe and hence has the potential to become a major wireless data market. Germany remains the single most important market in Europe based on the projected revenue opportunity, with France and the UK the next most important in the European market.

3.7 Australia

Through a combination of strong government support, an innovative and highly capable workforce, and strong R&D centers, Australia is an ideal location for wireless business, as many multinationals that have located in Australia can confirm. Multinationals like Alcatel, Cisco, Fujitsu, Nortel and Motorola have all based operations there to take advantage of Australia's competitive advantage. And with over 50 telecommunication carriers, many of whom are undertaking R&D with Australian universities and research organizations, it comes as no surprise that Australia is a recognized leader in wireless technology.

In 2001, the total number of subscribers reached 12 million across all networks with penetration levels of 58.8%. Notwithstanding these figures, competition slowed down in 2001, and the business strategy shifted from customer acquisition to customer selection and retention. One.Tel collapsed, and none of the major operators took advantage of Mobile Number Portability (MNP) to increase their market shares. This simply maintained the stagnant triopoly of Telstra, Optus and Vodafone, although even Vodafone reduced its competitive moves, effectively leaving the Telstra–Optus duopoly that was reflected in price increases soon after MNP was introduced. Virgin was the most innovative operator in terms of prices and customer service, while Hutchison has continued to try and acquire new young customers to its CDMA service without much success, and also pursues its ambitious 3G strategy.

3.8   Africa

According to the UN report released in October 2002, more cell phones have been activated in Africa in the past five years than fixed-line phones over the past century. The UN study said that of Africa's 2001 population of 816 million, one in 35 had a cell-phone, and one in 40 had a fixed-line phone. Internet use in Africa is increasing, although it still lags far behind Europe and North America. On average, one of every two people in North America and Europe use the Internet. One in 160 Africans used the Internet in 2001. With South Africa and northern Africa excluded, about one in 250 used the Internet in 2001. "With innovations such as wireless fidelity – commonly known as Wi-Fi – and other low-cost technologies and business models that are now being explored, we should aim to provide cheap, fast and eventually free access to the Internet," said UN Secretary General Kofi Annan.

In Africa, wars, years of corruption, and frustration with fixed-line systems is driving the growth in handsets and cellular networks across the continent, with South Africa leading the charge. In mid-2001, Nigeria – Africa's most populated nation at 125 million – possessed one of the lowest ratios of phones to citizens in the world: one line per 300 residents. In August 2001, things changed when two operators launched Nigeria's first nationwide wireless network. There was such pent-up demand that the number of subscribers exploded beyond all predictions in just a few short months. By the end of 2001, 450 000 subscribers had signed on for cellular services. By late spring 2002, the two operators had in excess of 900 000 subscribers each. Such explosive growth is revolutionizing how Nigerians communicate, and offers prospects for a better future where few prospects existed before.

Today, Nigeria boasts one of the five largest cellular networks in sub-Saharan Africa, a region that has lagged behind Asia, Europe and North America in wireless development. The other top five countries include South Africa and Kenya, with hundreds of thousands of new customers signing up for new cellular services (see Table 3.3).

The region is starting to catch up – and catch up rapidly – with the rest of the world in terms of market penetration for the cellular service. Having started nearly from scratch a few years ago, the region boasts more than 30 million wireless customers. That is no mean feat in a region beset by a wide range of economic, political and social challenges from abject poverty to government corruption. Intelecon Research and Consulting says the annual rate of growth for the entire African continent has been more than 82%, much faster than the 33% growth rate in the Americas. In

Table 3.3. Top sub-Saharan Africa mobile
markets (as of December 2001)

Country	Subscribers (in thousands)
South Africa	8888
Cote-d'lvoire	693
Kenya	610
Senegal	420
Nigeria	373
Uganda	316
Zimbabwe	314
Tanzania	309

Source: Pyramid Research.

many countries such as Cameroon, Kenya, Senegal and Tanzania, the annual cellular growth rates are running in excess of 300%.

Like their counterparts in Asia and Europe, African cellular operators use the "calling party pays" system for billing purposes. Customers pay only when making a call, and pay nothing when receiving one. The pre-paid system enables operators to by-pass costly, but inefficient, billing and payment systems – systems that would not work in national economies where transactions are conducted in cash. Another technology driving growth is short messaging services, or SMS. Exchanging messages by text rather than voice has proven to be a much more effective – and cheaper – means of communicating. Despite the steep growth curve of cellular networks, Africa still lags behind other regions of the world. Growth is largely confined to cities, while rural areas remain unconnected, but that, too, is changing rapidly.

3.9   The growth of WLAN

A fundamental challenge facing information technology (IT) decision makers is identifying and implementing the architectures, technologies and processes that reduce the total cost of ownership (TCO) of corporate networks – one way is the wireless LAN. Based on the results of this study, the present authors believe that if companies understand the true costs and economic benefits of wireless LAN solutions they will generate quicker returns on investments.

Currently, wireless LAN is not a replacement for the wired infrastructure, but it is a significant complement to what presently exists. Schools, manufacturing companies, hospitals and offices purchase these wireless LAN systems for two predominant reasons: they are seeking an increase both in user productivity and in IT team productivity.

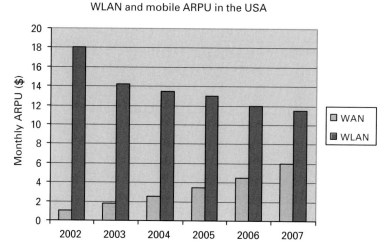

Figure 3.8. WLAN ARPU in the USA (source: Yankee Group).

Companies that integrate or upgrade with these systems will stretch IT resources further and be able to reap economic and business benefits sooner than companies taking a wait-and-see approach – especially now that a solid standard exists for wireless local area networking. As a result of this standard and of higher performing or higher speed systems, tomorrow's enterprises will be likely to have a mix of wireless and wired LAN systems. WLANs are also being adopted for home usage. As shown in Figure 3.8, it is forecast that the growth in WLAN usage will continue to increase for the foreseeable future.

As far as the growth in different geographical regions is concerned, regional differences are bound to stay; for example, mobile operators will have a larger role in Asia and Europe while independent providers with roaming agreements will survive in the USA. Consumer-dominated access in Europe and Asia will result in a higher emphasis on metered access while higher business use in the USA will lead to a predominance of flat-fee pricing.

In Europe and Asia, the density of population and cellular penetration is high; there is great reliance on public transportation, and the wireless markets are more consumer oriented. Hence, there will be a higher density of hotspots, and WLAN access serves as an extension to WWAN access. As such, mobile operators like Telia, Sonera, DoCoMo, Telefonica, Telenor, KDDI and BT are very active in this area. In Asia, some independent providers like MIS in Japan and Korea have started to appear, and free access is available at several airports and other hotspot locations.

In contrast, the USA has a higher penetration of laptop computers and PDAs, the highest Internet penetration, the highest 802.11 installations, and more advanced enterprise wireless data applications. So there is a larger demand for wireless data applications for business users and WLAN access will act as a substitute for fixed LAN access in some scenarios.

In this chapter, we have looked at the market, subscriber base, and technology evolution, the salient features of the markets of North America, South America, Europe and Asia. As is obvious from the discussion, each region has its own characteristics, its own challenges and potential for further growth, but as the next-generation wireless technology evolves, applications and services will become more global in nature and major carriers are already pursuing aggressive strategies to expand beyond their native borders. One does not have to live in Japan to enjoy i-mode; with DoCoMo's investments around the globe, i-mode technology is coming to more and more handsets outside Japan.

Here, we summarize the commonality and differences of the US, European and Japanese markets.

Systems and standards
Europe	United System (GSM)
USA and Japan	Multiple Systems (TDMA, CDMA, GSM, PDC, iDEN)

Digitization
Europe and Japan	Mostly completed (Europe), 100% in Japan
USA	AMPS remains

Billing model
Europe and Japan	Calling party pays (CPP)
USA	Non-CPP

Migration path to 3G
Europe	Single roadmap (GSM/GPRS $\rightarrow$ (EDGE) $\rightarrow$ WCDMA)
USA and Japan	Multiple roadmaps (cdmaONE $\rightarrow$ CDMA2000, GSM (GPRS $\rightarrow$ EDGE $\rightarrow$ WCDMA), PDC $\rightarrow$ WCDMA, TDMA $\rightarrow$ WCDMA)

While discussing all this, we briefly mentioned the "needs and expectations" of consumers. At the end of the day, everything boils down to "Are you satisfying the needs and expectations of the customers?". If not, you might as well be in some other business. Your contribution to the value chain needs to focus aggressively on the value that can be delivered to the end-user – the customer. Whether it is a consumer interested in games and bus schedules or an enterprise customer interested in checking the supply-chain and in being alerted of critical issues in the enterprise, it is important to focus on enhancing value. In the next chapter, we will delve into the needs and expectations of subscribers from both the personal and business sector point of view. We will use case studies to point out some key aspects that are critical for both sectors.

4 Subscriber needs and expectations

Over the past few years, the wireless phone has become an invaluable extension of daily human life. With the cost of voice services and phones being commoditized so rapidly, more and more people across the world are embracing wireless, even in regions where the technology has only recently been introduced. One of the key elements of this phenomenon is the value of the wireless device. It helps you stay connected (especially in case of emergencies) with people and information, and enhances personal and professional productivity (it can help you get things done in less time).

4.1 Personal subscribers

Consumer applications are designed for a different reason than enterprise applications. It is primarily to extend the brand (hence loyalty), extend the reach (especially in countries in Europe and South America, where wireless phones are the primary way for users to connect to the Internet), and generate revenue (transactions). Wireless Internet is all about "instant gratification" and "impulse transactions". Consumer companies can largely benefit from this new wave of application services. Wireless Internet is a prime candidate for promoting "social communities" – instant messaging, chat, auctions, etc. It is also great for providing information to the user on the move. Some of the key factors in keeping consumers satisfied are to focus on the applications and services from a consumer point of view, understand the customer and deliver the service based on their expectations and needs. The following criteria are very important for the consumer.

4.1.1 Cost effectiveness

Connection charges (especially circuit-switched networks) are still relatively high – another factor discouraging consumers from using wireless data. With the advent of "always on" networks, the situation will become a little bit better. The most effective applications, for which a consumer will be willing to pay a premium, are those where information is extremely time sensitive and where the benefit to the consumer of precise, timely information outweighs concerns about the cost of airtime (e.g. stock trades, emergency broadcasts). As the cost of handsets has decreased steadily, wireless penetration has grown rapidly and carriers will continue to provide subsidized handsets as we move into the data environment.

4.1.2   High-security perception

One of the biggest fears that consumers have in regards to mCommerce is security of personal data while making transactions using wireless devices. Although protocols are improving and carriers and application developers are designing more robust applications and services, security-conscious consumers need to be convinced that their personal data will not be compromised if they use a certain application. There are even more concerns about the possibility of unauthorized third-party interference, and customers will need to be convinced before they start making high-value transactions.

4.1.3   Extension of brand and customer loyalty

Consumers in general like to stick to a brand identity. If someone buys books regularly from amazon.com on the Web, they would not want to change to some other retailer if they are using a wireless device. Hence, it has become imperative for retailers and corporations to meet customers where they want them to meet and provide a user experience that is consistent across all access channels. This enhances customer loyalty and in turn directly affects the bottom line.

4.1.4   Privacy

Privacy is going to be the thorniest issues for the next few years. The technology that allows us access to rich, customized data also has the capability to collect unprecedented amounts of information on our daily lives. As such, it provides a ripe situation for hackers to play foul and for corporations and carriers to misuse this personal information. Carriers and application developers who can convince the consumers to trust them will benefit significantly as it will become one of the key marketing differentiators.

4.1.5   Personalization and customization

The mobile phone is a very personal device, so the service and the content delivered to the device needs to be customized appropriately as well. Unlike e-business applications, consumers expect a far greater degree of customization.

Table 4.1. Access to Fidelity Investments applications and services via various devices

Feature and capability	RIM 950 Handheld	Palm (wireless)	Internet-ready phone
Charts		×	
Intra stock screener		×	
Market indices	×	×	×
NetBenifits positions and holdings		×	
News		×	
Order status	×		×
Portfolio balances	×	×	×
Portfolio quote list		×	
Real-time quotes on demand	×	×	×
Top 10 gainers/losers/most active	×	×	×
Trading	×	×	×

4.1.6 Use of niche (idle) time

Mobile devices allow their users to do things during their spare time. People waiting onboard a plane or in an airport, or waiting for a bus or in a train, can use the idle time to send a message, play a game, buy a stock or bid for an expensive gift item. So, consumers would make greater use of applications, entertainment and services that help them kill time.

4.1.7 Ubiquitous access

As discussed in Chapter 2, one of the fundamental shifts in computing is universal access to information from a variety of devices on different networks using different interfaces and programming languages. Consumers do not want to get into the technology alphabet soup. They just want consistently reliable technology; they hardly care if they are being served by TDMA, GSM or CDMA, or if the interface is using WML, cHTML, or XHTML.

Customers demand that whatever content they desire to access be appropriately formatted according to their device capabilities and user preferences. As such, application developers and carriers need to worry about providing services on a whole range of devices that might use different network and user interfaces. For example, Fidelity Investments offer their trading and portfolio access services customized for RIM, the Palm family of devices and Internet-ready phones (see Table 4.1).

So, if you are targeting a consumer market, make sure you address all the variables associated with the market or else consumers will not adopt and stay with the application or service.

4.1.8   Location-based services

Mobile phones already offer the ability to give an approximate location based on cell-sector ID. A far more precise technology is available (as discussed in Chapter 7), and carriers are starting to deploy both network and phone-based technologies to enhance their infrastructure. Some systems like PHS, which have a higher concentration of cell sites for any geographical region, can actually pinpoint to a greater accuracy without the use of advanced network or home-based technologies. With location information, the market opens up for a range of new applications requiring location-specific information. It also makes information very relevant to the user in that instant of time and hence provides great value.

4.1.9   Messaging and presence-based services

These services are based on the consumers' physical environment and situation. For example, when a traveler arrives at the airport, a series of transactions take place in the same order, like checking in baggage, printing the boarding pass and luggage tags, authentication of identity, updating of frequent flier miles. A lot of these tasks can be automated to provide quicker service and convenience to the consumer. In addition, with the advent of technologies such as SIP (discussed in Chapter 7), it is possible to connect friends, family, and colleagues using a variety of channels, automatically. In future, people will tend to belong to multiple communities of interest and will want to project a different image of themselves and their availability to each community. The concept of calling a device such as an office phone or a mobile phone will transform itself into attempting to contact a specific individual or group of individuals. A person will want to project availability based on a variety of factors:

- accessibility of various forms of access,
- actual activities the user is involved in at any instant in time, and
- identity of the person attempting to contact somebody else.

     People will want an easy way to set boundaries on how they are contacted and how their availability is projected to different groups of people. Just as important, people will not want to be burdened with constantly having to provide specific input as to what is changing in their personal environment to keep their projections of availability current. One of the value propositions of any communications network will be how transparently the information enabling presence projections is managed. Both large and small changes in the physical location of a person may play an important role in determining desirable availability. People traveling a long distance from home may want to limit their contact to a small group from home while making themselves readily accessible for contact by people in the area where they are traveling. The fact that someone's mobile access switches to a location significantly away from home should be enough to trigger a different view of availability that is still customizable by the person involved, but without him or her specifically being forced to update anything.

4.2   Enterprise subscribers

The main goals of enterprise applications and services are as follows.
- To increase productivity by enabling faster, better, decision making and more efficient communications.
- To increase revenues and profitability by putting information to work at the point of need for individuals and operations.
- To improve customer/user satisfaction by improving communications and offering responsive services and accurate answers.
- To reduce cost and complexity of managing secure remote access accounts and services.

Any time the employee (user) is mobile or is in the field (meaning not near a wired connection), wireless Internet applications have an important role to play. Hand-held-based fields services applications can help users to be not only connected to their enterprise data but, because of that, to reduce time and cost to complete "a task". This has a direct impact on an organization's bottom line. For example, if the CPU of a critical server reaches above 90%, a system administrator in the field could be alerted of such a condition with the option to restart the machine, kill the process, or alert a service technician – all to be accomplished by a few taps on their wireless device. Even doing simple (yet extremely important) things like having access to email, calendar and address book is bound to boost productivity especially in large enterprises. Energy, utilities, financial services, consumer goods, logistics and retail are all prime candidates for the leverage of wireless Internet solutions and applications. Also, any company interfacing directly with consumers needs to pay attention to their CRM initiatives. By automating lots of daily tasks like expense reporting, dispatch and delivery, sales force automation, IT management, and access to corporate information databases, an organization can save both time and money and enhance productivity and efficiency.

So, what are some of the key needs and expectations for a busy enterprise user? Let us explore this a bit further.

4.2.1   Access to corporate PIM

Access to email, calendar, and address book is by far the single most important enterprise application for mobile executives and professionals. The value of instant access and delivery is immense. Wireless access to such information yields quicker information processing and response and thus directly and very tangibly affects employee productivity. Corporate users in a mobile environment have traditionally needed to access their company's server in order to use email, access the Internet or access a company intranet. Besides a PC, this method has required a GSM modem and connected mobile phone. Most corporate remote access email solutions are currently based either on Microsoft Exchange or Lotus Notes-Domino platforms. The alternative is to access emails forwarded from the corporate email account remotely using a Web browser. This is possible in some hotels and Internet cafes. Corporate solutions are being developed to allow users ubiquitous, instant access to personal management systems from a range of devices.

Instant access to email, corporate databases and ERP systems provide executives on the move with vital information for managerial decision making. Such information is usually available in simple text format, thus making it ideal for WAP and SMS. In Norway, Telenor Mobil has built an SMS-based application that allows corporate employees to request internal company information. This product includes phone numbers from the in-house directory, latest prices, product availability, internal news and customer information. The arrival of Intelligent Voice Response technology will enable users to give in-depth instructions over a mobile device while conducting another task.

4.2.2   Function or vertical applications

These applications are designed to be used by employees in a specific role (functional) or industry vertical. They typically involve interactive access to some enterprise information system, and the system is often an existing one that has been extended for mobile use rather than a system that was created specifically for mobile users.

Expense reporting

There are "road warriors" in every company, especially the sales and business development employees. They are constantly on the move. Sometimes it could be weeks before they are able to file their expense report and get reimbursed for all the costs they incur while being on the road. Sometimes these expenses are billable to the client. If the employees are able to file reports quickly, approval and processing could be done quickly. Faster processing means earlier reimbursements and billing to the client. The process works better with everyone involved.

Expense reporting can easily be integrated with a work-flow application to make it more attractive. Users can file their expense reports from anywhere, anytime, and not wait till they get back from their trips.

IT network alarm and monitoring

If you have been in IT supporting your corporate infrastructure, you would definitely appreciate the desire of keeping a check on your network, whether it is the performance of certain LAN segments, or intrusion detection, alarms and alerts, or summary reports. It would be nice if you could not only capture and extract information about your network, but also execute certain commands remotely. Sometimes

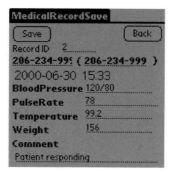

Figure 4.1. Medical application.

IT professionals who are on 24/7 duty, have to carry their laptops with them, boot their machines, and login to diagnose and fix problems. If a majority of these tasks (that involve only a few key strokes and not extensive debugging) could be accomplished remotely, the response time would increase and tangible cost savings could be realized. For corporations, the cost savings lie in

- inventory reduction,
- lower overhead costs,
- disappearance of an intermediary,
- lower procurement costs, and
- shorter product cycles.

Healthcare

Healthcare is an industry constantly in need of technology solutions to improve its services. Doctors, nurses and other medical staff always need ways in which they can serve their patients quickly while maintaining the quality of service. Applications like the one shown in Figure 4.1 can be used intelligently to store and process patient information in the hospital using handhelds (instead of paper). Similarly, medical guidelines can be accessed using wireless and handheld technology instead of browsing thick books. Wireless technologies can also be used by doctors to prescribe medication, and monitor patient condition remotely, especially under severe and emergency circumstances (high-bandwidth networks can not only deliver medical diagnostic information but can carry live video feeds from ambulances or other locations).

Retail

Retail industry is fiercely competitive. With the advent of the WorldWide Web, non-traditional companies like Amazon.com have emerged and have made brick-and-mortar companies like Barnes & Noble rethink their strategy. Competition is also driving profit margins down. The success of a retailer depends on inventory management, cost control and proactive customer service. To gain the competitive advantage, more and more retailers are turning to wireless Internet applications to enhance worker productivity, operational efficiencies, and anytime, anywhere customer service. On the sales floor and in the warehouse, wireless solutions can help in tracking materials and shipments from supplier and distributors to the customers, managing inventory, and in supporting point-of-sales activities. Because a lot of data can be collected in an automated fashion, analysis can be done much faster and the results can be used continuously to improve operations and customer service. Wireless Internet has allowed retailers to break away from their dependence of landline infrastructure for credit card processing. Wireless Internet applications are the next evolutionary step for retailers currently capturing data via batch data devices and fixed terminals. They are also a next step for arming mobile workers – from store managers to analysts – with information. This will lead to enhancing the customer buying experience, which is what every retailer aims for.

Transportation and logistics

The freight and trucking industry have been using wireless data applications for quite a while. Courier companies like FedEx and UPS have established a sophisticated wireless data network that has yielded productivity gains and increased customer service, which is worth in the region of billions of dollars. However, with wireless Internet solutions becoming more pervasive, it is no longer necessary to deploy expensive and proprietary wireless data networks to gain the competitive advantage. Applications can be easily built using the wireless Internet infrastructure to provide real-time communications. From the point of dispatch to the final destination, wireless Internet can be used to provide job status, dispatch information, proof of delivery, exception information, emergency notification and other critical data. Wireless asset tracking can enable managers to coordinate pickups and deliveries more accurately and to have a good sense of their assets at any given instant. Wireless Internet offers solutions for companies involved with aviation, courier delivery, freight trains, intelligent transportation systems, LTL trucking, maritime shipping, asset tracking and urban transport.

Field services

Everyday, thousands of field-service technicians and workers are dispatched to investigate, fix problems, and collect data from the field. Much of their day is spent traveling from one location to another. Often, they have to process lengthy paper work, which is detrimental to productivity and efficiency. Customers, too, dislike having to fill out laborious sheets of paper over and over again. Wireless Internet technology is a boon to field-services organizations. It can help not only in automating data entry but also expediting routing and dispatching, status reporting, and providing efficient access to corporate databases, customer billing, and vehicle location. The field-services market crosses multiple industries and vertical segments – manufacturing companies, third-party maintenance organizations, city governments, dealers and distributors, etc. For effective services, field-service personnel need real-time access to customer information, trouble tickets, product information, parts lists, schematics, dispatch schedules, maps and directions, weather conditions, restaurants, etc. With wireless Internet applications and services, all of the above can be accomplished. Since the data entry is automated in most instances, the time needed to resolve issues and to analyze and process data is drastically reduced, thus improving the overall productivity and customer service. Owing to efficient job scheduling and prioritization, billing cycles can be automated, thus reducing the operating costs, and increasing revenues. Field-services operations thrive on communications. With wireless Internet, cheap, reliable and secure solutions that can help realize immediate gains can be quickly built.

Utilities

Since the deregulation of the utility industry, the business environment in this sector has changed dramatically. Utility companies need to be better informed, and to be able to monitor their resources to stay competitive. Wireless-based applications are helping to make the leap by providing highly efficient, reliable data and information in real time at low cost. Remote meters in the field can be monitored and controlled from a central location. This allows problems to be detected (before they happen) remotely and maintenance crews to be assigned more efficiently. Wireless Internet will continue to play an increasingly important role in the development of this industry.

4.2.3 Sales force and field service

Mobile devices will enable sales and customer-service personnel immediate access to critical information concerning the customer they are traveling to visit, when this information is stored at central office. These devices will also give them the ability to make immediate transactions online. Benefits might include: order entry; product and spare parts availability confirmation; deal tracking; and "e-lerts" of business-critical information.

4.2.4   Enterprise resource planning

It seems logical to suggest that consumer mobile retailing principles should apply even more acutely in supply chain integration. An SAP application has already been developed for use on 3Com's Palm device. Much of a business's customer relationship management could be automated through wireless technology. An example of this is Orange's Wildfire. This is an automated interactive customer service tool that aims to improve customer service while reducing servicing costs.

The success of e-commerce CRM solutions with established ERP vendors such as SAP, Oracle and Baan joining the market looks set to be at least met by that in wireless data. SAP of Germany and Intentia of Sweden have announced links from ERP systems to WAP mobiles, allowing sales people, field engineers and technicians to login from mobile devices. Telia Mobile plans to use e-business alert software from Virginia and London-based Categoric Software for this kind of information. An unexpected event will trigger an SMS to be sent to a target group, giving them information about, say, a delay in the supply chain or a postponed onward flight. Faster information will give the recipients more control over what action they can take. Eventually, once WAP is more widely used, Telia Mobile expects the warning messages to become more interactive.

Sales force automation (SFA)

The next step is to strengthen the company's sales capabilities by a wireless device. A sales person can provide a faster response and contribute to better customer satisfaction by using wireless devices to search for product, inventory and customer information.

Figure 4.2 shows how the ratio of i-mode phones among DoCoMo corporate users is rapidly increasing. For example, DAT Japan, a Japanese delivery company using bikes, is using i-mode phones to increase business efficiency and communications from/to the control center and delivery persons.

Rugged devices

For most of the industrial applications, special-purpose rugged computing devices are needed. The devices and applications are often designed primarily to capture data rather than for data interaction. In some cases, an entire business process might have to be redesigned around the use of the technology. For example, rental car agents and postal workers routinely use their rugged wireless devices to capture customer information.

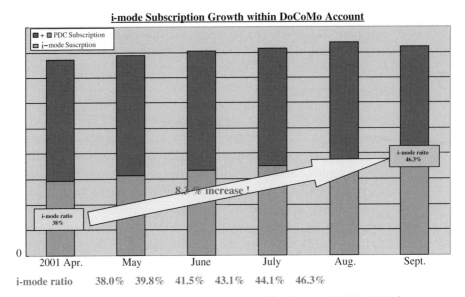

Figure 4.2. The growth of i-mode in the corporate market (source: NTT DoCoMo).

Security

The issue of security is an extremely important one for corporations embarking on implementing wireless data solutions for an enterprise. The coalescence of the computing and communications industries creates outstanding opportunities. However, any breach of information is potentially more harmful in the wireless world, as information is now interconnected and interdependent. Having strong security procedures and algorithms in place at various intermediary steps ensures that consumers feel comfortable embracing the information evolution.

Alerts-based enterprise

Enterprise event management and alerting (EM&A) systems are an indispensable part of an organization's technology infrastructure. They enable organizations to provide employees, partners and customers with interactive notification of time-critical events as they occur, thus allowing recipients to take rapid action to resolve issues that arise. The key requirements that an EM&A system must meet to deliver the benefits of true enterprise scalability fall into three major areas: architecture, deployment and communication. Enterprise alerting helps in "reducing time and response" to a situation in an enterprise, whether it is responding to a failed critical server or escalation of a supply-chain event breakdown, or maybe just a notification of delay of a shipment to a customer. Enterprise alerting notifies people of critical information and events and empowers them to capitalize on opportunities and respond to situations before they become problems. The results are real-time decisions that improve operational efficiency across the enterprise as well as enhancing speed, productivity, collaboration and customer loyalty.

The alert-based systems that support the enterprise should be independent of legacy systems or any specific application or database. It is also important that alerts are not restricted to any single standard or technology and there should be multi-tier escalation traps available.

- Reduced costs of delayed response. By accelerating the resolution of field-service requests and trouble tickets, InstantResponse can significantly reduce the costs of delayed response. These costs include lost revenues, lost enterprise productivity, lost customers and Service Level Agreement (SLA) violation penalties.
- Lower dispatch and escalation costs. By using real-time presence, availability and location information only to notify technicians who are present and available to respond, Instant Response eliminates the inefficiencies associated with slower notification methods like email, voice mail and paging. Automating the notification and escalation process can reduce or eliminate dispatch costs, enabling fewer dispatch staff to manage more distributed workers.
- Increased utilization and productivity of field technicians. By simultaneously enabling them to access the anchor trouble ticket and field-service application, other supporting applications and co-workers remotely and in real time, using IM-enabled mobile devices, the IRiS closed-loop response system empowers technicians to close more tickets in less time.
- Reduced rework costs. By ensuring that technicians are assigned only to resolve problems that they are equipped, qualified and available to solve, InstantResponse ensures that technicians can close tickets on the first attempt.
- Increased customer service and satisfaction. Faster ticket resolution, and real-time updates on resolution status, drive significantly higher levels of customer service and satisfaction.

Device and application management

As wireless devices and applications are becoming more commonplace in enterprises, it is becoming evident that IT needs to have a clear strategy and policy for device and application management. Similar to managing an installed base of desktop and laptop PCs, the total cost of ownership of an installed base of wireless computing devices is dominated by maintenance costs as opposed to the purchase cost of the devices. In reality, the cost of maintaining an installed base of wireless devices can be extraordinarily high without an effective wireless device management solution. Without a proper device management policy, information security is at high risk. We will discuss this issue in more detail in Chapter 9.

Use of WLANs

Mobile computers, personal digital assistants (PDAs), hand-held terminals, and bar-code scanners connected to WLANs are increasing the profitability of business operations. There are several types of businesses that have begun to use WLANs, in which the need for mobility and flexibility outweighs any sacrifice in data speeds.

- Hospitals – hand-held devices and portable computers are able to deliver patient information to doctors and nurses in real time, providing information that can save lives. In addition, administrative productivity is bolstered through the elimination of redundant paperwork.
- Warehouses – workers use wireless devices to deliver instant updates regarding supplies and inventories to centralized databases, providing more accurate information to the management team. Bar-code readers with wireless data links are used to enter the locations and to identify inventory, improving operational efficiency and reducing costs associated with physical inventory counts.
- Consulting and audit teams – small work groups can quickly establish private and integrated networks as they move from site to site.
- Dynamic environments, ad agencies, etc. – the cost of LAN ownership can be reduced significantly in dynamic environments with multiple additions and moves.
- Universities – many universities have implemented campus-wide wireless LAN technologies as students become increasingly PC dependent. New York University, for instance, provides each entering freshman with a laptop computer as part of his/her tuition. By implementing wireless LANs, colleges like NYU can provide accessible data to students, both as they travel around campus and when they are unable to access a wireline connection.
- Historic buildings, older buildings – many government and otherwise historic buildings have regulations against changing the building infrastructure, denying the possibility of pulling new wiring through the walls. Wireless LANs are, in these circumstances, the only possible form of data networking.
- Trade shows – WLANs provide quick and flexible set up capabilities that allow trade show personnel and convention centers constantly to change the network structure from one show to the next.
- Meeting rooms – real-time information allows management to make more informed decisions during meetings.
- Retail stores – WLANs enable frequent network reconfiguration.
- Restaurants and car rental agencies – real-time information transfer enables superior customer service. Examples include Avis and Hertz.
- Data backup – mission-critical applications can be implemented with a wireless backup in the case of malfunctions in the wired infrastructure.

In this chapter, we have reviewed the key issues that need to be addressed to satisfy subscriber needs and the expectations of wireless Internet applications and services. In the next chapter, we will look at the wireless value chain, the players involved, and what part they play in keeping the ecosystem healthy and competitive. We will also use models to analyze the wireless value chain and try to learn from past history.

5 The wireless value chain

The world's two fastest growing technology sectors – wireless communications and the Internet – are changing rapidly and dramatically. As we have discussed in previous chapters, not only are these two highly dynamic markets changing, they are also converging. Taken together, the changes pose significant challenges for both wireless and wireline service providers. But challenge also brings opportunity. Wireless operators are facing a substantial margin squeeze for their mobile voice services. Although the number of subscribers is increasing and traffic and overall usage are up, growing competition is driving prices down. To remain competitive and minimize "churn", wireless operators must differentiate the services that they offer, and retain high-quality, revenue-generating subscribers.

The Internet explosion is changing the way consumers behave – from social communication, to making purchases online, to the way we receive news. And now, to meet the growing demands of mobile consumers, the Internet revolution is going wireless.

The convergence of these two technologies creates lucrative new revenue opportunities for wireless operators, service providers and content portals. This new open-market environment opens the door for new services, new partnerships and a new revenue model that will help operators retain their current end-users, attract new subscribers, and improve their bottom line.

The introduction of wireless Internet services is allowing operators dramatically to expand both their service and their revenue models. The operator's traditional revenue mix (primarily flat, monthly subscription fees) is changing in the Internet-driven world. In this new environment, operators are also deriving revenues from subscriptions to data service and content, e-commerce, advertising, advanced network services such as Virtual Private Networks (VPNs), and Quality of Service (QoS) guarantees.

Significant growth in annual service revenues is expected, as narrowband subscriber growth slows and higher-priced broadband services begin to experience considerable subscriber growth. That means enormous opportunity for wireless operators. Media companies and wireless service providers can also expect new revenue opportunities in mobile commerce, advertising and content, resulting from changing consumer behavior.

The new wireless Internet value chain reflects the emergence of new competitive forces and a new economic environment. Historically, operators 'owned' the value chain because they owned and maintained the network, and acted as the service provider to offer the only available service – voice. But with the Internet revolution, users now demand powerful applications and services. To meet these changing expectations, the wireless value chain has been expanding and evolving. With the emergence of the Internet, many new services, service providers and business models have also emerged. The new wireless Internet value chain incorporates new players who have the expertise needed to implement wireless Internet services. The operator no longer owns the entire value chain and is in partnership with other content and service providers to the subscriber base.

The right combination of network elements, services, applications and professional services will ensure the successful implementation of a wireless Internet solution for all parties involved. Partnerships need to be established between wireless operators, service providers, equipment vendors, industry, service bureaus, systems integrators, terminal vendors, end-users and academia, among others. With each party contributing its expertise towards a successful implementation of wireless Internet, risk is reduced and innovative applications are delivered on time.

Implementing the new wireless Internet value chain requires a paradigm shift by wireless operators – a shift that involves new and different players and elements. As wireless operators plan their migration, it is imperative that they engage the right partners with the expertise to ensure the successful implementation of an end-to-end wireless Internet solution.

In the past few years, the wireless value chain has changed significantly. The convergence of the computing and communications industries has blended the capabilities and value of some players while diminishing the value proposition of others. The wireless value chain is an extraordinary environment of competitive dynamics, rapid innovation, and continuous re-alignment. We will now discuss the main segments in the value chain and also try to understand the complex relationships that exist. We will focus on the fundamental principles of value-chain analysis and the value it adds to assessing relationships. Then, these principles will be used to project light onto the wireless industries, especially the device and infrastructure value chains. We will use Porter's Five Force analysis and Charles Fine's double-helix models to analyze the value chain.

5.1   The wireless value chain

The wireless value chain includes the following major players.
• Investment community

- Venture capitalists
- Corporate ventures
- Equipment
  - Parts and components providers
  - Device manufacturers
  - Device OS providers
  - Communications infrastructure providers
  - Computing infrastructure providers
- Networks
  - Operators (including MVNOs)
- Software
  - Middleware providers
  - WASPs
  - Application developers
  - System integrators
- Services
  - Content providers
  - Portals
  - Content aggregators
  - Application aggregators
- End-users
  - Consumers
  - Mobile business users

The industry value chain shown in Figure 5.1 is a map of the major segments within the wireless industry. These players work with each other to conduct exchanges. A description of each value-chain segment is listed in Table 5.1 along with the services they provide and the individual value they add to the chain.

Before the arrival of wireless data services, the value-chain structure consisted of end-users who bought products and services from the network operators who operate and maintain the network. Network operators purchase devices from device manufacturers and infrastructure from infrastructure providers. The chain stopped here as mobile voice was the only service available.

Today additional links are forming; both voice and data services are being offered and new players are entering the value chain to support these niche markets. Application providers and content providers have partnered with network operators to design and develop solutions for consumers and enterprises. Application providers are forming information relationships with both infrastructure and device manufactures and may soon provide enough incentives and pressure to open up the closed software architectures of the devices and infrastructure components. Further, as data become more feasible for mobile operators and standards become accepted for new applications, content providers will be likely to accelerate personalized and customizable content.

Table 5.1. Description of value-chain segment

Equipment	Components	Antenna
		Amplifiers
		Display
		Transceivers
		Signal processors
		Integrated circuits (ASIC)
		Analog–digital converters
		LCD screens
		Batteries
		GPS
		Bluetooth components
		Scanners
		Camera
	Infrastructure	Cell towers
		Node systems
		Operations and management
		Systems
		Base stations
		Routers and switches
		Servers
		Billing systems
	Devices	Phones/smartphones
		PDAs
		Pagers
		Laptops
		Telemetric devices
	Operators	Data-only carriers
		National carriers with roaming agreements
		Paging networks
		Wireless LAN networks
Software	Software components	Operating systems
		Gateways
		PIM Applications
		Management and security systems
		Location based services
		Caching
		Compression
		Payment systems
		Speech systems
	Enterprise	Application and service hosting
		mCommerce services
		Professional services
		Software products
	Content	portals
		Voice Portals
		WISPs

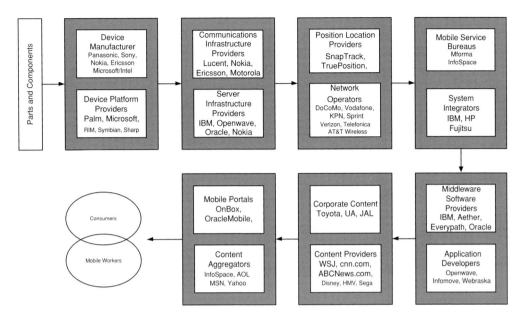

Figure 5.1. The wireless value chain

We will now review each of the value chain segments in some more detail and list some of the major players in each segment.

5.1.1 Device manufacturers

Device manufacturers and technology platform (operating system – OS) providers are critical in the value chain. Consumers generally identify very well with a particular handset brand, rather than a network operator. These devices are very personal in nature and in many cultures have become the equivalent of fashion accessories, especially among teenagers. Mobile phone sales have been growing steadily, with over 400 million units sold in 2001 and about 498 million sold in 2002.

It is not that Nokia is releasing 20 new handsets every couple of quarters, nor is it the upcoming multimedia messaging service (MMS) standard. Not even the onslaught of new and weirder-looking phones from Motorola, or those 'must-have' snazzy color screen phones. What is interesting is how non-wireless industry players are making inroads into the wireless scene, thus making the already fierce competition more intense. Chip-maker AMD, for example, recently bought Austin-based Alchemy Semiconductor, which designs processors for personal digital assistants (PDAs) and wireless devices. Also, the archrival of Microsoft, Sun Microsystems, recently joined with Ericsson and Apple to develop music and video for mobile phones. But what is most exciting is the battle between software giant Microsoft and the no. 1 mobile phone manufacturer Nokia to grab the bigger slice of the emerging smartphones market.

Table 5.2. Major handset manufacturers

Handset manufacturer	Market share (%)
Nokia	35.6
Motorola	15.7
Samsung	9.5
Siemens	8.4
Sony/Ericsson	5.4
Others	25.6

Source: Gartner, Dataquest, Deutsche Bank.

Nokia, the Finnish giant, supplies one of every three phones sold in the world with about 37% of the global market share. For four years it was the market leader in the handset business. Now, it faces challenges and the wrath of the software giant, which is spinning its SmartPhone 2002 operating system (OS) against Nokia's Symbian that was formed by Nokia, Motorola, Ericsson and Psion. Microsoft wants to get its new SmartPhone 2002 OS into 25% of mobile phones worldwide by 2005, says its chief executive officer Steve Ballmer. With no market share whatsoever now, the Redmond giant seems very determined to get into the small screens by securing key alliances with personal computer/mobile chip maker Intel and circuit maker Texas Instruments (TI) to create all kinds of smartphones.

Meanwhile, in a move to counter the new competitor, Nokia is also working with TI to offer its smartphone blueprints to other handset makers to produce Internet-ready phones. Nokia seems to have the upper hand to rally more support from major handset makers, while Microsoft has managed to get only Samsung to use its software.

Whether or not Microsoft will bite into Nokia's share of the market remains to be seen. Many industry analysts are predicting that the software company may not be successful because of the lack of supportive handset makers. Table 5.2 lists the major handset manufacturers and their respective market share as of the fourth quarter of 2002.

5.1.2   Operating system providers

With the convergence of computing and communications devices, there has been a symbiotic effect on the device operating system industry. Operating systems such as that from Palm, which used to be primarily for computing tasks, are now capable of handling the communication aspects of the device. The major operating systems in the market are from Symbian, Palm, RIM, and Microsoft. Linux is also gradually showing up on some PDAs and, given its popularity, it is likely to become a major player in the sector. In addition, there are some other proprietary platforms available

Table 5.3. Major wireless operating systems and solutions

Device operating systems	Currrent solutions	Upcoming products
Palm	Kyocera QCP 6035, Handspring Visor Palm Mobile Internet Kit Samsung smartphone	Nokia smartphone Motorola smartphone
Symbian	Ericsson R380 Nokia Communicator Siemens smartphone Kenwood smartphone Nokia smartphone SANYO communicator	Psion communicator Panasonic, Sony, Phillips and Motorola are licensees
RIM	RIM 780	
Linux	Sharp Zaurus SL-C700 Sharp Zaurus SL-A300 AML M7100	Samsung Yopy
WinCE	Sagem Pocket PC/phone Sony CMD-Z5 Benefon Q Mitsubishi Trium smartphone Mitsubishi Mondo Sendo Z100 smartphone Samsung SGH-N350 HP Pocket PC phone	Samsung GSM smartphone Siemens smartphone
Proprietary	Sprint Neopoint Motorola V100 Motorola Accompli 008/003 communicators	

from Motorola and Neopoint. Table 5.3 provides a sample of solutions in the market at the time of writing.

As the industry is making a transition from thin-client handsets to the thick-client environment, the industry is going to see a disruption in the value chain as desktop players get into the mobile application and services market. Already, IBM, Microsoft, Sybase and Oracle offer scaled-down version of their flagship products for mobile phones and handhelds. With thick-client, consumers can interact with information independent of the network by downloading and storing information and thus drawing more value from their handsets. Both consumers (games, entertainment) and corporates (enterprise applications) will benefit from this transition. This shift is also becoming a battle ground for two emerging technologies for devices: J2ME from SUN and BREW from Qualcomm. The next section contains a discussion of some of the differences and benefits of these two rival technologies.

J2ME vs BREW

With the announcement of BREW, Qualcomm immediately became the low-cost alternative for phone vendors to Web-enable their phones. The BREW software will be included free of charge with Qualcomm's CDMA chipsets. With Qualcomm controlling 90% of the CDMA chipset market, this software is guaranteed to be available at least to a large subsection of wireless vendors. In a world where the average consumer pays under $100 for a mobile phone, CDMA vendors and carriers will feel some pressure to opt for the BREW standard, where the real cost will be the price of designing the user interface (UI). We anticipate that other CDMA handset manufacturers beside the few listed are planning to incorporate BREW, but have just not announced their intent publicly. Those manufacturers may not feel any urgency in issuing such an announcement because BREW is already considered part of the chipset at this point. Handset manufacturers who have announced their support for BREW are Denso, Hyundai, Kyocera, LG Telecom and Samsung.

Sun's J2ME platform, though not as versatile as BREW, will benefit from a first-mover advantage. It is already in use on many i-mode phones in Japan, and Sun hopes to have dozens of Java phones available worldwide. Sun will benefit significantly from its partnership with Openwave, which plans to make J2ME a standard feature of its browsers. However, Sun is still pursuing agreements with individual phone manufacturers to license J2ME. Motorola will sell the first J2ME-enabled phone in the USA; it will run on Nextel's iDEN network. Some of the companies that plan to include J2ME as a stand-alone feature are Nokia, Siemens, Motorola and Palm. One might ask why vendors would choose to license only J2ME, rather than have it bundled with the Openwave browser. One of the reasons is clearly that the Openwave browser will not include J2ME software for several months, so there is a time-to-market issue. The other answer is that some companies, most notably Nokia, prefer to use their own proprietary browser. For that reason, we expect there will be a significant market for J2ME as a stand-alone product. Handset manufacturers who have announced their support for J2ME are Ericsson, Fujitsu, Hitachi, J-Phone, LG Telecom, Mitsubishi, Motorola, NEC, Nokia, Palm, Panasonic, Samsung, Sharp, Siemens, Sony and Symbian.

5.1.3   Infrastructure providers

Infrastructure providers furnish the necessary components to complete the big picture. These providers range from suppliers of the wireless base and switching stations (Nokia, Ericsson, Motorola and Lucent) to providers of public-key infrastructure for encryption and security (Verisign and Certicom). Some players specialize in niches like personalization, performance measurement, business intelligence, and so on. The major players in this space are listed in Table 5.4 by category.

Table 5.4. Infrastructure providers and their specializations

Specialization	Vendor	
Communications – wireless specific	• Nokia • Alcatel • Ericsson • Motorola	• Lucent • Qualcomm • Nortel • Siemens
Communications – gateways	• Phone.com • Nokia • CMG • Siemens • IBM	• Ericsson • Dr. Materna • Alcatel • Palm • Infinite Mobility
Position location vendors	• Cambridge Positioning Systems, Ltd. • Cell-Loc, Inc. • Ericsson • Harris Communication • Airbiquity • Lockheed Sanders	• Lucent • Nokia • Nortel • SigmaOne • SiRF • SnapTrack (Qualcomm) • TruePosition
Computing	• Oracle • Microsoft • Sybase • HP	• SUN • Netscape (AOL) • IBM
Smart cards	• Schlumberger • GemPlus • Mondex	
Caching	• Akamai • Inktomi	
Business intelligence	• Business Objects • Cognos • Brio	• IBM • Oracle • Microsoft
Personalization	• Art Group Technology • Broadvision • Yodlee	• Net Perceptions • Blaze Software
Others	• Verisign (digital certificates) • F-Secure (anti-virus)	• Certicom (digital certificates) • Visa, Mastercard (online payment)

5.1.4 Network operators

Network operators provide and manage the link between an application and the device or user through their networks. Network operators are always looking for ways to increase the average revenue per user (ARPU), and the advent of wireless Internet will benefit them the most. Some of the early adopters of the technology are using wireless Internet applications and services to differentiate themselves from the competition and enter new markets with new business models. Wireless Internet provides them with the opportunity continually to expand and improve their networks and provide a wide range of services to their customers. Table 5.5 provides the list of operators with licenses in various countries and their coverage obligations.

5.1.5 The rise of MVNOs

When Sprint PCS announced a deal with Virgin Mobile, a new breed of operators called mobile virtual network operators was born. The idea is that companies with strong brand recognition like Virgin can capitalize on their customers' loyalty to provide wireless services, including reselling airtime that they purchase from the traditional wireless carriers and rebrand under their own names. Several other players who have strong global or regional brand recognition like American Express, Walmart and Schwab are considering entering the lucrative but largely untested market. Not every business is suited to launching a virtual mobile service. Forrester Research says a successful MVNO candidate needs four attributes:
- brand power to keep customers loyal,
- a way to control customer acquisition costs,
- a core business attribute that can be transferred to the mobile service, such as content or a loyalty rewards program, and
- the organizational commitment to spend what it must and take the necessary risks.

   The most indispensable partnership for US MVNOs – teaming with a network operator to carry voice and data traffic – is also the most difficult one to set up. Striking deals with US network operators is problematic for several reasons: besides struggles over the wholesale price of network time, US carriers want partial ownership of the new venture. They do not want simply to rent out their networks and let someone else maximize their value. If the MVNO is successful, the carrier expects to share in the upside. Carriers are also afraid that MVNOs will steal their customers. MVNOs are planning to target smaller segments and under-penetrated portions of the market such as low-income people and youth (research firm Yankee Group pegs wireless penetration among US teenagers and young adults at 35%, slightly below the nation's overall penetration of 50%). Because the carriers have not successfully reached these segments, they are more open to partnerships that address the under-penetrated segments.

Table 5.5. Network operators around the world and their 3G plans

COUNTRY	LICENCE DETAILS (US$)	LICENCE HOLDERS	TIMETABLE	Comments
Australia	Six licenses awarded through auction in March 2001 for a total of $580m	Telestra, C&W Optus, Hutchinson, Vodafone Pacific, CKW Wireless, 3G Investment	Licenses started from October 2002	License duration is 15 year
Austria	Six licenses awarded in November 2000 for a total of $618m	Mobilcom, max.mobil, Connect Austria, Tele.ring, 3G Mobile communications, Hutchison 3G Austria	Mobilkom Austria launched 3G network in September 2002 (first UMTS network in Europe)	Cost/head of adult population = $90. License duration is 20 years
Belgium	Three licenses awarded in March 2001 for a total of $413m. (Meant to issue four licenses, but cut back to three after lack of interest.)	Belgacom Mobile, Mobistar, KPN Orange	30% of population to be covered by March 2002, 40% by March 2003, 50% by 2004 and 85% by 2006	All initial 3G launches got delayed by at least a year
Denmark	Four licenses awarded in September 2001 for a total of $472m	TDC Mobile, Telia Mobile, Orange, Hi3G	Operators are obliged to cover 30% of the population by the end of 2004 and 80% by the end of 2008	Hi3G planning to launch services in 2003
Finland	Four licenses awarded in March 1999 for free	Sonera, Radiolinja, Telia, Suomen 3G	Commercial operations meant to start by January 2002	All launches got delayed, some indefinitely
France	Two licenses awarded in July 2001, two more licenses to be offered. If all four licenses are sold government should get $2.23bn plus a percentage of operators' revenues	France Telecom and Cegetel	80% of population within eight years for voice services. For data 144 kbps: 20% by 2003, 50% by 2009	Launches delayed untill 2004
Germany	Six licenses awarded in August 2000 for a total of $46.11 billion	D2 Vodafone, T-Mobile, E-Plus, Viag Interkom, Group 3G, Mobilcom Multimedia	25% of population to be covered by end of 2003 and 50% coverage in three to five years of launch	Cost/head was $657. All launches delayed. Carriers struggling to survive
Israel	Three licenses awarded in December 2001 for a total of $157.1m	Partner , Cellcom, Pelephone	Partner expected to start operating system by 2003	Cost/head was $3.64

(cont.)

Table 5.5. (*cont.*)

COUNTRY	LICENCE DETAILS (US$)	LICENCE HOLDERS	TIMETABLE	Comments
Italy	Five licenses awarded in October 2000 for $10.04bn	Omnitel Pronto Italia, Ipse, Wind, Andala, Telecom Italia Mobile	Operators are to offer 3G services to major cities by the second half of 2004.	IPSE is out of business
Japan	Three licenses granted in June 2000, with no fees	NTT DoCoMo, J-Phone, KDDI	NTT DoCoMo launched 3G (FOMA) in Oct 2001. J-Phone launching their 3G network in Dec 2002. KDDI launched services, based on CDMA2000 1X technology, on 1 April 2002	One of the few countries where both WCDMA and CDMA2000 1X networks are up and running and attracting consumers
Netherlands	Five licenses awarded in July 2000 for a total of $2.5bn	Libertel, KPN Mobile, Dutchtone, Telfort, 3G Blue	Services to be launched by end of 2003	France Telecom's Dutchtone and Deutsche Telekom unit Ben reached an agreement to share parts of their infrastructure in December
New Zealand	Four licenses awarded in January 2001 for a total of $51.4m	NZ Telecom, Clear Communications, Telstra Saturn and Vodafone.	–	–
Norway	Four licenses awarded in November 2000 for a total of $92m	Telenor Mobil, Netcom, Tele2Norge (Broadband Mobile originally held a licence but gave it back to the government)	90% coverage within five years of launch	Cost/head was $17
Portugal	Four licenses awarded in December 2000 for a total of $342.4m	Telecommunicacoes Moveis Nacionais, Telecel, Optimus, OniWay	40% by 2003, 60% by 2005	Launches delayed by one year
Singapore	Three licenses awarded in April 2001 for a total of $173.4m	SingTel Mobile, MobileOne Asia, StarHub	Provisional deadline of 31 December 2004 for all licensed 3G operators to have nationwide network. This can be reviewed by the Singapore's InfoComm Development Authority (IDA)	Cost/head was $40

Table 5.5. (*cont.*)

COUNTRY	LICENCE DETAILS (US$)	LICENCE HOLDERS	TIMETABLE	Comments
South Korea	Three licenses awarded at $1.billion each	KTICOM, SK Telecom, LG Telecom	License winners launched their services in 2002. Government is urging companies to provided full-fledged service by 2003	As of mid-2003, had the highest number of 3G users in the world. CDMA2000 1X network accounted for most of them
Spain	Four licenses awarded in March 2000 for a total of $446.5m	Telefónica Moviles, Airtel, Amena, Xfera	All cities by mid 2002 (with >250k inhabitants)	Launches delayed due to equipment problems
Sweden	Four licenses awarded in December 2000. Government got total of $107 000 from 10 beauty contest entrants, and will get 0.15% of the annual revenues of license winners	Europolitan, Tele2, Orange Sweden, Hi3G Access Group	First network opened in December 2001, 99.9% of Swedes to have access by 2003	Europolitan and Orange have networks up and running. Cost/head was $0.006
Switzerland	Four licenses awarded in December 2000 for a total of $120m	Swisscom Mobile, Orange Communications, dSpeed, Team 3G	50% of population must be covered by 2004	Cost/head was $19
Taiwan	Five licenses awarded in February 2002 for a total of $1.4bn	Chungwha Telecom, Taiwan Cellular, Taiwan PCS Network, Eastern Broadband Telecommunications, Far EasTone	50% of population within three years of construction of 250 base stations	The end of Taiwan's auction brought relief to officials who had begun to worry that prices were climbing too high
UK	Five licenses awarded in April 2000 for a total of $35.361bn	Hutchison 3G, Vodafone, BT 3G, One2One, Orange	80% of population to be covered by 2007	Newcomer Hutchison 3G started their 3G network on 3.3.3. Cost/head was $576

(*cont.*)

Table 5.5. (*cont.*)

COUNTRY	LICENCE DETAILS (US$)	LICENCE HOLDERS	TIMETABLE	Comments
US	Auction for PCS licenses that can be updated to 3G held in January 2001, raising $16bn for the US treasury. The spectrum was originally bought by Nextwave in the mid-1990s, but was reauctioned by the government after Nextwave filed for bankruptcy. A second auction for licenses in the 700 MHz band of the spectrum is also planned, but this is already in use by broadcast television stations, and as a solution for sharing the space is sought, the auctions have been postponed, possibly until September 2004.	The biggest lots of spectrum in the January 2001 re-auction were bought by Verizon Wireless, AT&T Wireless, Cingular Wireless and T-Mobile	Verizon and Sprint have launched nationwide services based on CDMA2000 1X technology covering, AT&T Wireless planning to deploy WCDMA services in 2004	

(Source: Financial Times, Company Data, CSFB Research, cellular-new.com, UMTS at the Crossroads – Chetan Sharma, Sunil Jain)

5.2   Aggregators – content/application/services

Aggregators are essentially service providers that collect content from a variety of sources and repackage available data from a single source. With this method, users do not have to browse from site to site. In addition, thanks to personalization, the user is presented with only the content from various information types and sources that interests him or her. The portal information can be stored or collected in real time. Portals from the Internet world were among the first movers in this space. Firms such as Yahoo!, Lycos and MSN are partnering with or buying firms that have expertise in wireless delivery to extend their valuable content to mobile devices.

Aggregators can either specialize in one or more content categories or they can be generalists. Some aggregators, such as Charles Schwab, provide content related to business and financial markets; others, such as CNN.com, specialize in providing news across multiple disciplines and sectors. Some mobile portals also provide personal information management applications such as email, calendar, instant messaging, and so on. The key to successful delivery of content is personalization and localization.

Sensing the growing market opportunity, many mobile operators are also jumping in the mix to become content aggregators. They have partnered with some top-tier content providers and aggregators to build their own wireless portals. BT Cellnet with its Genie, Telia with its MyDOF (My Department of the Future), Sonera with its Zed, and Deutsche Telekom's T-Mobil with its T-D1@T-Online were among the first movers in Europe. In the USA, AT&T and Sprint PCS were among the first movers, and in Japan, NTT DoCoMo's i-mode service has been a big hit with consumers.

The concept of aggregation has swiftly migrated from content to applications. Instead of dealing with hundreds of application developers, carriers want to deal with a set number of players, preferably with application aggregators like Openwave and InfoSpace who are already working with the carriers on a variety of applications and services like PIM, chat, etc. So, if somebody comes up with a unique exciting application, the carriers' preference is for that application developer to work with their application aggregator to minimize integration headaches.

5.2.1   Middleware solution providers

This segment of the value chain emerged as a result of the convergence of the wireless and Internet industries. In 1999, one of the biggest challenges in the wireless Internet arena took advantage of aggregated content and transformed it for disparate devices that have different capabilities for different users who have differing preferences and needs. Middleware software is the glue that binds the communication and computing worlds together and thus is a very important component of the value chain. The earlier incarnation of the middleware technology was a standalone product suite that talked to the application server to transform content on the move. The transformation was done either at the gateway level at the carrier or enterprise, or within the premises of the corporation. A new breed of companies called the Wireless Application Service Providers emerged to address the need for quick set up of middleware to transcode legacy or Web content for wireless devices. Middleware software is generally based on XML at its core for transcoding existing or new content into a variety of formats such as WML, VoiceXML, HDML, HTML, Web Clipping, and so on. These products connect to application servers, databases, and various gateways, such as WAP, UPLink (from Openwave), Palm.Net, SMS, and VoiceXML to push the content onto the devices. With time, application server providers such as IBM, BEA, Oracle, and Microsoft, have incorporated transcoding and middleware functions into their server offerings. Even vertically focused players like Siebel and SAP provide a middleware piece along with their core offering that adapts the content to the various devices that are interacting with it.

5.2.2   The vanishing of WASPs

During the early days of wireless Internet solutions, in 1998–2000, there was an instantaneous deluge of companies claiming to be wireless ASPs. The market had the need for a solution that could take corporate content and solutions to wireless devices in a short period of time. Larger players such as Charles Schwab did not have the trained resources to launch and implement such a project in a short space of time. So, they turned to WASPs to connect their customer data to the wireless world through a WASP-based model. Aether Systems provided the infrastructure for RAD (rapid application deployment). With the bursting of the economic bubble, the need for 'time-to-market' solutions and services went away for most part, as did the need for WASPs, especially for larger players like Schwab, Citibank and the like. WASPs still play an important part in the wireless ecosystem, offering services to small-to-medium sized companies who do not want the headache of dealing with wireless protocols and infrastructure. Larger companies, who are always wary of their customer data, very quickly brought the WASP technology in-house. Many of the established WASPs quickly adapted to the licensing-model, which is more tenable for large corporations. Some smaller players like NetMorf, 2Roam, and Thin Air Apps could not adapt quickly enough to the changing market dynamics and disappeared.

5.3   Enterprise solution providers

Companies in the enterprise solution space are continuously evolving; the service and product offerings are changing fast to align with market needs. In the late 1990s, it was difficult to categorize such vendors as a product company, a services company, a hosting service, or something else. But, after the economic burst in 2000–2002, firms have gravitated towards certain areas of expertise. There are two categories of players in this field: adaptive technology players and vertically integrated wireless applications and services.

Adaptive technology players are used for dynamically converting existing corporate applications or for delivery to browser-based and proprietary devices. The solutions offered by these players offer the flexibility of enabling the delivery of existing ERP, CRM, and Intranet-hosted systems to the wireless device.

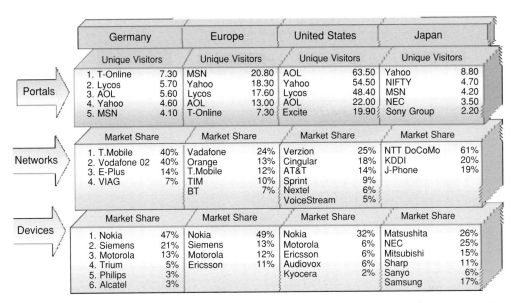

Figure 5.2. A look at the value chain across different geographies (source: Yankee Group, 2002).

Players who provide vertical integration of applications and services are used to tailoring wireless solutions from scratch, sourced directly from corporate data. Vertical applications (and services) offer the benefit of rapid deployment and a full end-to-end solution that works out of the box. What they might lack in flexibility, they make up for in terms of the provision of industrial strength solutions for a specific need. In the context of application services, the corporate customer also benefits from minimum up-front investment and resource commitment.

So, far we have discussed several segments of the value chain and the important players in each segment. Figure 5.2 compares the top players in the three key market segments: Portals, Networks, and Devices.

5.4   Where is the value in the wireless value chain?

As industry value chains continue to morph, some links in the value chain are in danger of becoming commodities, which severely limits the opportunity to create significant value. These areas are also less feasible for new investments. Over the past decade, so much money has been invested in several areas of the value chain that many of the companies can only hope to recover their investments instead of worrying about capturing added value. For example, in the case of major carriers like Sprint PCS and others, who have invested heavily to build network "pipes", are forced into a fierce battle over pricing so that the pure voice-based ARPU is coming

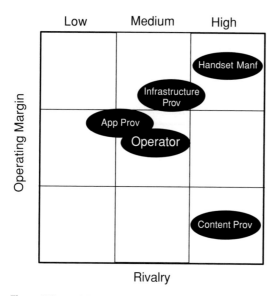

Figure 5.3. Positioning in the wireless value chain.

down. Every player in the chain has to add some value or else, with declining margins, there are no opportunities to add value in their wireless offering. The same is true for content companies such as Yahoo! and AOL as information like news, weather, financial and stock trading is readily available from several sources. Content needs to be unique and tailored towards individual needs and preferences.

For the foreseeable future, the value in the wireless value chain will lie in applications and software (see Figure 5.3). There is an almost endless number of both industry-specific and general-business applications that could become part of the daily lives of tens of millions of workers in the next few years, and many of these applications are still being developed or are yet to be developed. Companies that provide wireless applications will become the ultimate winners because they have the ability to create a differentiated product that adds value to the operators' offering or that streamlines business processes. As was the case in the PC market, it was not the manufacturers of hardware that ultimately captured the value; it was the software providers and applications developers.

In this section, we have reviewed the wireless value chain. In the next section we will take a closer look at factors affecting the value chain.

5.5   Analyzing the forces influencing the value chain

The wireless industry has experienced rapid change almost continually since its commercialization in 1983, continuing into the rapid growth of the 1990s and into the twentyfirst century. These patterns of change will continue to persist and intensify.

Table 5.6. Wireless industry segments and their clockspeeds

Device	Fast	Hardware (Nokia, Ericsson, Morotola, Palm)	Fast
		Operating System (Palm, WinCE, Psion)	Slow
		Applications	Medium
Access provider (AT&T Wireless, NTT DoCoMo, Vodafone, Cingular)	Fast	Standards (GSM, CDMA, 3G)	Slow
		Equipment (Lucent, Nortel)	Medium
		Network design (IP, Sonet, ATM)	Slow
		Product and service packaging	Fast
Content	Fast		

It is important to understand how and why these changes take place so a foundational framework can be laid out to assess and discuss the dynamics of buyer and seller in a rapidly changing landscape. For each segment in the value chain, it is important to understand the speed at which things are changing within that ecosystem. To analyze this we will use the concept of 'ClockSpeed' introduced by Professor Charles Fine at Sloan School of Management, MIT, in his landmark book *ClockSpeed: Winning Industry Control in the Age of Temporary Advantage*. The ClockSpeed concept seeks to understand the various rates of evolutions called 'clockspeeds' in any given industry. Each industry evolves at a different rate, depending in some way on its product clockspeed, process clockspeed, and organization clockspeed. For a more thorough coverage of the concept, please refer to the text. The discussion of Fine's model in this chapter also draws on the work done by Scott Constance and Jeff Gower. Table 5.6 lists the important segments of the wireless value chain and summarizes the clockspeeds within those segments.

Our goal in this section is to provide an assessment of industry change. We have broken this down into the following steps:
- assessment of entry of new players,
- analysis of competition in the market, and
- the wireless industry's competition intensity and desire to innovate.

We will be using Charles Fine's double-helix model and Porter's Five Force analysis to discuss the above.

5.5.1 Five Force analysis

So what forces are influencing the value chain today? To help us start this analysis we will look at Porter's 'Five Force' analysis[1]. This tool allows us to look at supplier

---

[1] For an introduction to Five Forces principles, please refer *Competitive Strategy, Techniques for Analyzing Industries and Competitors* by Michael E. Porter.

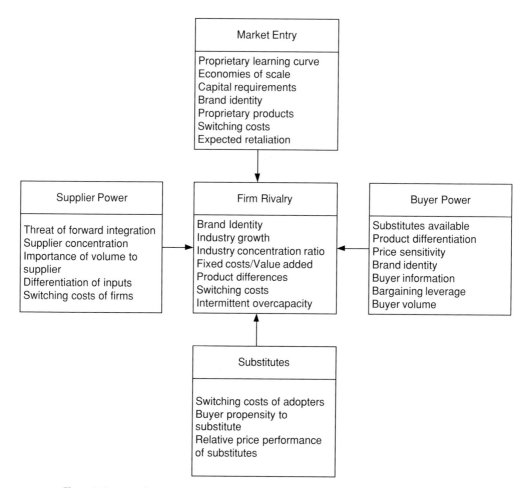

Figure 5.4. Porter's Five Force analysis.

power, firm rivalry, buyer power, the effect of substitutes, and the ease or difficulty of market entry. Five Competitive Forces analysis was refined by Michael Porter, one of the major figures in the field of strategic management. The five forces are as follows.

• The threat of new entrants into an industry or a market served by a specific company.
• The bargaining power of suppliers.
• The bargaining power of customers.
• The threat of substitute products or services.
• The intensity of the rivalry among existing firms.

As can be seen in Figure 5.4, there are many quantitative and qualitative measures or metrics to consider when assessing the value chain with the "Five Force" analysis. Projecting this onto the wireless value chain and simplifying the analysis, this type of analysis helps us to do the following.

Market Entry	Entry
Operator	Hard
Infrastructure Prov	Difficult
Device Manf	Difficult
Application Prov	Easy
Content Provider	Easy

Supplier	Power
Operator	Strong
Infrastructure Prov	Intense
Device Manf	Intense
Application Prov	Weak
Content Provider	Strong

Competition	Rivalry
Operator	Intense
Infrastructure Prov	Intense
Device Manf	Intense
Application Prov	Weak
Content Provider	Strong

Buyer Power	Power
Operator	Strong
Infrastructure Prov	Weak
Device Manf	Strong
Application Prov	Strong
Content Provider	Strong

Substitutes	Number
Voice/Data Network	Many
Network Comps	Few
Cell Phones	Many
Application/Content	Many

Figure 5.5. Wireless industry using Porter's model.

- Identify the actual activities performed by business units.
- Analyze the value created by these individual activities.
- Examine how linkages and flows to external buyers and suppliers build value as successive processes occur.
- Map the exchanges of flows into and out of the organization.
- Identify activities that are key to the success of the strategy.
- Understand resource allocations with a view to allocating resources in accordance with the contributions of the task to strategic direction.

We can measure the forces that each player feels within the value chain (see Figure 5.5). Market entry is measured either as hard (difficult to enter) or easy (low barriers to entry). Buyer power is weak, strong, or intense, based on their ability to demand products at particular price. Substitutes can be many or few, based on the number of potential alternative products or services offered in the market. Supplier power ranges from weak, to strong, to intense, based on the suppliers' ability to extract additional rents. And finally, competition is weak, strong, or intense, based on rivalry among firms. Overall it gives us an understanding of the types of force each player within the value chain is feeling.

Two apparent conclusions can be drawn. First, network operators, device manufactures and infrastructure providers are experiencing tremendous pressures from suppliers and buyers, and competition from rivals and substitute products is intense and increasing. Second, the newer players, application developers and content providers are in much less competitive markets, but they are dependent either on the buyers to help develop the newer markets or on the suppliers who control the interface to products on which these applications operate.

5.5.2   Clockspeed in network markets

Fine's double-helix model illuminates the degree to which periods of vertical and horizontal industry structure determine the fates of products, companies and industries. Fine's model helps participants to prepare for shifts in industry structure and hedge against the potential losses caused by these shifts. Fine emphasizes the notion of temporary advantage – participants must realize that one era's critical capabilities may become commodity capabilities in the next era.

Locating a participant's offering within Fine's double helix (Figure 5.6) is a prerequisite for understanding likely forthcoming changes in industry structure and making optimal preparations for these changes. As we have stated before, two drivers

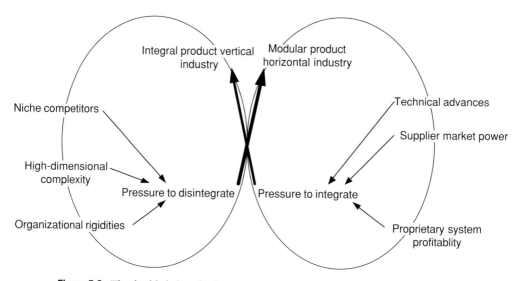

Figure 5.6. The double helix of industry structure.

of clockspeed are competitive intensity and technological innovation. When the industry structure is vertical, and product architectures integral, the competitive intensity is low.

Firms enjoying the rewards of slack competition will eventually confront the potential entry of competitors into niches of their integral architectures. The difficulty of keeping ahead of competition across many dimensions of technology and the organizational complexity of large organizations and complex products will put pressure on vertical firms to disintegrate, especially in the presence of outsourcing offers from potential suppliers.

The acquiescence of vertical firms to these pressures will foster the dynamics of transition, in which the previously vertical firm becomes dependent upon its new suppliers. Continuance of these dynamics results in complete modularity, as participants specialize in the production of horizontally applicable components. These horizontal firms depend on upstream suppliers for factor inputs and on downstream customers to absorb their offerings.

We have come to see that industries, products and firms cycle through the double helix, shifting from vertical/integral to horizontal/modular and potentially back through the helix. Further, it has become apparent that specific submarkets or product groups within the overall industry may operate at different clockspeeds, thereby contributing more or less to changes in buyer and seller behavior.

Vertical industry structure

In vertical markets competitive advantage comes from economies of scale in the fabrication of components; control over delivery, quality, and rates of technical change; reduced vulnerability to holdup by suppliers; and a quicker information flow. Vertical markets have limited direct competition and competitive threats do not exist. However, a trade-off exists; the vertical nature of the market reduces the competition for complementary products and drives slower adoption.

Transition from vertical to horizontal

The transition from a vertical market to a horizontal market increases the competitiveness of the market. Niche competitors provide incentives for firms to give up pieces of production, which increases entry and supplier power. Higher-dimensional complexity limits the economy of scale and increases potential vulnerability to holdup, while organizational rigidities decrease the transparency of information. All this increases the pressure to disintegrate, and increases both competition and adoption.

Horizontal industry structure

A horizontal industry is highly competitive. Competitors enter the market freely, driving down profits and competing on costs. Firms seek to use their small-differentiated advantage to push other competitors out of the market. As prices drop through price wars, adoption accelerates and the market continues to expand from new entrants.

Transition from horizontal to vertical

As adoption slows, firms identify technical advantages in one subsystem and gain competitive advantage over their many competitors. This market power encourages bundling with other subsystems to increase control and add more value. Further increases in market power in one subsystem encourage engineering integration with other subsystems to develop proprietary integration solutions. Competition begins to decrease as suppliers are squeezed out of the market and larger firms regain vertical control.

In summary, vertical market structure benefits firms who can extract greater rents, decreases competition along the chain and lowers the adoption of the networked product. As we shift to horizontal market structure, competition and adoption increase as complementary products increase the benefits that can be realized through standardization and interoperability among components. Buyers favor horizontal markets with greater competition while firms favor vertical market structures with competitive advantage and decreased, or little to no, competition.

5.5.3   Disturbances throughout the value chain

Although we consider that all firms act uniformly, it is important to remember that each firm within a segment can influence the value chain and therefore disrupt the traditional way exchanges occur. Firms within a segment that is exchanging cash indicate that one of the two segments are a seller and the other is a buyer. These players develop a link either out of necessity, strategic intent, or for learning. Figure 5.7 captures the key reason that each wireless value-chain participant is linked to the other in today's market environment.

In the following sections we will examine two specific value chains: the device value chain and the infrastructure value chain. In these discussions we will highlight the dynamics that these chain have experienced in the past and how the forces that make up clockspeed have contributed to these dynamics. We will then conclude the analysis with a look at how these two chains are influencing each other and how firms within different segments of the chain have and will respond.

5.5.4   Device value chain

As shown in Figure 5.8, the device value chain contains a number of players. Buyers purchase cell-phones from either VARs or network operators who in turn buy the phones or devices from manufacturers. These manufacturers then purchase software from application developers, circuit components from circuit board

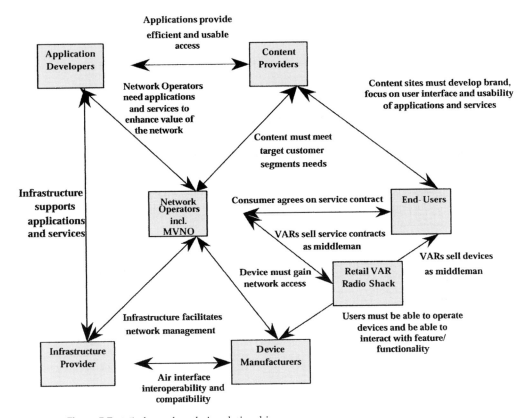

Figure 5.7. Wireless value-chain relationship.

component manufacturers and non-silica products from non-circuit board component manufactures.

Over the past decade the device value chain has seen the greatest increase in clockspeed. In 1983 the market for mobile devices was limited to a few players like Motorola, Nokia and Ericsson who produced the integral devices in a vertical industry structure as shown in Figure 5.9. The rapid growth of mobile subscribers has stressed the capabilities of different value-chain segments. In the early 1990s, till 1995, the market for voice-enabled cell-phones grew at a dramatic rate. As shown on the double helix, a shift began to occur in the industry structure. Vertical/integral device manufacturers scrambled to meet production levels between 1992 and 1993 as both demand and complexity accelerated. This increased the amount of outsourcing and the capacity dependence of these companies.

In 1994, delays of DSP chips and Flash memory translated into a production slowdown and a major shortage of devices to the market. Since this time demand has continued to grow, the complexity of the phone architecture has increased, and devices have shrunk in size. The average number of components per phone increased from 250 to 900, the size of the phone has decreased between 40–60%, and

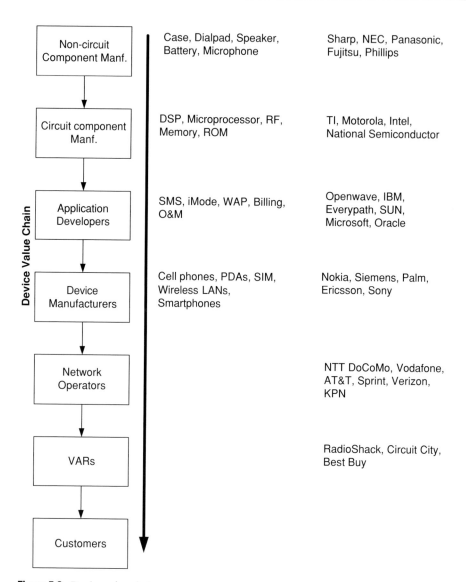

Figure 5.8. Device value chain.

the life-cycle of the phone has moved from 24 months to 17 months. All this has pushed the industry structure out of the vertical loop and into the horizontal loop. The modularity of mobile devices continues today as disparate chips consolidate into single fabs, non-silica components become transferable, SIM cards allow the user to transfer personal information between devices, and operating systems open up to application developers.

Texas Instruments (TI) is becoming a major player in the handset chip market, supplying baseband chipsets that control mobile phones. The latest phones with color screens can run sophisticated applications, and use more powerful and more

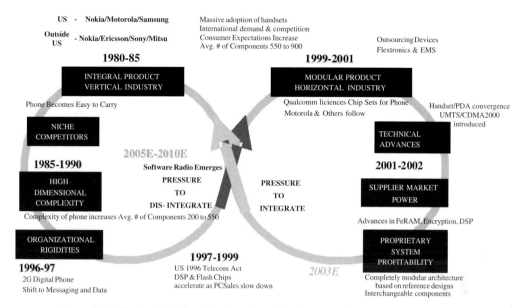

Figure 5.9. The double helix of device value-chain dynamics (source: Scott Constance and Jeff Gower, 2001).

expensive baseband chips, offering prospects for better margins and higher revenues for semiconductor suppliers. According to TI's chairman and CEO Tom Engibous, TI's baseband chips and the surrounding electronics will eventually account for as much as 65% of the cost of building a mobile phone compared with 30% in phones available in 2002. Given the lucrative nature of the business, Intel, Motorola and Qualcomm are wrestling to gain marketshare.

Also, Microsoft has teamed with Intel to work on a template for a high-end phone that they will license to cell-phone makers. The template will include software from Microsoft and chips from Intel, and aims to extend the Intel monopoly from the PC market to the cell-phone industry. As high-end cell-phones become increasingly complex, it is becoming expensive for a single player to design the units from scratch. As such, they are buying ready-made chunks of technology from semiconductor makers, software companies, and even competitors.

Will the modularity and horizontal segmentation of the mobile device industry continue? Recent pressure from tightening capital markets has forced a number of competitors to continue outsourcing or divest portions of their operations. The answer is that it will depend on network operators, application developers, and device manufacturers' ability to invest appropriate value, apply pressure, and create incentives between each other.

5.5.5   Infrastructure value chain

The infrastructure value chain also contains a number of players. Buyers of voice and data services purchase network access services from network operators who in turn purchase switches, gateways, modems and other components from infrastructure providers. These manufacturers then purchase base-station components, cell switching components, and PSTN/Internet components from suppliers. Unlike the device value chain, the infrastructure value chain has not seen a dramatic increase in clockspeed over the past decade; however, more recently pressures have started to increase and incentives are becoming more attractive as a number of new entrants and existing players pull the value chain towards a horizontal/modular industry structure.

Back in 1983 the market for infrastructure components grew as network operators slowly began to expand their geographic coverage within northern Europe, the USA and Japan. Limited in the most part to the same early entrants – Motorola, Nokia, Ericsson and NTT – networks were developed in test markets and eventually expanded to larger metropolitan areas.

Rapid growth of mobile subscribers between 1990 and 1995 began to justify a mass-market approach in wireless as customer demand grew outside the traditional emergency services and elite global enterprise consumer segments. Also during this period the clockspeed of air-interface standardization began to speed up. What was once a TDMA-eight, PDC- and GSM-dominated digital platform (between 1994 and 1996) suddenly expanded into five major platforms with eight to ten popular variations being supported by 15–20 large network operators. The most influential of these was CDMA IS-95, which emerged as the next-generation air interface and promised a lower cost to the subscriber.

The creation of new air-interface standards had an unusual effect on the wireless industry. What should have resulted in a decrease in clockspeed as we moved away from a single standard actually increased clockspeed. "Spider webs" of value-chain partnerships (along air interfaces, mobile operating systems and mobile application protocols) and "influence networks" have been created to push standardization of various platforms.

Infrastructure providers who had developed the interfaces through large investments in R&D have now shifted their investments into creating spider webs of supporters throughout the value chain. Firms have either scattered their bets towards many potential outcomes or have focused their investments, trying to push specific outcomes and create mass adoption.

More recently, a new set of forces has contributed towards driving the industry structure towards the horizontal/modular market. During 2000–2001, over 150 billion dollars were spent on spectral licenses for next-generation 3G services. Along with the drying up of the capital markets, this has resulted in delays of projected 3G rollouts. This shift in the economy has forced network operators to reconsider their investment strategies.

No longer can operators hope for a positive NPV over a period of five to ten years. Clockspeed is becoming too fast and resulting in greater complexity, expensive coordination, and insufficient return periods. Network operators are beginning to

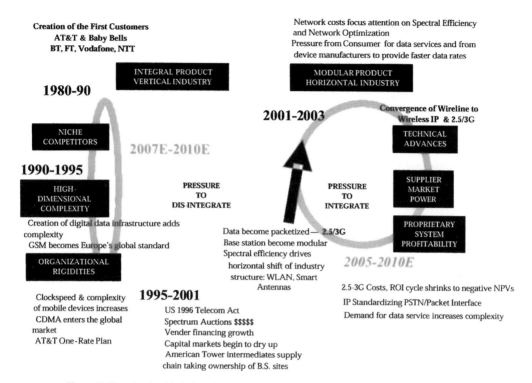

Figure 5.10. The double helix of the infrastructure chain (source: Scott Constance and Jeff Gower, 2001).

pressure infrastructure providers to create more flexible, scalable and updateable systems. No longer will the economics and adoption growth allow for "throw away" base stations, switches and gateways. Already we have seen companies like American Tower intermediate the value chain by taking ownership and management of large portions of the base-station towers in the USA. In Europe, competitors are following suit.

There are also new "Mobile Virtual Network Operators" (MVNOs) like Virgin who are beginning to lease spectrum from traditional network operators in Europe who are experiencing financial difficulty. Other niche suppliers have focused their attention on spectral efficiency. Mercury Computer System is pushing for modularization of the base station. Their goal is to create a network of wireless-networked base stations with embedded adaptive processors. This would allow for continuous updates of the latest optimization and signal processing software. The company boasts that its multi-user detection (MUD) application can increase capacity and distance of signals by over 60%. This could cut down the number of base stations in a typical area by between 25 and 40%.

Other spectral efficiency providers are marketing additional solutions. Dr Martin Cooper founded ArrayComm in 1997 and has developed smart antennas that will decrease the amount of inefficient signal transmission by reducing interference.

ArrayComm's products also promise dramatic performance increases compared with current technology.

Another disruption to the wireless WAN industry is being propagated by wireless LAN technologies like 802.11 (Wi-Fi) by providing cheap alternative to 3G for localized high-speed data access. Although Wi-Fi is a complementary technology to 3G, most carriers today are perceiving it as a threat to their 3G ambitions. As such, they have been now openly embracing Wi-Fi as one of the services they want to offer to their customers. VoiceStream's acquisition of bankrupt Wi-Fi service provider Mobilestar was a step in that direction. Other players like AT&T, Sprint PCS and NTT DoCoMo have also been experimenting with the technology and service offerings. We will deal with the 3G vs. 802.11 debate in more detail in Chapter 10.

In Figure 5.10, we see that the double helix of the infrastructure industry has so far sustained its vertical/integral architecture. Infrastructure providers have been able to ignore niche competitors and their efforts to increase the spectral efficiency. The recent pressure from tightening capital markets has also had an effect on the infrastructure market. Additionally, the pressure from the huge costs of moving to 3G applications and services is beginning to shift the industry structure.

5.5.6   Disruptive phenomena

How do you tell the difference between a disruptive technology and a false positive? Dr Clayton Christensen – a noted Harvard professor and author of *The Innovator's Dilemma* – explained his litmus test: "Does it compete against non-consumption? Does it target customers that are delighted to have a crummy product? Does it help customers do something they wanted to do but didn't have the means for or is it predicated on wanting to get something done that consumers haven't prioritized?" If the answer is mostly yes, technology is a good candidate for creating disruption.

5.6   Conclusion

In this chapter we have looked at all the critical segments of the wireless value chain. We listed the major players in each segment. In the second half of the chapter, we took an in-depth look at the factors that impact on the value chain using two different analysis models. As in any industry, it is important for players within the value chain constantly to evaluate their role in the ecosystem and how they are being supported, and what they can do to support the ecosystem. Ignoring fundamentals of business is always disastrous and this is especially true in the extremely competitive industry of wireless.

In the next two chapters, we will take a look at global wireless technologies, from systems and architectures to network and access technologies to software technologies.

6 Global wireless technologies: systems and architectures

We have come a long way from the early 1980s when the first-generation analog cellular networks were deployed; cellular phones used to weigh more than a pound, looked ugly, and were considered the toys of only the elite. Fast forward to the present. There are already several successful deployments of third-generation (3G) digital wireless networks, the sleek cellular phones weigh less than a few ounces and are considered indispensable tools for common man worldwide. In this chapter, we will look at the technologies impacting the wireless world and lay out a framework of technologies that will become the basis of our discussions in future chapters. We will cover not only some of the core wireless technologies but also the computing and communications technologies that impact on the wireless world. Detailed discussion of these technologies is beyond the scope of the book; however, we provide a list of references for anyone interested in further research.

6.1 The pervasive computing landscape

One of the most fundamental transition that is in progress in the information world is the evolution of pervasive computing (also known as ubiquitous computing). Pervasive computing in a nutshell means access to any information, anytime, anywhere using any device and any network. It means empowering the end-user with unprecedented access to information and the ability to act on it. This transition has been fostered by the ever-growing convergence of the computing and communications industries with consumer electronics and content and services. The ferocious pace with which innovations are being introduced to the market is breathtaking. Different hardware, software, network and peripheral technologies are coming together to form the pervasive computing ecosystem where information is readily available, transactions can be carried out with the touch of the button, enterprise data are extended beyond the shackles of the desktop to be accessed by people in the field, and the time between your thought and action has decreased to mere seconds. As depicted in Figure 6.1, the needs of pico-, micro-, and macro-environments is serviced

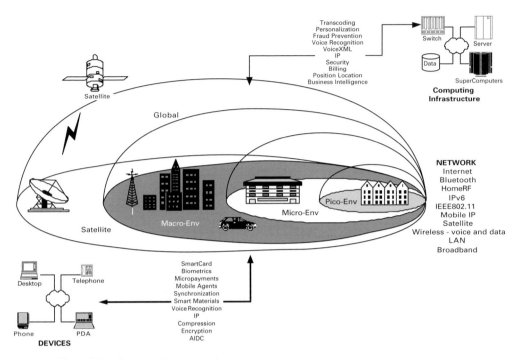

Figure 6.1. The pervasive computing ecosystem.

by different technologies and it is this interfacing of various computing and communications technologies that is making the concept of pervasive computing a reality. Pervasive computing has a far-reaching impact not only on design and development of devices, infrastructure, applications and services, but also on the cost savings and time efficiencies of an individual or a corporation. Pervasive computing is not only impacting on the existing legacy and Internet applications and services, but is also enabling a whole new suite of applications and services. This fosters innovation and competition – the backbone of progress. We will be discussing many of the technologies listed in Figure 6.1 later in the section.

Let us discuss an example to appreciate the impact of pervasive computing on the daily lives of professionals in the field. Every city has a Utilities department whose responsibility is to ensure that all basic infrastructures (electricity, water, transportation) are up and running, all the time. Typically, this is done through staff called field and service engineers. These engineers go out into the field (responding to public calls or otherwise), investigate the problem, and take notes in the field on paper forms. Once they get back to their offices, the information is either entered into a computer database or some more forms are filled out before the information makes it way into the databases. Typically, a supervisor reviews the report produced from the database and analyzes the problem by geography, so that the issues within a geography can be grouped and handled at one time. After review, based on availability, he or she assigns an expert to take a look into the issue or problem. This expert picks up the paper report and goes to the site, does further investigation, and fixes the problem. On his or her return to the office, another paper copy report is filed, which the supervisor reviews before the issue is considered closed. As, you can imagine, the process of solving these problems from start to finish is very tedious, highly inefficient, and it can take days, even months, to get small issues resolved. The tracking of the process gets in the way of solving real problems, quickly and in a cost-effective manner.

If these field engineers, experts and supervisors were armed with wireless and GPS-enabled PDAs, which had an application installed on these devices to meet their needs, things would be so much better. Field engineers could file their reports, while still on the ground, and their supervisor could program business rules into the application that intelligently assign tasks to experts in the field, in real time. Experts could get instant messages on their PDAs regarding work assignments based on their level of expertise and geographical coordinates. Once the problem is solved, experts could file their reports on their PDAs as well. These data are then entered into the back-end database servers, automatically. The task that took days and months could now be completed in a matter of hours, even minutes. More problems could be solved much more quickly. There would be no need for loads of paperwork or confusion. Time and cost savings to the department, city and, in turn, residents could be immense. For large and remote geographical areas like some parts of the USA and Canada, it is economically impractical to install a wireless data infrastructure. In those instances, it makes sense to get access to the same enterprise information via voice using interface technologies such as VoiceXML and SALT (Speech Application Language Tags). Users can use a simple ubiquitous device – the telephone, to interact with the system via natural voice and just by using a few words can access and update data in real time. So, as we stated earlier, the same back-end information can be accessed as well as acted upon using a variety of different interfaces and end-devices depending on the users' preferences or situation and the ecosystem should be ready to provide that flexibility to the end-user. In a few years' time, this key aspect will become so ingrained in our daily lives that we will not even notice its existence; we will just expect it to be omnipresent.

The benefits of automatic data entry have far-reaching effects on industries like retail, utilities, medical, sales, customer service and emergency. Table 6.1 lists some of the applications and service areas where pervasive computing is already truly transforming the landscape.

As is evident from the applications and services listed is Table 6.1, pervasive computing impacts on various aspects of every industry and just like Internet-centric computing transformed client–server computing so dramatically, pervasive computing is the natural evolution to interconnect devices, computing power, information and people. And wireless is at the heart of pervasive computing to make all this

Table 6.1. Sample of applications impacted by pervasive computing

Application area	Applications	
Broadcast	• News • Weather advisory	• Travel advisory • Advertising
E-commerce/ E-business	• Shopping for books, jewelry, etc. • Auctions • Dynamic distribution • Customer service • Stock trading • Banking • Location-based services • Dispatch and delivery • Field services	• Exception and business intelligence reports and data messaging • Customer relationship management • Enterprise resource planning • Supply chain management • Point of sale • Package pickup, delivery and tracking
Entertainment	• Soap opera updates • Interactive multi-user games • Horoscopes • Puzzles, quizzes	• Inventory control • Credit/Smart card verification • Chat
Information	• News, sports • Weather • Travel • Movies, restaurants	• Music • Gambling, betting • Lottery tickets • Traffic, airline schedules
Intranets	• Sales force automation • Access to Intranet information like emails, calendar, contact list • Contracts signing workflow • Remote network diagnostics, analysis and recovery	• Exchange rates • Yellow pages • Movie rental • Work flow automation • Video conferencing, shared white board
Medical	• Medical guidelines lookup • Doctors writing prescription	• Expense reports • Asset tracking
Personal	• Health care monitoring • Access to information like email, calendar, contact information • Tracking of pets and children	• Remote diagnosis • Emergency services: dispatch and location • Myportal consisting of links that are important to me
Telemetry	• Engine and vehicle statistics • Burglar or fire alarm reporting • Vending machine status • Weather statistics like temperature, wind velocity and direction, rainfall rate, barometric pressure	• Home and appliance networking • Stream statistics • Vehicular traffic statistics • Field measurements: gas, electric, water meters

possible. In the rest of this chapter, we will take a look at systems, architectures, software and network-access technologies. We will then discuss a wide spectrum of technologies that enable the wireless ecosystem: we will have a tutorial on systems and architectures; we will discuss the network and access technologies in this chapter; and in the next chapter we will discuss the role of various software components that have made world of wireless Internet possible and that provide the impetus for future growth and evolution.

6.2 Systems and architectures

In this section, we will briefly describe the systems and architectures of major cellular systems in the world. A detailed study of the cellular systems is beyond the scope of the book. Please refer to the References for further research. To help the reader to understand the architecture of a wireless network, we will describe each component at a high level and how they fit together. Figure 6.2 shows a basic wireless network with each of its major components. Let us first get familiarized with some of the terminology.

- Authentication center (AuC). An AuC is responsible for authentication and validation of services for each mobile device attempting to use the network.
- Base-station controller (BSC). A BSC controls a cluster of cell towers. A BSC is responsible for setting up a voice or data call with the mobile terminal and managing the hand-over when the mobile unit makes the transition from one cell tower boundary to another without disruption of service. For the purposes of our discussion, a BSC can also be referred to as a base station.
- Cell tower. A cell tower is the site of a cellular telephone transmission facility. Wireless coverage of any city is generally divided into rough hexagonal boundaries with one cell tower to cover each region. Depending on the usage and geography, this hexagonal shape could vary.
- Data rates. Data rates refer to the speed with which data can be transferred from point A to point B. In a wireless context, it is the speed with which a wireless device can communicate with the wireless network.
- Equipment identity register (EIR). An EIR stores and checks the status of mobile identity and electronic serial numbers.
- Home location register (HLR). An HLR keeps track of information regarding the subscriber. For example, the HLR keeps a record of the last time the mobile unit (cell-phone, or Palm, for example) was registered on the network. Note: mobile units register with wireless network every few seconds to identify their location – this helps in quicker call set up when BSCs have to find the mobile device.
- Mobile identity number (MIN) and electronic serial number (ESN). All phone equipment used in a wireless network carries these identification numbers. MIN and ESN are used for verification authentication and billing purposes.

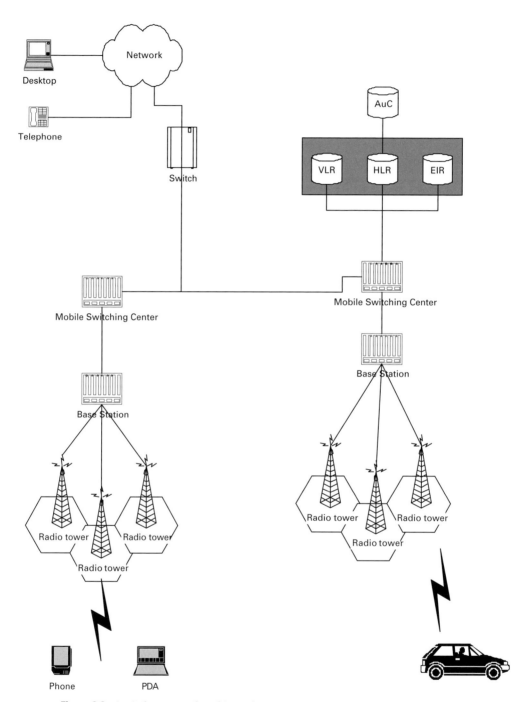

Figure 6.2. A wireless network and its major components.

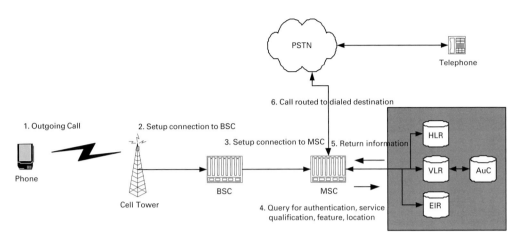

Figure 6.3. Call set up process in wireless networks.

- Mobile switching centers (MSCs). An MSC connects all the base stations together to pass communication signals and messages to and from subscribers operating on the network. An MSC is connected to a visitor location register (VLR); a home location register (HLR), authentication center (AuC) and an equipment identity register (EIR) and other MSCs.
- Operations and maintenance center (OMC). An OMC is also connected to the network to provide functions like billing, network management, customer care and service provisioning.
- Radio frequency (RF) transceiver. A transceiver is a combination transmitter/receiver in a single package.
- Visitor location register (VLR). A VLR records information about mobile units that have roamed into their network from other networks – that is, it tracks visitors. For example, if your cell-phone is registered to operate with AT&T's network in Seattle, and you want to make a call in New York, the VLR registers details about the mobile and its plan in New York.

To set up a call, a wireless device's radio frequency transceiver sends a message request to the nearest base station. The base station recognizes the call signal and routes it to the MSC. MSC queries the HLR or VLR (depending on the original registration of the mobile; if it is local, HLR will be queried, otherwise VLR), EIR and AuC for location, service qualification and features, and authentication. Depending on its destination, the call is routed either to another base station (or the same one, if the recipient is in the same area) or to a traditional wireline via the public switched telephone network (PSTN) or to an Internet device via the Internet. The call can be either a voice or data call. Figure 6.3 shows the message flow for a call originating from a wireless device to a LAN phone. For the reverse call set up, a LAN phone attempts to connect to the wireless device.

Since the early 1980s, wireless technologies in the three major geographical areas (Japan, Europe and the USA) have evolved differently. With the advent of 2G and 2.5G network access technologies such as CDMA, GSM, GPRS and EDGE, voice and data networks began to converge to provide a richer application and services environment to the user. With 3G, wireless data rates have graduated to LAN-like access rates to provide high bandwidth and QoS (quality of service).

6.3 Packet data

Wireless packet data is the data transmission service using a packet-switched network for wireless access. The typical examples of packet data cellular systems currently available are GPRS, CDPD, and PDC-P.

The term packet-switched transmission describes the type of network in which relatively small units of data, called packets, are routed through a network based on the destination address contained within each packet. Breaking communication into packets allows the same data path to be shared among many users in a network. This type of communication between sender and receiver is known as connectionless (rather than dedicated) communication. Most traffic over the Internet uses packet switching, and the Internet is basically a connectionless network. Users are generally charged based on the number of packets sent.

Circuit-switched transmission describes a type of network in which a physical path is obtained for, and dedicated to, a single connection between two end points in the network for the duration of the connection (thus, it is called dedicated communication). For example, the ordinary voice phone service is circuit switched. The wireless operator reserves a specific physical path to the number you are calling for the duration of your call. During that time, no one else can use the physical lines involved. Users are generally charged for the duration of the connection time.

GPRS introduces the novel concept of the device being always connected to the network, meaning that the device and terminal can get instant IP connectivity. The network capacity is used only when data are actually transmitted. Because of high available speeds, GPRS, along with EDGE, provides a nice transition for operators who are not keen on investing in 3G infrastructure up front.

Figure 6.4 shows a typical GPRS network configuration.

PDC-P is wireless packet data using the Japanese second-generation system, PDC. PDC-P provides the DoPa and i-mode service of NTT DoCoMo and allows for burst transmission speeds up to 9.6 kbps for i-mode and 28.8 kbps for DoPa.

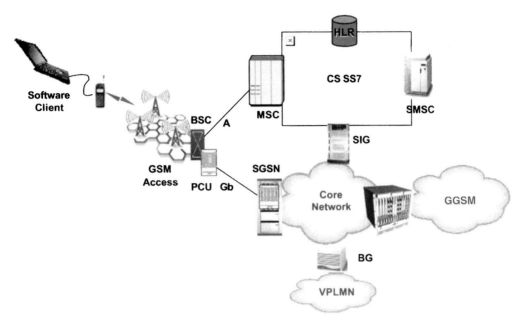

Figure 6.4. GPRS network configuration (source: Nortel). SGSN: Serving GPRS Support Node; GGSM: Gateway GPRS Support Node; PCU: Packet Control Unit.

Table 6.2. Packet switching vs circuit switching

Packet switching	Circuit switching
Efficient for short- to large-burst transmissions	Good for very large transmissions (for example, a 1 Mb data file)
Call set up and termination are almost immediate	There is a delay in call set up and termination
Carries broadcast capabilities	Enables point-to-point (P-to-P) connection
One logon at power up	Logon (call) for every transmission
Airlink is generally secure	Airlink is not secure

6.3.1   Packet switching vs circuit switching

Some of the key differences between packet- and circuit-based transmissions are highlighted in Table 6.2. The various types of wireless data networks are presented in the following sections. Let us discuss why the transition from a circuit-switched world to packet data is so important.

6.3.2   Packet data is economical

From a service point of view, wireless packet data has significant merits over circuit switched data for small data transmissions. The tariff can be minimized owing to

Table 6.3. Sample billing for DoCoMo content services

Application/service	Billing (Japanese yen)
20 character text	0.9
Look at news	14.7
Restaurant information	23.1
Mobile banking (check account)	9.7
Airline (check seats)	24.6

packet-volume billing rather than connection-time billing of circuit-switched data. That is why packet data is sometimes also called connectionless communication.

It is widely recognized in Japan that the technical reason for i-mode's success is packet-data transmission because users can enjoy economical tariffs. The packet tariff for i-mode is only 0.3 yen for 128 bytes. Table 6.3 shows some average packet tarrifs for typical content and applications (1$ = 120 yen, February 2003).

In spite of cheap tariff systems, the average monthly income by packet data (i-mode) per user was over 1540 yen as of March 2002 (total ARPU 8480 yen). The packet tariff for the FOMA service (DoCoMo W-CDMA) is cheaper than i-mode and it costs only 0.2 yen for 128 bytes.

6.3.3   Packet data is "always on"

Users feel that they are always connected to the network, so it is called "always on". This feature gives users freedom of care on the tariff and on the time of the communication. For the mobile user, "always on" communications not only deliver valuable wireless data and voice services to mobile subscribers, but also fundamentally change the way people interact with one another in all aspects of their daily lives. "Always on" services and applications are controlled and customized by the mobile user such that users can tailor these services to their constantly changing needs. This is one of the main reasons why RIM has been so successful with its Blackberry service. For the wireless operator, the "always on" network delivers two critical benefits: a reduction in costs and an increase in time-to-market with new revenue-generating services so that operators can deliver "always on" voice and data services that their customers demand.

By becoming simply another part of everyday life, "always on" communications are giving rise to a whole new era in human culture. Thanks to these innovative communications capabilities, "always on" defines an era in which everyday life is easier, more productive and with far fewer hassles than ever before. After all, that is surely the basic idea of communications! As more and more wireless operators around the world deploy "always on" networks, the already-apparent shift in human culture will become even more pronounced. Consider the following real life scenario. A traveling business executive, en route to the airport, is alerted by her mobile communications device that she has a high-priority message. Clicking a button, she sees a text message appear instantly on the device's display screen; it is from the airline, informing her that her flight departure has been delayed by two hours. She decides to take advantage of the delay to get some work done. She is due to give a speech at a conference next week, and she had asked her assistant to draft some text for her. In the airport business lounge, she uses her mobile device to access first the corporate intranet and then the speech draft. She places a call to her assistant and, using a document-sharing feature, they go through the draft together. As her assistant types in her suggested changes on an office PC, the executive sees the revisions in real time, as they appear on the screen of her mobile device.

6.3.4   Packet data is high capacity

In addition to the above features, packet transmission can increase the actual traffic capacity for randomly generated calls.

Here is one example. Several years ago (before i-mode), DoCoMo experienced huge simultaneous voice calls in downtown Tokyo every Friday evening, especially after 5 p.m. These calls were short-duration communications between friends, family and colleagues who needed to talk about the plan for the night or to decide on meeting places. As they all tried to make the calls at the same moment, the network was sometimes overloaded and very often calls could not be completed. This became a social problem at that time and DoCoMo received many complaints from the users.

Today, most of this voice traffic is replaced by i-mode mail and DoCoMo rarely receives such complaints from users, thanks to the smaller bandwidth requirement of packet data than a voice call.

It should be noted that packet data has merits and demerits. Packet data is economical for randomly generated and also relatively small traffic signals such as short text and web access. For real-time communication such as a video conference and a video downloading service, packet data is not appropriate, since some packet data systems can not confirm the transmission speed in case multiple users share the same transmission bandwidth.

6.3.5   All-IP network

It is important to understand the meaning and implications of the all-IP network. Does it play at the transport, service or application level? The ultimate goal for adopting IP as a unifying protocol is to converge to a single protocol at the application layer. For example, the architectural principles for the all-IP UMTS network clearly states that the UMTS core network shall be independent of the underlying transport mechanism (see Figure 6.5). For the IP transport layer, layer 2 options are ATM, PPP, or MPLS. Therefore, wireless operators have several options with respect to implementing the initial packet infrastructure, as long as the core can be evolved to

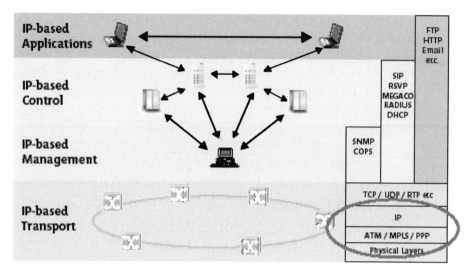

Figure 6.5. All-IP 3G architecture (source: Nortel).

support the high-bandwidth requirements of the future. IP already exists in the core of today's carrier-grade transport networks, but it is primarily being carried over ATM. ATM offers the asynchronous nature of packet switching, i.e. data are sent only as needed while requiring connections to be set up before any data are actually sent, as with circuit switching.

To enable the successful deployment of next-generation wireless Internet services, today's wireless operators must plan for a graceful migration to a pervasive IP infrastructure based on IP version 6 (IPv6) as an underlying strategy to enable these services and reduce deployment costs. Wireless operators must study their networks and plan the evolution of their core infrastructure carefully to ensure a seamless migration to IPv6 in the future.

The packet core is the most important piece of these networks. The reliability, scalability and versatility of its components will ensure the future success of the entire network. In the near future the packet core will need to support both voice and data traffic over IP as the network evolves. Given the efficiency of the packet-data network, a service provider will realize operational cost savings associated with leased transport, as well as additional revenue generated via personalized IP services. However, the initial costs of 3G spectrum and infrastructure will be considerable.

Currently there are several cellular systems in operation around the globe. They are both analog and digital systems. The first generation (1G) of wireless technologies, such as AMPS in North America, TACS in Europe, and NMT in Japan, was based on analog transmissions. As analog systems are being replaced and will be no longer in operation in some countries such as Japan, we will not touch on analog systems in this book.

The second generation (2G) of technologies, such as TDMA and CDMA in the Americas, PDC in Japan, and GSM in Europe and Asia, is digital in nature and provides improved system performance and security. In addition, there are two micro-cellular systems: DECT and PHS. Most new deployments are digital, with a few notable exceptions, such as those in South America and the USA. Even though digital networks are widely deployed, analog networks still service over 40% of cellular phones in the USA.

Apart from these distinct generations of technologies, there are additional technologies, which are referred as 2.5G wireless technologies. These technologies are High-Speed Circuit-Switched Data (HSCSD), General Packet Radio Service (GPRS), and Enhanced Data Rates for Global Evolution (EDGE). They enhance the existing digital technologies and provide a nice transition to the 3G technologies. GPRS-based systems are already being deployed on some wireless networks. Available rates for HSCSD are up to 38.4 kbps, those for GPRS are up to 144 kbps, and theoretical data rates for EDGE hover around 384 kbps.[1] So, network operators have an option of deploying GPRS and EDGE before they invest in expensive 3G infrastructure. The third generation (3G) of technologies includes UMTS and CDMA2000 technologies.

6.4   2G cellular systems (GSM, cdmaONE, PDC and TDMA)

As is evident from recent developments in the wireless industry, the rise of cellular service has been phenomenal. Every year, millions of new subscribers are enjoying the benefits of wireless services. With figures like these, the success of the cellular phone service rivals that of the car in the twentieth century. Table 6.4 provides a comparison of the four major cellular systems.

6.4.1   Digital cordless-based public systems (DECT, PHS)

As these two cordless systems (DECT, PHS) were developed based on digital cordless technology, they both employ TDMA-TDD access in the 1.9 GHz band. The system concepts of these systems are very similar, i.e. to bring an indoor digital cordless phone to outdoor public coverage. The output power of an outdoor base station is limited to several hundred milliwatts. As a result, the coverage area of a base station is so small (several hundred meters in radius) compared with that of cellular systems (0.5 km–1.5 km radius) that DECT/PHS operators have to install huge numbers of base stations outdoors.

---

[1] Practical data rates depend on the class of devices, the number of timeslots allocated to data, and the number of simultaneous users on the network.

Table 6.4. Comparison of 2G standards

	GSM	cdmaONE(IS-95)	PDC	TDMA (IS-136)
Frequency band	900 MHz, 1.8 GHz, 1.9 GHz	800 MHz, 1.8 GHz	800 MHz, 1.5 GHz	800 MHz, 1.9 GHz
Frequency channel separation	200 kHz	1.25 MHz chip rate/1.2288 Mcps	25 kHz (interleave)	30 kHz (interleave)
Modulation	GMSK	QPSK (FW) OQPSK (RV)	$\pi/4$ sift GMSK	$\pi/4$ sift QPSK
Access method	TDMA 8 ch (16 ch/half-rate)	DS-CDMA	TDMA 3 ch (6 ch/half rate)	TDMA 3 ch
Duplex	FDD	FDD	FDD	FDD
Transmission rate	270 kbps	max. 14.4 kbps (circuit switch)	42 kbps	48.6 kbps
Coding method	Full-rate RPE-LTP. Half-rate, enhanced-half rate (option)	QCELP EVERC	11.2 kbps VCELP (full-rate) 5.6 kbps PSI-CELP	VSELP ACELP
Delay equalization	Yes	No	No	yes
Standardization	ETSI	TIA (IS-95)	ARIB (RCR STD-27)	IS-136, UWC-136
Main service areas	Europe, Asia, North America	Americas, Korea, Japan	Japan	Americas
Main operators	DT, FT, Telefonica, TIM, KPN, AT&T Wireless	Sprint-PCS. KT, KDDI, TeleSP	NTT DoCoMo, J-phone	AT&T Wireless
Number of countries of operation	197	58	1	22

The network architecture of micro-cellular systems uses the ISDN (integrated service digital network) network, while most of cellular systems use dedicated network from PSTN/ISDN. So, PHS/DECT is considered "wireless" ISDN service.

From a technical point of view, these micro-cellular systems still have the strength to compete with 2.5G/3G cellular systems. As they both employ 32 kbps-ADPCM coding schemes per radio carrier, they can provide 32 kbps data transmission capability. Some operators provide 64/128 kbps data service by multiplexing the radio carriers. Also, in some regions such as China and South America, DECT/PHS has been widely introduced as a WLL (wireless local loop) system thanks to its economical installation. WLL is a powerful solution for building up telecom infrastructures quickly instead of installing a cable to each home.

Figure 6.6. FOMA (3G) concept phones from NTT DoCoMo (source: NTT DoCoMo).

However, from a marketing point of view, DECT/PHS operators are still struggling to get more subscribers because of the relatively small area coverage, as mentioned above. In Japan, the total number of PHS subscribers has been hanging around five million for several years.

NTT DoCoMo is considering its PHS system for data centric wireless service and offering a PCMCIA type PHS card (p-in compact) for data users. These systems are believed to be kept for the future as comparative digital systems to 3G (see Figure 6.6).

6.5   Wireless LAN, home networks

Wireless LAN technology such as IEEE 802.11x or Wi-Fi (Wireless Fidelity) has recently become an accepted part of the corporate infrastructure. Corporations such as Microsoft have wired their large campuses with Wi-Fi technology to provide network access to corporate data from anywhere in the campus. Wireless LANs are applicable to all industries, and are especially suited for in-building networking, providing a convenient way to access the network through a simple use of a PCMCIA card for laptops and handhelds and access points, which connect the devices to the

LAN. It is fairly easy to set up an 802.11 network although appropriate attention must be paid to some of the common installation and configuration problems such as security, interference, multipath and interoperability. Industries such as retail, hospitality, warehousing, small business offices, and healthcare can really benefit from a quick, convenient way of setting up a reliable network. For instance, in hospitals doctors and nurses must maintain accurate patient records to ensure proper care. Most of the time, they use a paper-based system, which is prone to errors, requires a lot of time and training, and is just not reliable. Using wireless LAN PCMCIA cards, doctors can very conveniently use their PDAs, laptops or tablet PCs to enter the data directly into the database in real time with a few taps on the screen. This way doctors and nurses can spend more time with their patients rather than tackling paperwork.

Recently some cellular operators have shown strong interest in the feasibility of wireless LAN as a business. T-Mobile (VoiceStream), the North American unit of (T-Mobile International) Deutsche Telecom, has announced the launch of a wireless LAN service based on the now-defunct Mobilestar assests. NTT DoCoMo formally started doing technical feasibility tests in March 2002 to look for new revenue sources. The following is a brief description of various 802.11 standards.

- 802.11b. This is a wireless technology for transmitting data at 11 Mbps across short distances of about 300 feet. The technology operates within the unlicensed 2.4 GHz industrial band, the same wavelength used by microwave ovens and cordless phones. Given its rapidly decreasing price points, 802.11b will continue to dominate the landscape for WLAN chipsets for the next couple of years.
- 802.11a. This is an emerging standard that operates at 5 GHz and supports data rates of 54 Mbps over distances of approximately 50–100 feet. Although the 802.11a product was not available until the fourth quarter of 2001, some customers have already decided to postpone 802.11b purchases and jump directly to 802.11a should it prove a sufficiently robust platform. Although 802.11a has been ratified in North America, the standard has not been fully approved as an international standard, as Japan and some European countries are holding out on ratification. There are some concerns about the performance of 802.11a chipsets relative to 802.11b. Cost and backward compatibility will continue to be an inhibiting force for the adoption of 802.11a for some time. 802.11a, as a standalone, is not backwards compatible to the 2.4 GHz networks and thus requires the deployment of separate 802.11a access points.
- Despite some drawbacks 802.11a has some very compelling advantages over 802.11b and will eventually become a widespread standard, particularly in combination with .11a/11b or .11g chipsets used for enterprise deployments. 802.11a also uses OFDM but in the 5 GHz spectrum and provides eight RF channels as opposed to three for 802.11b and 802.11g in the 2.4 GHz spectrum. Although the range of 802.11a is arguably less than that of 802.11b, some tests of the system suggest that the system still outperforms 802.11b on an absolute throughput basis at all ranges

(in obstruction-less environments). 802.11a tiers down from 54 Mbps to 48, 36, 24 and 12 Mbps at the perimeter of the network, while 802.11b tiers down from 11 Mbps to 5.5 Mbps and 2 Mbps at the perimeter of the network.

- 802.11g. This is the newest of the ratified 802.11 standards; 802.11g provides 802.11a-like speed but is backwards compatible with .11a.
- 802.11i. This is a security enhancement to 802.11 networks.
- 802.1x. This is also a protocol for enterprise-type Cisco, 3Com, or other servers that provide hardware-driven encryption for SOHOs and enterprise environments.
- 802.11e. This is a QoS standard for video, which prioritizes the order of audio and video and supports peer-to-peer networking and multimedia. Companies such as Sharewave/NETGEAR have been large proponents of 802.11e in the past, citing the ability to provide DVD-quality video via an 802.11b connection at 7.2 Mbps. The advent of 802.11a and the considerably higher bandwidth for .11a at 5 GHz has further enhanced the possibility of multiple multimedia streams within a home or SOHO environment. Carriers worldwide are becoming interested in the QoS capabilities of 802.11e and are participating in the standards process.
- 802.11f. This is an enhancement to 802.11b that supports interaccess-point communication and enhanced network management capabilities.
- 802.11h. This is an improvement to 802.11a that could theoretically help 802.11a and HiperLAN coexist in Europe. The 802.11h protocol is supplementary to the MAC layer to comply with European regulations for 5 GHz WLANs – European radio regulations for the 5 GHz band require products to have transmission power control (TPC) and dynamic frequency selection (DFS) to minimize transmission output and interference with radar and other systems. The 802.11h standard was expected to be finalized by the end of 2002, although with the recent move from many European countries to permit the use of 802.11a in its current form, there is a question as to whether or not 802.11h will ultimately be necessary for the commercial adoption of 5 GHz WLANs in Europe.

Table 6.5 compares the main WLAN standards.

6.5.1   IEEE 802.11x

A host of major communications and PC OEMs, including Cisco, Motorola, Siemens, Nokia, IBM and Intel are making wireless LANs a critical piece of their overall broadband product strategies. Locations already wired for Ethernet access can be turned into WLAN simply by plugging a network access point (AP) into an existing jack, leveraging existing Ethernet infrastructure. Connecting via wireless eliminates the need to rewire buildings with multiple fiber, Ethernet or cable access points.

IEEE 802.11 or Wi-Fi is aimed at medium-range, higher data rate applications. The standard specifies a 2.4 GHz operating frequency using two forms of spread spectrum modulation: frequency hopping (FHSS) and direct sequence (DSSS). IEEE 802.11b is a data rate extension of the initial 802.11 DSSS, providing operation up to 11

Table 6.5. WLAN standards

Standard	Frequency	Maximum range	Practical data rate (theoretical)
IEEE 802.11b	2.4 GHz	30–100 m indoors 100–500 m outdoors	5 Mbit/s (11)
IEEE 802.11a	5 GHz	Same as above	32–38 Mbit/s (54)
IEEE 802.11g	2.4 GHz	Same as above	20 Mbit/s (22)
HiperLAN/2 (European Standard	5 GHz	Same as above	32–38 Mbit/s
Bluetooth[1]	2.4 GHz	10–100 m	Asymmetric 721 kbit/s downstream and 57.6 kbit/s upstream; symmetric 432.6 kbit/s

[1] Though not a practical WLAN technology, several Bluetooth vendors have been trying to sell Bluetooth products in the WLAN market with little success. As discussed in other sections of this book, Bluetooth is more suitable for PAN scenarios rather than WLAN.

Mbps. Using 802.11a, data rates of up to 54 Mbps can be obtained using orthogonal frequency division multiplexing (OFDM) modulation in the 5GHz frequency band. IEEE 802.11g is aimed at providing interoperability between the a and b versions of the standard. The usage of IEEE 802.11 is growing so fast that now the wireless Ethernet card is a standard feature on newer laptops. Access point kits are available from leading manufacturers such as Intel, Cisco and Lucent. A lot of consumers are using IEEE 802.11 for home networking as well. There are also efforts to provide interface between wireless WAN technologies and IEEE 802.11. A Seattle-based company called RadioFrame is attempting to build a radio that acts as such an interface.

WLAN configurations include both peer-to-peer functionality and client/server networking, offering fully distributed data connectivity.

(1) Peer-to-peer. A peer-to-peer network is a WLAN in its most basic form. Two PCs equipped with wireless adapter cards form a simple peer-to-peer network, enabling the PCs to share resources. This type of network requires no administration or preconfiguration, but also bypasses the central server, inhibiting client/server sharing. In *ad hoc* mode, multiple PCs can communicate directly with each other; no access point is needed.

Applications include:
- collaborative work groups,
- small/branch offices sharing resources,
- remote control of another PC,
- games for two or more players,
- demos.

(2) Client and access point. Client and access point networks allows for extended range capabilities; they are also able to benefit from server resources, as the access point (AP) is connected to the wired backbone. The number of users supported

by this type of network varies by technology and by the nature and number of the transmissions involved. Generally, client and access point networks can support between 15 and 50 users.

(3) Multiple access points. Although coverage ranges in size from product to product and by differing environments, WLAN systems are inherently scalable. As APs have limited range, large facilities such as warehouses and college campuses often find it necessary to install multiple access points, creating large access zones. APs, like cell sites in cellular telephony applications, support roaming and AP-to-AP handoff. Large facilities requiring multiple access points deploy them in much the same way as their cellular counterparts, creating overlapping cells for constant connectivity to the network. As network usage increases, additional APs can be easily deployed.

There is significant confusion surrounding wireless standards. In addition to 802.11b, there is 802.11a, a faster technology beginning to ship into channels, and 802.11g, a standard recently ratified by the IEEE. The newest of the ratified 802.11 standards, 802.11g, provides 802.11a-like speed but is backwards compatible with 802.11a. It is expected for 802.11g to be generally slow to ramp industry wide and most of the initial uptake on 802.11g to occur in 2003/2004, probably in 802.11a-combo chipsets.

WLANs operate currently in three different unlicensed frequency ranges: 902 MHz, 2.4 GHz, and 5 GHz. These are called the ISM (Industrial, Scientific and Medical) bands, each with different characteristics and advantages. The basic trade-off with the frequencies involves range versus data rate – the higher the frequency, the higher the data rate, but the smaller the range, and vice versa. The 5 Hz unlicensed band has more than three times the spectrum as the 2.4 GHz band available for 802.11a WLAN access, allowing 802.11a to provide an average of eight channels versus three channels for 802.11b. But because of the shorter range of transmission, 802.11a networks in the 5 GHz band may require more access points in a standard network in comparison with 2.4 GHz 802.11b (see Figure 6.7).

6.5.2   HiperLAN

HiperLAN began in Europe as a specification (EN 300 652) ratified in 1996 by the ETSI (European Telecommunications Standards Institute) BRAN (Broadband Radio Access Network) organization. The current version, HiperLAN/1, operates in the 5 GHz radio band at up to 24 Mbps. The standard uses the same frequency band in Japan and the USA. HiperLAN uses GMSK modulation and it also provides QoS support for various data needs.

6.5.3   UltraWideBand (UWB)

UWB, the newest entrant to the WLAN marketplace, is a low-cost, low-power method of delivering broadband connectivity. The technology uses a wide, flat signal that

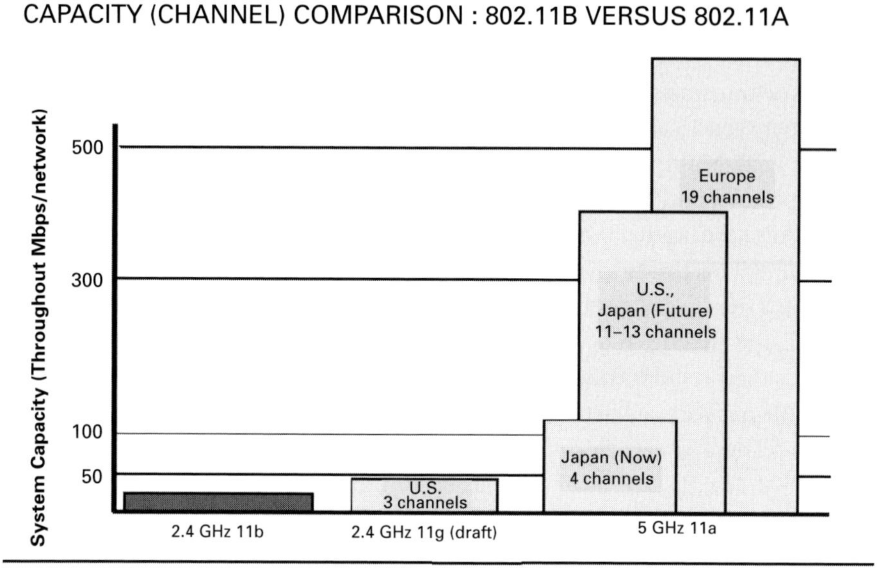

CAPACITY (CHANNEL) COMPARISON : 802.11B VERSUS 802.11A

Figure 6.7. Comparison of 802.11a and 802.11b. (source: Atheros Communications).

pulses rapidly. Initially established as a military application, the technology is now being looked at for its commercial applications, with the potential to achieve 500 Mbps at a fraction of the cost of an 802.11 network. UWB uses milliwatts of power to transmit, and gains throughput by spreading the transmission across the entire frequency band. Additional cost savings are gained through the reduction of the traditional radio functions, like an oscillator, filter and mixer. Major backers of UWB include Intel, Sony, Siemens, WorldCom and Marconi. Other UWB companies include XtremeSpectrum and Time Domain. UWB is discussed in more detail in Chapter 13.

An additional possibility for the future generation of 802.11 wireless LANs comes from an earlier attempt to establish a commercial wireless industry based on line-of-sight technology known as the Multipoint Microwave Distribution System, or MMDS. Major service companies like AT&T, Sprint, WorldCom, XO, Winstar and Teligent deployed MMDS systems and failed miserabley. Experts contend that MMDS technology failed in part because it required the receiver to be within sight of the transmitter, and also because it operated in more costly licensed frequency bands (~3.5 GHz). Learning from the mistakes of the original MMDS business plans and from the success of 802.11 operating in unlicensed frequency bands, companies ranging from start-ups to established OEMS are working on the development of the next generation of WLAN technology, often dubbed "4G". As a takeoff of 3G, 4G intends to expand the LAN to the WAN but maintain the current compelling cost points of deploying WAN architecture. Major backers, including IBM, Nokia, and private companies such as Navini Networks, are pulling together ways to build nationwide 802.11 deployments. We will discuss more about Mesh Networks in Chapter 13. Additionally, Smart antenna/multibeam technology has been considered as a method for transmitting 802.11b-type speeds across long distances. Companies such as Etherlinx are attempting to develop software-designed radio technology that, like Mesh, utilizes repeater antennas on the outside of homes.

Theoretically, each individual repeater would communicate with a central antenna (likely at 3.5 GHz), and would use 802.11b to deploy the signal within the home. To facilitate the growth of the combination of MAN (metro area network) and LAN networks, a unifying standard known as 802.16 is being developed by the IEEE Working Group on Broadband Wireless Access Standards. 802.16 is intended more specifically for wireless metropolitan area networks, with WLAN as an important future network extension/component. The IEEE is working to finalize 802.16 for WMAN + WLAN operation between 2 GHz and 11 GHz, with an enhanced physical layer and a MAC layer of existing MAN standards. Given the reduced prospects for nationwide 3G deployments in the United States and Europe in the near term, the push for less expensive MAN and LAN technologies may begin to gain momentum in the next few years.

6.6   PAN

Personal area network or PAN technologies are aimed at short-range, high-speed applications. Although some vendors would like you to think of these as wireless LAN technologies, the strength of PAN technologies such as Bluetooth lies in their capability to provide a low-power, cheap option to replace wires that connect various devices, appliances and computing systems. These technologies can also be used for small-range data exchange and transaction-execution applications. For example, a passenger could check-in using their Bluetooth-enabled PDA or phone into airline terminals or trains and get security clearance based on their ID information, which securely interacts with an authorized terminal server. The goal of Bluetooth SIG is to drive the cost of a Bluetooth chip to less than $5 so that all new devices are Bluetooth-enabled.

6.6.1   Bluetooth

Bluetooth is a low-power radio technology that replaces the cables and infrared links for distances up to 10 m and data rates up to 1 Mbps. Bluetooth uses FHSS in the 2.4 GHz frequency band. More than a thousand companies worldwide support and/or are working on standardizing the technology. Key applications include synchronization and exchange of data, electronic cash/wallet, telematics, authentication and validation, and replacement of cables. Some of the manufacturers are also positioning Bluetooth as a wireless WAN technology, although it is difficult to compete with the robustness and pervasiveness of IEEE 802.11. Industry groups are also working on coming up with interoperability specifications (to work with 802.11 and HomeRF).

6.6.2   AIDC/RFID

Automatic identification and data capture or AIDC technologies allow for very short-range data exchange. Operations such as logistics, supply chain, transportation, shipping and warehousing benefit a great deal from technologies such as radio frequency ID (RFID), barcodes, and mechanical and inductive flags.

6.7   Evolution scenarios towards 3G

It was towards the end of 1985 that the TG (Task Group) 8/1 of the ITU-R started its first activity on 3G radio access technologies. In the standardization bodies, 3G was first called FPLMTS (Future Public Land Mobile Telecommunication Systems) and then the name was changed to IMT-2000. IMT stands for International Mobile Telecommunication and 2000 refers to the year 2000 when the service was expected to start and also to 2000 MHz frequency.

After 14 years, at TG 8/1's 18th meeting in November 1999, IMT-2000 radio interface specification recommendation was finally completed as IMT.RSPC and was certified officially at the RA (Radio Assembly)-2000 meeting in May 2000.

The original motivation for 3G systems in the 1980s was to have a single global standard wireless system since there are multiples of 2G systems as discussed above. Japan in particular was a very enthusiastic supporter for the concept of a single global standard wireless system since the PDC system (2G system in Japan) is operated only in Japan. In the 1980s, there was no Internet service, and DSP and device technology was still immature so it was natural that the basic concept of 3G was to focus on voice services.

The five principal and original concepts of 3G were:
(1) advanced commonality of system design on a global basis,
(2) service commonality within the IMT-2000 network and with the fixed network,
(3) high-quality service (same as fixed wireless),
(4) small mobile terminals available around the world,
(5) global roaming capability.
It is interesting to note that the terminology "wireless data" did not exist at the time, but it is now the mainstream service for current 2.5G and 3G systems.

During the 1990s, many countries introduced and started operation of 2G systems, and 2G operators in the world have made tremendous investment to build their 2G networks. Also, cellular service was accepted enthusiastically by the public and the success of GSM especially in Europe and Asia was phenomenal.

It was seen that if current 2G users continue to use current 2G services in the twentyfirst century, operators will need to collect debts caused by their investment. So the concept of migration from 2G to 3G architecture was proposed. It was a natural transition and a quite reasonable proposal.

The concept of migration scenario from 2G to 3G was proposed for TG8/1 in 1996. In the process of standardization, a "network family concept" was proposed. The concept is that the core network of 3G should be based on current 2G core networks and flexible connection between radio access modes and core network can be achieved based on operator needs. As a result, 3G systems and architectures are designed as an advancement of 2G core network platforms. Finally two system architectures for 3G were standardized: one is GSM enhanced (GSM-MAP) and another is ANSI-41 enhanced. In addition, IP-based network architecture is under study for future solutions.

Concerning radio access technologies, the story was different from the network issue. In September 1998 (which was the deadline for the ITU-R proposal on radio access technologies), ITU-R received a total of 16 candidate specifications (10 for terrestrial and six for satellite) from around the world. Out of 10 proposals for terrestrial systems, CDMA accounted for eight and TDMA for two. Since then, intensive effort at 3G-PP/3G-PP2 has been made to finalize IMT-2000 radio specification recommendations and finally at TG8/1's 18th meeting in November 1999, IMT.RSPC was agreed upon. The outline of IMT-2000 is shown in Figure 6.8.

The OHG (Operator Harmonization Group), which aims to reduce the number of multiple standard specifications of 3G, proposed the concept shown in Figure 6.9.

As a conclusion, the following ITU recommendations were made by the OHP:

(1) CDMA technology can be considered as a set of packages with three different modes of DS (direct spread), MC (multi carrier) and TDD;

(2) TDD technology (layer-1) temporally includes both UTRA TDD and CWTS (Chinese proposal) and will be discussed for unification in 3G-PP.

6.7.1   Migration strategies of 2G operators

Based on these understandings of "family concept" and the OHP, some 2G operators have already established and announced their own migration scenarios towards 3G. These scenarios are very important for these operators since the scenarios will decide the investment policies for the next 10 years. Here are some case studies.

GSM operators

It is a relatively simple decision for current GSM operators to go to a GSM-enhanced (GSM-MAP) network with DS-CDMA (W-CDMA) air interface. This considers

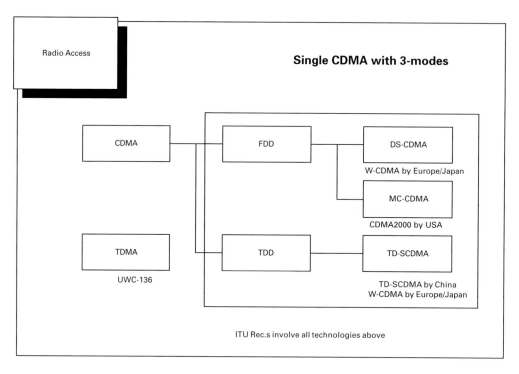

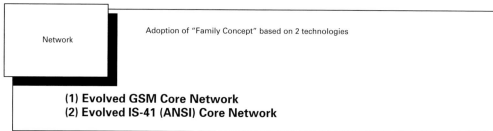

Figure 6.8. Concept of IMT-2000 radio access systems (for terrestrial).

maximizing the usage of existing GSM infrastructures and minimizing additional investment cost, and hence seems to be an optimum solution.

However, some GSM operators, especially in Europe, are having financial difficulties jumping into this scenario as in some European countries it was extremely expensive to get the 3G frequency license, and this has damaged their financial condition. So, some of the operators, in order to cope with the rapidly increasing demand for a wireless data service, are introducing GPRS (GSM packet network) as a short-term solution and the W-CDMA system as a long-term solution. They also considered GPRS as a good testing infrastructure for forthcoming 3G services and applications.

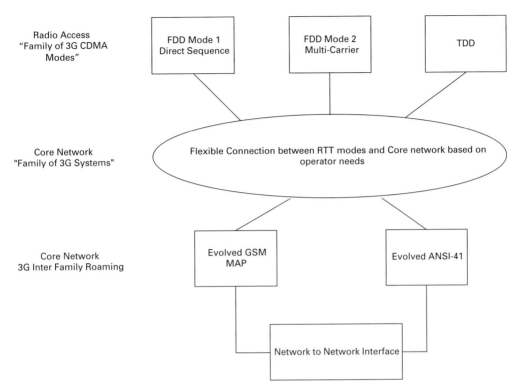

Figure 6.9. Three-mode proposal on 3 G by the OHP.

cdmaONE operators

Some cdmaONE operators have announced that they are going to the ANSI-41 enhanced network with CDMA2000 1X air interface. CDMA2000 1X technology is intensively studied at 3G-PP2 in detail as the continuous road map of the cdmaONE system. In the USA especially, most of the IMT-2000 frequencies of the 2 GHz band are already occupied as the national PCS band; the cdmaONE system has been widely introduced. For the air interface, the CDMA2000 system with maximum compatibility with cdmaONE will be used. For the network, they can upgrade the existing ANSI-41 network infrastructure with relatively small investment cost.

CDMA2000 1xEV-DO (HDR; high data rate) is a wireless packet system with a maximum 2.4 Mbps data rate for the down-link channel; cdmaONE operators can attach an HDR system to the existing cdmaONE infrastructure. KDDI (Japan) announced the start of their HDR service in 2002.

TDMA (IS-136) operators

TDMA (IS-136) operators need a different strategy from GSM or cdmaONE operators. For example, AT&T Wireless, one of the major TDMA operators in the

USA, with more than 17 million TDMA users, has announced its network migration scenario as follows.

Step 1. Introduce GSM/GPRS systems nationwide.

The GSM/GPRS infrastructure overlays the configuration of the existing TDMA network nationwide.

Theoretical maximum data speed is 115 kbps.

Step 2. Enhance GSM/GPRS air interface to EGDE (enhanced data rates for GSM evolution).

EGDE and GPRS differ only in modulation technology (8-PSK) and encoding scheme for FEC (Forward Error Correction).

Theoretical maximum data speed is 384 kbps.

Step 3. Introduce GSM-enhanced (GSM-MAP) network with DS_CDMA (WCDMA) air interface.

Theoretical maximum data speed is 2 Mbps.

For newer networks, Step 2 can be achieved by upgrading just the software, not the hardware or infrastructures.

PDC operators

PDC operators (NTT DoCoMo and J-phone in Japan) have to implement an overlay-type 3G network in addition to existing PDC infrastructures since there is no compatibility with PDC and 3G standard. The two companies both have selected a GSM-enhanced (GSM-MAP) network with DS_CDMA (WCDMA) air interface for 3G.

The strategies for these operators to go to GSM-based WCDMA are slightly different. For NTT DoCoMo, the frequency shortage problem (mainly in the Tokyo area) has been a serious headache for a long time, especially after the explosive increase of cellular subscribers after 1995 (about 10 million per year). So, DoCoMo spent billions of yen in investment for R&D of WCDMA technology and core networks over more than ten years. Also, the great successes of the wireless data service of i-mode has made DoCoMo confident of future high-speed wireless multimedia business. DoCoMo launched a WCDMA commercial test service in June 2001 and followed it with a commercial service known as FOMA (Freedom of Mobile multimedia Access) in October 2001.

For J-phone, which is a strategic unit of the Vodafone global group, compatibility of 3G infrastructure with European GSM/future 3G operators must be considered as the main priority. J-phone announced the launch of its WCDMA commercial service in mid-2002.

6.8   Devices and terminals

One of the key factors in the acceleration of the mobile society is the gradual miniaturization of communication terminals and devices and their convergence with computing technologies. The progress made in the past decade in terms of size, weight, functionality and marketing appeal is just phenomenal. For teenagers, telephone terminals have become akin to fashion accessories. The corporate information and data that used to fill a briefcase can now be conveniently packed into a small phone or PDA – essentially giving you a mobile office on your palm. One of the main factors in promoting a mobile society is the small, light-weight terminal. To create a small device, development of both hardware and software is indispensable.

From the hardware point of view, the following items need to be taken care of:
- battery
- display
- antenna
- filters
- LSI and packaging
- RF amplifiers.

From the software point of view, the following items are critical to the usability and functionality of the device:
- speech coding technology
- contents programming language
- human interface and character input scheme.

Let us briefly discuss the hardware components.

6.8.1 Battery

The battery is the key device for achieving a small, light-weight terminal. Presently, the lithium-ion battery is most commonly used thanks to its light weight and its high-capacity performance. Another reason is that the 3.6 V operation per cell of a lithium-ion battery is appropriate for cellular system. With this, typical 2G cellular/microcellular terminals have several hundred hours of stand-by time, which has greatly increased the ease of daily usage. The development race on battery technology for mobile systems is so intensive and fast that R&D on battery technology is usually promoted under very confidential and strict information control and is not widely available.

6.8.2 Display

In March 1999, the first color-display cell-phone was released in Japan and, since then, the color variations have improved dramatically from 256 colors to 65 000 colors. Introduction of a color-display device to i-mode has made a great impact on its sales. Also, content providers compete to provide the most compelling color content. Display has become a very critical factor for the wireless data market.

The display of a cellular phone is usually made of a Liquid Crystal Display (LCD). Power consumption of TFD-LCDs has been reduced to less than 10% of that of

10 years ago owing to the continuous effort of device engineers. For example, introduction of reflection-type color LCDs was a milestone in 1997 as it made the consumption power 1/7, weight 1/2 and thickness 1/3 of its predecessors. As a result, the ratio of energy consumption of the LCD part in a cell-phone total is now reduced to less than 1%. About 33% of the total consumption energy is used for the trans/receiver part and 66% is used for the control circuit part in a cell-phone.

6.8.3 Filters and LSIs

The RF-band filter is one of the key elements of a cell-phone since good RF filter characteristics can improve the frequency efficiency of the total system. Recent developments of SAW (Surface Acoustic Wave) filters gave that function wider band width, higher frequency band and contributed greatly to achieving a small and compact device. Multi-media SOC (Silicon/System On a Chip) is a combination of signal processing and LSI technology. Thanks to SOC, the multi-media function of a cell-phone, such as the real-time TV phone and video clipping service of 3G, can be achieved on a single chip.

6.8.4 Antenna

Antenna research has a long history – since Marconi, more than 100 years ago. The design philosophy of an antenna for a cellular system is basically the same as for other fixed radio systems, but the following three issues should be considered for cellular systems:
- strong requirement for a small and compact physical shape
- sturdiness (to protect it from consumer abuse)
- variable dataspeed requirements.

These three factors make the optimum antenna design of a cell-phone very complicated. Also, antenna configuration is greatly affected by overall system design. For example, the GSM system employs no diversity reception, but the PDC system does employ that method. In addition, non-unified allocation of 2G/3G frequencies on different continents requires multifrequency-band function for the terminal.

As a device becomes smaller, the effect of the human body on antenna performance becomes larger. However, performance tests on the human body require much time and labor. Some intensive research studies are focused on analyzing how the human body affects the antenna performance and how such an effect is reduced by optimum antenna design.

Over the past five years, the convergence of the computing and communications industries has given rise to very powerful software for phones and PDAs, so that one can not only talk over these devices, but also use them as a computing devices. Let us discuss some of the key software aspects of such devices.

6.8.5   Speech coding technology

This will be discussed in Chapter 7.

6.8.6   Contents programming language

This will be discussed in Chapter 7.

6.8.7   Human interface and character input scheme

This will be discussed in Chapter 7.

6.9   Smart cards

The smart card is an intelligent card which obtains some important information such as authentication, security, subscriber information, billing information, phonebook and so on. SIM (Subscriber Identity Module) is a typical example. It was developed for GSM by ETSI and it contains information such as authentication key, restricted outgoing number, phonebook, received SMS messages and also advanced value-added service features. The most important advantage of the SIM is that it helps in national/international roaming. In these cases, a GSM user doesn't need to bring their cellular phone itself, but only a SIM card. When a person arrives at a destination area or country, they can just insert a SIM card into a rental GSM terminal and then all the billing information is automatically routed to their home bank account. A technical study on cross-infrastructure roaming capability (such as GSM/cdmaONE or GPPS/WCDMA) has been promoted by many operators.

At ITU-T, UIM (User Identity Module) was also standardized for IMT-2000. UIM has similar concepts to the SIM, and the UIM is an IC-card with an inbuilt CPU. UIM has improved its security against electrical and mechanical attacks by a third party. Combining UIM function and PIN (Personal Identification Number) can increase its security robustness to extremely high levels.

Smart card technologies such as SIM and UIM are considered powerful tools for future service application enhancement and also for e-commerce applications. Standardization bodies such as GSMA, ETSI, TIA, 3G-PP, etc., are collaborating intensively with each other for future enhancement of smart card capability.

In this chapter we have discussed the various cellular systems around the world and what kind of evolution strategies are likely to be adopted by carriers in the next few years, along with various network and access technologies. We also took a deeper look into the components that make up devices and terminals. In the next chapter, we will discuss the software technologies that are important for the wireless world.

7 Global wireless technologies: network, access, and software

7.1 Network and access technologies

In Chapter 6 we discussed various cellular systems around the world and what kind of evolution strategies carriers are likely to be adopting in the next few years. In this chapter we will discuss the key network and access technologies and software technologies impacting on the wireless world. We will first take a look at the wireless network architecture and then discuss various wireless WAN, LAN and PAN technologies. This discussion is presented to make the reader familiar with the spectrum of wireless technologies. For a more comprehensive discussion, please refer to the References, where we have listed some useful texts for detailed technological discussions.

7.2 Position location

In 1996, the FCC in the USA, via a series of orders, had mandated all wireless carriers to provide automatic location identification (ALI) as part of phase II E911 (Enhanced 911 for emergency services), with the implementation starting on 1 October 2001. In the past four years or so, the FCC has adjusted the requirements to better suit the reality of position-location solutions. These solutions can be broadly divided into two main categories: network based and handset based. Techniques such as Angle of Arrival (AOA), Time Difference of Arrival (TDOA), and MultiPath Fingerprinting (MPF) are the commonly used network-based solutions, while GPS or network-assisted GPS are the primary handset-based solutions. Both solutions have their advantages and disadvantages, short-term and long-term.

7.2.1 Network-based solutions

A TDOA system works by situating location receivers at three sites, with each site having an accurate timing source. When a signal is transmitted from a mobile terminal, it propagates at approximately 300 m/μs to each antenna site. When the cell site receives the signal, it is time stamped. By examining the differences in time stamps, TDOA can determine a mobile terminal's location by computing intersecting hyperbolic lines. To work correctly, the base stations must be time-synchronized to better than 100 ns.

Also called Direction of Arrival, AOA has several versions. Small-aperture direction finding is the most common version and requires a complex antenna array at two or more cell-site locations. The array consists of four to twelve antennas in a horizontal line. The antennas work together to determine the angle (relative to the cell site) from which a cellular call originated. When at least two sites can determine the angles of arrival of a given call, the caller's location can be determined from the point of intersection of projected lines drawn out from the cell sites.

Multipath is a phenomenon of a wireless RF signal bouncing off solid objects like buildings, towers and so on. While multipath deteriorates TDOA and AOA's accuracy in locating handsets, this technique utilizes the multipath characteristics (measures the phase, the timing, and the amplitude path of all the RF signals from a single caller) of an RF environment to locate a handset. Multipath characteristic patterns within a block are analyzed and stored as fingerprints in the database. When a handset transmits its RF waveforms, algorithms try to match these RF characteristics to the fingerprints in a database, thus identifying the block where the call came from.

7.2.2 Handset-based solutions

The most commonly considered handset-centric option is the Global Positioning System (GPS). It takes advantage of the multi-billion-dollar investment the US government has made to establish a satellite infrastructure for location determination. For several years, 24 satellites have been operational, and they have provided accurate, continuous, worldwide, three-dimensional position and velocity information at no charge. The basis of GPS is triangulation from satellites. To triangulate, a GPS receiver in the handset measures distance using the travel time of radio signals from the satellites. Since satellites already know their location, a precise location of the GPS receiver can be computed. Mathematically, four satellites are needed to determine exact position, but using some error correction and digital signal processing techniques the number of satellites required for accurate measurements can even be reduced to one (three satellites are required initially, and one or two thereafter). The network-driven GPS method places a minimal GPS front end in the handset and lets the wireless infrastructure equipment handle all the calculation and position determination. Autonomous GPS technology is based on a complete GPS subsystem in a handset. Wireless carriers and handset vendors could place all location-determination functionality inside the phone, making virtually no changes in wireless infrastructure.

The FCC adopted the following revised ALI accuracy standards for phase II location accuracy and reliability:
- for handset-based solutions: 50 m for 67% of calls, 150 m for 95% of calls, and
- for network-based solutions: 100 m for 67% of calls, 300 m for 95% of calls.

The FCC required wireless carriers to report their plans for implementing E911 phase II, including the technology they planned to use to provide caller location by November 2000. The top US wireless carriers – Verizon, Cingular, AT&T Wireless, Sprint PCS and Nextel Communications – account for roughly 76 million subscribers. In reviewing their plans and progress made for E911 deployment, apart from Sprint PCS, these carriers are way behind in their implementation schedule. For some, the service will not start until later in 2003. The position location industry in the USA is very complex. Since there is such diversity in wireless standards (AMPS, TDMA, CDMA, GSM, GPRS, iDEN), wireless carriers have been struggling for the past few years to come up with a strategy that will suit both short-term needs and long-term goals. It is pretty clear that in the next five years or so, GPS-based solutions will become the norm for the industry, but it is the short-term compliance that is causing the carriers to split hairs. The majority of the carriers are being forced to consider a dual-implementation strategy: network-based solutions for their antique AMPS and second-generation networks and GPS-based solutions for GSM and next-generation networks. For GPS-based solutions, since it is a handset-based solution, the existing handsets in the market needs to be recycled and for the existing handsets, the only way you can locate is by using network-based solutions, which means infrastructure costs in millions, if not billions.

Another conundrum is recovery costs. Carriers want to see their investments recovered as quickly as possible; they would ideally like the state to fund the roll-out. There are still several states that are unclear on cost-recovery schemes. NENA (National Emergency Number Association), APCO, and other public-interest groups have been arguing on behalf of consumers for a speedy implementation of position-location solutions so that they can start providing E911 services – the domino that was supposed to enthuse the industry with all sorts of applications and services. The issue is very evident at AT&T Wireless, which has filed for a waiver as it concluded that the network-based technology (E-OTD, Enhanced Observed Time Difference) that it has tested is not meeting FCC ALI accuracy requirements for its TDMA network; AT&T is proposing Mobile-Assisted Network Location System (MNLS) technology for its TDMA networks. The most significant problem facing this technology is its accuracy. AT&T gave an expected MNLS accuracy estimate of 250 m for 67% of the calls and 750 m for 95% of the calls, which is short of the FCC requirement. According to the FCC definition, MNLS is a network solution because it does not require modifications to legacy handsets and, thus, it is subjected to the less stringent accuracy requirements of 100 m for 67% of the calls and 300 m for 95% of the calls. So, even though the solution is less expensive and has some nice features, it probably will not make the grade.

For handset-based solutions, handset manufacturers are wary of production capacity risks. There are no rules for the evolution of technology in the USA and there are different technical innovations occurring simultaneously. Manufacturers are asking customers (carriers) to prioritize, but nothing has been forthcoming according to Nokia. Nokia believes that it would take two years from an order date to get to 50% activation of ALI-capable handsets for a single carrier and, for that to happen, GPS needs to be available into the lowest-end handsets to meet the numbers. Also, since accuracy requirements are statistical, there is no general agreement amongst carriers regarding accuracy of the various technologies, because solution accuracy varies depending on the topography – accuracy of a solution is quite different in the plain fields of Topeka, Kansas, to the urban canyons of Manhattan, NY. Table 7.1 lists some pros and cons of each solution.

There are also issues regarding interoperability. In this situation there is a need to establish a global forum to address the complexity and multiplicity of current solutions and the market situation. With this purpose in mind, Motorola, Nokia and Ericsson established the Location Interoperability Forum (LIF: www.locationforum.org) in September 2000. The LIF's purpose is to define and promote – through the global standard bodies and specification organizations – a common and ubiquitous location services solution.

The forum's goal is therefore to define and promote an interoperable location services solution that is open, simple and secure. This solution should use appliances and internet-based applications to obtain location information from the wireless networks independent of their air interfaces and positioning methods. In Japan and Europe, applications and services based on position location are already on the market. We will be discussing some of key issues with implementation of position location applications and services and some potential solutions in Chapter 9.

7.3   Fiberless optical system

Wireless is not the only area where telecommunications have made some amazing advances in recent times. Breakthroughs in optical communications have also had a tremendous impact on the way data are transmitted from point A to point B, especially WDM (Wavelength Division Multiplexing) technology. NEC announced 10.9 Tbit/s transmission speed with 40 Gbit/s per carrier (multiplied by 273 carriers) at the OFC 2001 conference. It is imperative that WDM will be the basis for multi-media broadband networks in the twenty-first century. Also, the development of optical devices such as SOA (Semiconductor Optical Amplifier), VCSEL (Vertical Cavity Surface Emitting Laser) and GaAs on Si is enormous. Using these technology advances, the deployment of optical network is only going to grow.

It is natural to apply the enormous optical technology progress to another telecom area – wireless. Terabeam, a Seattle-based company provides one example. Terabeam

Table 7.1. Comparison of various position-location technologies

ALI	Pros	Cons
TDOA	• Provides location for all handsets • Can utilize the RF information to provide network design and monitoring services • Indoor coverage is similar to indoor RF coverage	• Requires extremely precise time synchronization • Requires TDOA systems at 80–100% of the cell sites, thus increasing the overall cost of the system • Requires at least three cell sites for accurate location determination • Multipath propagation effect (problem resulting from bouncing wireless signals) deteriorates the accuracy of the system • This technique is able to track location only at the start of the call • For CDMA systems, because of the near–far problem, the signal-to-noise ratio (SNR) at the neighboring base stations decreases and leads to inaccuracies in position-location estimation
AOA	• Provides location for all handsets • Antennas could be used to improve capacity, reduce interference, and better system performance • Indoor coverage similar to indoor RF coverage • Can track the call in progress	• Requires at least two cell sites for accurate position-location determination • Requires AOA systems at about 100% of the cell sites thus increasing overall cost of the system • Requires highly expensive antenna beams that need to be kept in a calibrated state all the time • Since AOA tracks on the voice channel, additional processing is required to query MSC (Mobile Switching Center) to get the information about the user (MIN) • For CDMA systems, suffers from the same near–far problem as discussed in the TDOA section • No privacy, carrier can track the user all the time (when the phone is on)
MPF	• Single cell site can be used to locate and track multipath rays from a handset thus decreasing the capital and operating costs (as compared with TDOA and AOA solutions) • RF characteristics analysis could be used for network design and optimization • Works very well in urban and dense areas • Can track the call in progress if on the same voice channel (no handoffs)	• Requires several iterations to build the fingerprint database. The fingerprints would change with weather or construction layout • Antenna beams need to be kept calibrated • No way of tracking hard handoffs • Stationary handsets harder to track (in comparison with moving handsets) • Poor performance in rural areas • No privacy. Carrier can track the user all the time (when the phone is on)

Table 7.1. (*cont.*)

ALI	Pros	Cons
GPS	• Does not require expensive network modifications and infrastructure. The cost of putting GPS in a phone is declining fast, thus making it a very attractive long-term solution for carriers • Much higher accuracy possible • Provides privacy. Location could be available on demand	• No location of unmodified phones (FCC mandate requires location to be provided for all handsets and there are over 61 million handsets without GPS capability) • GPS enabled phones will affect battery consumption • Up to 2–3 minutes of initial warm-up time • Poor location determination in buildings and other shadowed environments • Need more memory and software for handset

has developed a fiberless optic transmission system and deployed it in US cities for test and evaluation. The system can support a gigabit Ethernet data port based on IEEE 802.3 and a maximum link distance of 2 km theoretically.

The key features of the Terabeam system are quick installation and alignment owing to small and lightweight optical transmitters. Also, the antenna has an automated tracking system to compensate for building movement and earthquakes. As the installation work for fiber optics in a metropolitan area such as New York and London is becoming very expensive, this may provide the solution to having high-capacity backhaul networks in timely manner (see www.terabeam.com). Fixed wireless solutions also provide alternatives to ground fibers in many of the developing countries, which don't have the necessary infrastructure.

Fixed wireless reminds us of a legacy technology of the millimeter wave transmission system, which was intensively researched and developed more than 30 years ago all over the world. At AT&T and NTT labs 20 years ago, the research project on the millimeter wave transmission system was terminated and most of the researchers shifted their research focus to the then-unknown but challenging field of fiber optical transmission. Needless to say, the focus and resources invested at that time paved the way for the current fiber optics industry.

7.4   Smart antennas

Antennas are a key component of the wireless network infrastructure. Their role is to establish radio transmission between base stations and wireless devices. These antennas can be configured such that one antenna can serve a geographical radius (360°) or it can also be sectorized to serve two or more segments within the same geographical radius. This helps in optimization, interference reduction, spectrum utilization, capacity enhancements and performance improvement. In recent years, with the help of "smart antennas", the performance of wireless systems has further improved. The better the utilization of the spectrum, the greater the number of users that can be served within the same geography with less amount of interference. Recently, Arraycomm – a company that pioneered the concept of smart antennas – has demonstrated the use of smart antennas for improved wireless Internet access.

Smart antennas use sophisticated signal processing and computing techniques rapidly to optimize receptions and transmissions from multiple users at the same time. This results in the wireless system's ability to serve up many more connections in any given area and with a given amount of radio spectrum. Arraycomm claims that its I-Burst technology can provide up to 40 times more capacity than 3G systems will provide in the future. Arraycomm's technology has been successfully used in base stations by Kyocera in its PHS wireless network since 1998. It has resulted in improved coverage and network capacity.

7.5   Wireless WAN

Please refer to Section 6.2 for a discussion on wireless WAN networks and different technologies.

7.6   IP-based technologies

With the advances in communication technology, Internet Protocol (IP)-based telephony using VoIP enabling voice services on a data network is becoming commonplace. In general, this means delivery of voice traffic in digital form bundled in discrete packages rather than in the traditional circuit-switched protocols of the public switched telephone network (PSTN). VoIP voice and data services can be employed to provide a more cost-effective, efficient and flexible way of building networks. These technologies are based on open standards and provide for the separation of functions such as call control and switching. The distributed nature of VoIP allows innovation and enables service providers to compete for different parts of the network continuum, while at the same time interoperable standards ensure that the overall network model remains consistent. The next-generation converged networks allow many different communications systems to interoperate.

Media gateways are critical internet working elements that translate between networks having differing standards. This provides conversion of streamed media formats such as voice or video and manages the transfer of information between the different networks. The key element that is used to support VoIP is the set of standards that have evolved over the years to address the various problems associated with voice traffic. Protocols such as H.323, Session Initiation Protocol (SIP), and

Media Gateway Control Protocol (MGCP) are used for call set up and media gateway control by providing an interface for the media path conversion between the legacy circuit network and clients and between the packet network and clients. Let us look at some IP-based protocols that are impacting the next-generation networks.

7.6.1 Session Initiation Protocol (SIP)

SIP, as described in Internet Engineering Task Force (IETF) RFC 2543, is a text-based protocol that leverages the power of the Internet by borrowing such common elements as the format of Hypertext Transfer Protocol (HTTP), Domain Naming System (DNS), and email-style addressing. It provides the necessary protocol elements to create services such as call forwarding, call diversion, personal mobility, calling and called party authentication, terminal capabilities negotiation, and multicast conferencing.

7.6.2 IPv6

IPv6 or IPng (IP next generation) is the next version of the current Ipv4 protocol and is designed to solve the problem of running out of IP addresses. Ipv4 uses 32-bit numbers, which allows for only four billion distinct network addresses. Because of the proliferation of devices and the pervasiveness of the Internet, we are quickly running out of those addresses. IPv6 uses 128-bit addresses, which allow for about 340 trillion trillion trillion unique addresses, enough to assign an IP address to every known thing on earth. Ipv6 also allows for QoS and strong packet-level encryption features. Vendors like Cisco and SUN are already shipping IPv6 compliant products, but it will be a while before we are able to complete the transition from IPv4 to IPv6.

One of the reasons for moving to IPv6 is to improve router performance. IPv4 has a variable-length header and variable-length payload. This is flexible and minimizes address overhead but means that a router never quite knows what to expect. IPv6 defines a standard header size of 40 bytes. The 40 bytes are divided down into eight fields rather than the 14 fields used in IPv4. The defined header size and reduced number of fields means router latency can be reduced and packet-shaping/traffic-shaping protocols can be more rigorously applied.

7.6.3 Multi-Protocol Label Switching (MPLS)

MPLS is then used to break the packet stream into fixed-length cells, grouping packets within an IP session into a single flow, which can be tagged to optimize router throughput. (A definition of a flow is a sequence of packets treated identically through a "possibly complex" routing function – the idea is to pass down long-lived flows, for example, multi-media rich media streams to be switched by hardware.)

7.6.4   Reservation Protocol (RSVP)

RSVP sits at the edge of the network (for instance in Windows 2000) and defines several levels of service: high quality/application driven for applications that can declare their resource requirements – for example MP4-encoded "declarative content"; medium quality where the application identifies the type of traffic flow needed, for example isochronous or non-isochronous but letting the network determine priority; low quality network driven – basic latency bounds and a minimum bandwidth guarantee and best effort.

7.6.5   Differentiated Services (Diffserv)

Diffserv provides a coarse-grained but simple way to categorize and prioritize network traffic. It is used to define four levels of service – platinum, gold, silver, bronze – and is used as a mechanism for grouping traffic flows sharing similar QoS attributes. MPLS and Diffserv used together will typically reduce queuing delay from 20–30 ms to 5 ms.

Delay is, however, only part of the story. An additional parameter is delay variability (also known as jitter). Delay variability is a consequence of packet loss triggering send-again protocols and is therefore related to the provisioning of buffer bandwidth. As traffic has become more asynchronous (increasingly bursty), buffer bandwidth has become increasingly hard to dimension. Essentially, bursty traffic can be accommodated by over-provisioning buffer bandwidth – a 2 Mbyte buffer at 60% usage delivers a 4×10^{-2} packet drop rate, and a 64 Mbyte buffer reduces the drop rate to 3×10^{-6}. Note that UDP (User Datagram Protocol) could be used to hide the packet loss (UDP allows packets to drop while TCP/IP requires dropped packets to be sent again), but dropped packets are bad news for differentially encoded rich media.

And there is the snag – packet-routed networks promise greater bandwidth efficiency but need to deliver similar dynamic range to existing network topologies, i.e. to support real-time rich media, packet-shaping protocols have to deliver an order of magnitude improvement on existing Internet latency performance. The jury is still out as to where this is achievable. Even if it is, bandwidth efficiency will be little better than existing circuit-switched networks.

7.7   Voice

Voice-based access has a very close synergistic relationship with wireless applications and services. As discussed before, voice not only provides a channel for information access when wireless access might not be possible or is convenient.

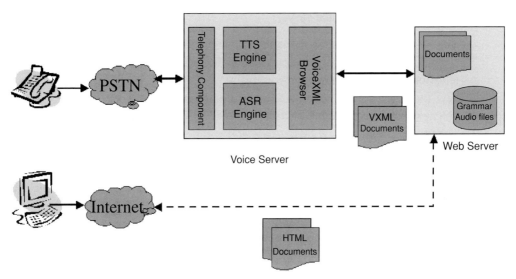

Figure 7.1. Voice Web system.

Voice and wireless Internet-based technologies can also work together to provide a multimodal framework to the user to provide a much better and richer user interface. Also, voice can be used as a biometric input for user authentication and validation. In the past couple of years, with the advent of VoiceXML and the recent release of its 2.0 specification, VoiceXML has become the *de facto* standard for voice-enabling applications and services. VoiceXML represents the next generation in interactive-voice response applications (IVR). VoiceXML is to voice applications what the Internet was to client–server computing. It allows us to be unbound by the constraints of the proprietary IVR programming languages and speech systems. Using an open architecture, one could mix-and-match best of breed solutions and not get locked down by "a" technology or vendor – it is a familiar story. In addition to VoiceXML, speech recognition accuracy has increased tremendously over the past couple of years, and of course with more processing power at hand, one could run more sophisticated algorithms quickly thus improving the chances of accurate recognition.

A voice Web system (Figure 7.1) consists of several components, which are discussed in the next subsections.

7.7.1   Telephony network

This can be a PSTN (regular analog) line or lines coming through a PBX system, ISDN lines or VoIP network. The telephony network is connected to the VoiceXML gateway. The telephones can be regular phones or IP phones if connected to the VoIP network.

7.7.2   VoiceXML gateway (voice server)

The VoiceXML browser is the key component that requests the VoiceXML documents, interprets them, and controls the dialog flow. It also controls speech and telephony resources. These resources include Automated Speech Recognition (ASR), Text-to-Speech (TTS), play/record audio, and telephony network interface.

The telephony component is responsible for all the telephony features like Dual Tone Multi-Frequency (DTMF) extraction and detection, call placing, call transfer and call termination. Typically, a single instance of VoiceXML browser has an instance of the ASR and TTS engine. If there is a need to grab user input, it passes control to the ASR engine, and for generating speech output it passes the request to the TTS engine.

The ASR engine is responsible for recognizing the user utterance and converting it into text, which is forwarded to the VoiceXML browser. The TTS engine is responsible for generating the speech output from a text that is sent to the engine by the VoiceXML browser. If the audio output is a prerecorded audio, the VoiceXML browser forwards the raw data to the telephony component.

7.7.3   Web/application server

This is typically a Web server that runs the application logic, and may contain a database or interface to an external database or transaction server.

7.7.4   Internet-style network

This is a TCP/IP-based packet network that connects the application server and voice server via HTTP.

7.7.5   VoiceXML documents

These define the voice user interaction and dialog flow control.

7.7.6   Grammar files

These files define the valid commands that are allowed during the voice interaction. Grammar can be defined at the development stage or generated dynamically at the run time.

7.7.7   Audio files

These are prerecorded audio files that are played back, or the recordings of the user input.

Let us look at where speech-enabling an application makes sense and, more impor-tantly, where it does not. Speech applications can be used when:
- voice is the most convenient mode of device input (example scenario: driving in a car),
- navigation is complex and commands are embedded deep into the menu structure,
- users have a physical disability.
    Speech applications are not very useful when:
- the work environment is very noisy,
- the user is in a situation of talking with devices and people at the same time (will lead to SR problems),
- it is easier to accomplish the task using other means of device input – keyboard, mouse, etc.,
- content returned from the server is large and visually complex,
- the task requires the user to compare data items,
- information presented to the user is personal and confidential.

However, speech recognition by its very nature is an imperfect technology. What does that mean? When interacting with a computer or handheld, the way we interact with these devices basically remains the same for almost all users. This is not the case with voice. There are large variations in languages, dialects and accents. Our voice could vary if we have a cold or are too tired. In addition, today, the conversations for most part have to be directed to have good recognition accuracy. If the conversation is free flowing, the computer has not only to identify the utterance but also the context (e.g. "Could you write up a letter to Mr Wright to give him the right directions?"). An average human would get the right writes to Mr Wright, but would a computer? If it's Natural Language Understanding-enabled, it has a good chance. This can work in a demo environment, which has had some neural network training, but to get it to the level where we can expect it to be part of most voice applications is probably not going to happen until the end of 2004. These advances will also tie in with the progress made on the multi-modal browsing front. It has the potential of increasing the usage of wireless applications tremendously. It would mean that we would also require speech recognition (SR) on the device as opposed to SR on the server (which invariably is the case today). Most of the multi-modal applications do not really need NLU since the interaction is directed but NLU can make multi-modal browsing so much powerful (start imaging yourself talking to devices) – think agents!

VoiceXML 1.0 was released in March 2000, it became a W3C standard later in that year and then VoiceXML 2.0 was released in October 2002. VoiceXML 2.0 standardizes several key elements and introduces some new tags as well. There is some final compromise work going on amongst various players. Several markup languages have already become a standard, such as Speech Recognition Markup Language, N-gram Grammar Markup Language, and Speech Synthesis Markup Language. Of particular importance is the work being done to standardize multimodal access using

Multimodal Dialog Markup Language (MDML). Multimodal access allows the user to employ both speech and screen tap to interact with applications and services. For example, the user could input his or her request for getting directions to a certain destination by speaking to a device (phone, PDA, autoPC) and the directions appear on the screen in text and graphics. The SALT (Speech Application Language Tags) consortium was launched in October 2001 to work in area of multimodal access. Advances in multimodal technology are discussed further in Chapter 13.

With advances in speech recognition and synthesis technologies and standardization of the family of voice standards, voice is going to become a more commonly used interface for interaction with devices and information. This will be more evident in car-ridden societies like North America where, according to IBM, users spend over 500 M hours per week in automobiles.

7.8   Telematics

Telematics is a new terminology combining "telecommunication" and "informatics" and it refers to the services and systems of the automobiles with wireless interfaces for information communication. The purpose of telematics is to send a variety of information from the car and also to receive the location-related information for the driver and passengers. With this capability, we can expect a lot of applications and new business opportunities.

The telematics industry is on the fast track for anticipated growth. By the middle of the present decade, the vast majority of new cars sold are expected to be telematics-ready. Steady growth in telematics deployment is anticipated over the next several years, generating billions of dollars in equipment and service revenues.

In an emergency situation, a telematics system automatically notifies a response center when an airbag deploys. This is especially important when the driver is unable manually to contact a response center. In some telematics systems, automatic notification to the response center also occurs in a roll-over accident – common on the fast-moving autobahns in Germany.

With telematics systems, drivers can request information and quickly receive help in emergencies or problem situations. Telematics systems in the future will extend these capabilities even further and bring greater levels of intelligent transportation to all drivers. Services available today include:

- vehicle management services – empowering consumers to unlock car doors remotely if keys are accidentally locked inside,
- car-theft notification and tracking services that are triggered by an embedded alarm system in the vehicle,
- convenience voice services – offering users mobile yellow-pages inquiries,
- convenience data services – enabling motorists to access real-time stock quotes and current news reports.

A car user can get location-related information automatically. For example, when a user talks to the navigation system "Are there any good Italian restaurant around here?", the telematics on-board terminal will typically respond by voice "Yes, the recommendation is the Tully's at 1110, 4th street, Seattle" and follows with "The special menu of the week is Lasagna. Do you want to be navigated from here?".

Here is another example. No one likes to be stuck in traffic jams. Major cities around the world are notorious for traffic jams. The damage done by traffic jams on social and economical activity is huge. When a car is stuck in a traffic jam, the telematics system can automatically find an alternative route to the destination and show the route on the terminal display. As the system constantly measures the average speed of the car, it is easy to get a grip of the traffic. GPS can be a tool to get the location information.

Another application is to inform the driver about the real-time status of the car – such as brake oil or tires or distance to the nearest gas station. As this function is expected to increase the security of the car, a security company can dispatch a person to the troubled car when the company gets an alarm signal from the car.

Recently, some wireless carriers have been actively working on developing telematics applications and services. The background is that as the increase of subscribers and ARPU are approaching saturation, so carriers need to create new markets. Also, the introduction of high-speed digital cellular systems such as GPRS and 3G can provide the ideal infrastructures for telematics. Car companies are aiming to use telematics as a tool to add value to the automobile and consequently accelerate sales.

Some telecom vendors such as Motorola are very positive about developing telematics. In spite of aggressive development worldwide, the market demand forecast for telematics is still not clear. The main reason for this is that there has been no great success up to now and also the market demand strongly depends on each country.

As the USA is such a car-centric country, the market opportunity is very positive. According to the Yankee Group, there are likely to have been over 1 billion wireless devices worldwide by 2003, and more than $50 billion of commerce transactions in the USA will be wireless; the telematics industry will be a $5 billion industry in 2005 in equipment and service revenues (source: The Strategis Group).

Compared with the USA, the average usage time of a car in a day is reported to be only 27 minutes in Japan. So, we need to investigate in detail the respective market requirements. In the USA, telematics is focused on business use in the car, such as a moving office, while in Japan it is more focused on entertainment use in the car, such as infotainment (information + entertainment).

In Japan, telematics is called ITS (Intelligent Transport System) and the active R&D of ITS has been promoted by the ITS promotion committee, which was established in 1999. The committee is a joint project with more than 100 companies including government organizations, and the main promoters are Toyota and the NTT group. The concept of the ITS project is to develop

- an advanced car-navigation system for traffic jam solutions,
- an automatic charge collection system for toll roads,
- a sensor system for traffic safety support,
- an automatic driving system.

Denso, the fourth-largest maker of car parts in the world and a main engine supplier for Toyota, is now concentrating its R&D on ITS. Denso has the top position in customer satisfaction in the US market for high-end car navigation systems. Denso also announced a joint development with Microsoft on a next-generation car-navigation system in March 1999. This project aims to combine CD, DVD, a car-navigation system and a cellular system into one with a Windows CE-based OS. It seems the project is aiming at promoting PC function in a car environment.

NTT DoCoMo has a different approach to ITS/Telematics. DoCoMo announced a joint development project with Nissan in February 2002. The purpose of this project is to promote FOMA (the WCDMA system of DoCoMo) as a platform of ITS/telematics and to accelerate its application. Dr Tachikawa, CEO, sometimes refers to the possibility of doubling the cellular business market size by introducing FOMA into not only human-to-human communication, but also in machine-to-machine (human) communication. Machine-to-machine (human) communication using the 2.5G/3G infrastructure will give us very promising business opportunities and new ecosystems.

7.9   Biometrics

The business case for utilizing wireless biometric technology is to decrease managed security costs and to provide the highest level of security while allowing the user to maintain his or her own password. There are a myriad of offerings for wireless security available to the enterprise customer. PKI is an important issue, and there are some techniques that are better than others. However, it is important to step back and look at the problems that wireless security must solve. Authenticating the user to the device, and the device to the network, is more challenging than it first seems. Digital signatures are now binding and can be held up in a court of law; yet the technology and the legislation backing the technology are not clear. Biometrics offers great promise, especially for voice authentication.

Responding to a biometric market possibly worth $15 billion by 2005, Fujitsu Microelectronics America (FMA) unveiled a fingerprint ID system for cell phones and PDAs. Users are authenticated by using a 1.28 cm × 0.20 cm sensor working together with sophisticated algorithms creating unique profiles that are compared against a fingerprint database. Systems like the company's Sweep Sensor are replacing passwords and PINs. The biometric signature technology built into these new fingerprint-based verification systems provides a new form of highly reliable, secure user authentication for unlocking and enabling secure communication over mobile devices.

Biometrics is defined as a unique, measurable characteristic or trait of a human being for automatically recognizing or verifying identity. Let us look at the types of security techniques available.

1. What do you have? ID, photo and so on.
2. What do you know? Passwords, PINs.
3. Who are you? Biometrics.

Humans are unique, and so are their physical and behavioral traits. By successfully extracting the measurable information from these traits, an individual can be mapped with that extracted information and hence it can be used for verification and identification purposes. Although biometrics has been around for a while, it has been only recently that it has started gaining some traction with the computing and communications industry for consumer applications.

First let us look at some numbers. According to a report released recently by the International Biometric Group,

- biometric revenues are expected to grow from $399 million in 2000 to $1.9 billion by 2005, and
- revenues attributable to large-scale public sector biometric usage, currently 70% of the biometric market, will drop to under 30% by 2005.

Finger-scan and biometric middleware will emerge as two critical technologies for the desktop, together comprising approximately 40% of the biometric market by 2005.

All biometric systems essentially operate in the same fashion. First, a biometric system captures a sample of the biometric characteristic. Then unique features are extracted and converted into mathematical code. Depending upon the needs and technology, several samples could be taken to build the confidence level of the initial data. These data are stored as the biometric template for that person. When identity needs checking, the person interacts with the biometric system. Features are extracted and compared with the stored information for validation.

Types of biometric identification systems available today are listed in Table 7.2, and Figure 7.2 provides accuracy versus cost analysis for various biometric technologies.

Based on cost, accuracy, and ease of use, voice- and fingerprint-based solutions are going to be more prevalent in commercial applications while face and eye print-based solutions will be used for things such as public security (at airports, malls, stadiums and highly secure facilities such as nuclear installations).

7.9.1 False rejection and false acceptance

The efficiency of a biometric system is measured by its accuracy for identifying authorized individuals and rejecting unauthorized people. The false rejection rate

Table 7.2. Various biometric technologies

Biometric technology	Definition	Comments
Face	The system analyzes the unique shape, pattern, and positioning of facial features	This approach could be preferable where man–machine interaction needs to be minimal, for example, under disability or harsh climatic conditions
Finger scanning	It uses analysis of minute points like finger image ridge endings, or bifurcations (branches made by ridges)	Automated fingerprint identification systems (AFIS) are already being installed on a variety of devices like desktops and wireless phones (Sagem)
Hand geometry	The three-dimensional images captured by hand geometry systems are unique and can be highly accurate	This technique has been used with considerable success at high-profile events like the Olympics and places like airports
Finger geometry	Finger geometry is much less of a dominant market force than finger imaging, but is particularly well proven in the area of physical access control, regulating the movement of people within secure areas	This technique is suited to large-scale high-volume applications like a secure area or airport check in
Iris recognition	Iris recognition is one of two biometric techniques that focus on the eye; the other being retina scanning. The iris is the colored ring of textured tissue that surrounds the pupil of the eye. Because of the unique characteristics found in the iris, information capture about it provides over 200 independent variables that can be compared – making it a highly accurate biometric system	Because of this accuracy, it is a good candidate for network access authentication
Palm	Palm biometrics is similar to finger scanning	An abundance of minute data are found in the palm, making this technology useful for criminal and civil applications. Palm systems are often integrated with AFIS technology to provide law enforcement with a complete crime detection kit

Table 7.2. (*cont.*)

Biometric technology	Definition	Comments
Retina	The retina is the layer of blood vessels situated at the back of the eye. The biometric technique used to capture data from the retina is often thought to be the most convenient for the end-users. An end-user must focus on a green dot while the system uses a harmless beam of light to capture the unique retina characteristics	Retina biometric systems are often considered to be some of the most unbeatable security systems
Signature	Signature biometrics are primarily concerned with the study of dynamic characteristics: the movement of the pen during the signing process rather than the static image of the signature. Many aspects of the signature in motion can be studied, such as pen pressure, the sound the pen makes against paper or the angle of the pen. This study of signatures is very much a behavioral biometric	Tablet-based systems are widely deployed in the United States and elsewhere
Voice	Voice biometrics combines the physical and behavioral characteristics of an individual. This technology is different from speech recognition. Voice biometrics goes beyond speech, and identifies the speaker of the words. Techniques like measurement of change of frequency between phonemes can be used	Voice-based speaker verification has applications in wireless networks, voice mail, interactive voice responses (IVR), calling cards and so on

(FRR) is defined as number of instances an authorized individual is falsely rejected by the system. The false acceptance rate (FAR) refers to the number of instances a non-authorized individual is falsely accepted by the system. Biometric systems often allow the FRR and FAR variables to be configured (see Figure 7.3). Typical values of FRR and FAR are close to 1%.

In March 1999, the Human Authentication API, Biometric API (BAPI), and BioAPI specification groups joined to combine their efforts into the BioAPI organization. This organization consists of representatives from various technology, health care, finance and government sectors.

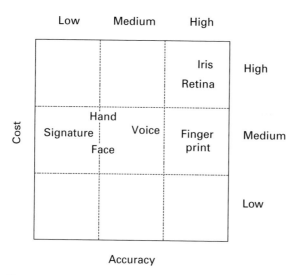

Figure 7.2. Cost analysis for various biometric technologies.

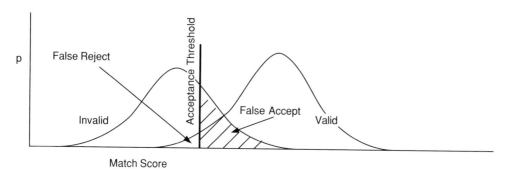

Figure 7.3. Error trade-offs.

Biometrics systems are not immune from false rejections or identifications as the performance varies according to sample quality and the environment in which the biometric samples (fingerprint, voice, face scans) are being gathered. For example, fingerprint recognition could be affected by angle and pressure of placement, cold, dry or oily fingers, high or low humidity, cuts or other effects on the finger, etc.; voiceprints can be affected by variations in background noise, a voice affected by illness and cold, quality of device capture, etc.; an iris scan can be affected by movement of the head or eye, glasses or colored contact lenses.

The debate of liberty vs security will continue to be a fierce one. The Public's perception of privacy vs security is changing and it is more willing to embrace biometrics for both commercial and non-commercial applications.

With the discussion of technologies that impact various subscriber environments, the pervasive computing ecosystem we discussed previously is beginning to take shape. It is clear that wireless WAN, LAN and PAN technologies need to work together in an integrated fashion to provide a seamless user experience to the subscriber. The end-user is less worried about the underlying technology and more interested in the applications and services that are built on top of those base technologies and services. As such, software becomes critical; it becomes the catalyst to enable compelling and useful applications and services to become a reality. In the next section, we will take a look at which software technologies are having an impact on the wireless ecosystem and why.

7.10   Software: catalyst to wireless Internet

By now we have familiarized users with various technologies that are important to the wireless industry. So far we have discussed wireless systems and architectures, and the various network and access technologies in some detail. We now turn the discussion to one of the critical elements of the ecosystem – the software. Again, this discussion is presented only to familiarize users with important software technologies and not as a comprehensive text on the subject. For readers with further interest, we have provided suggested reading in the References.

7.11   SMS

The Short Message Service (SMS) is the ability to send and receive text messages to and from mobile telephones. The text can contain words or numbers or an alphanumeric combination. SMS was created as part of the GSM Phase 1 standard. Each short message is up to 160 characters is length when Latin alphabets are used, and 70 characters in length when non-Latin alphabets such as Arabic and Chinese are used. The vast majority of SMS usage is accounted for by consumer applications. It is not uncommon to find 90% of a network operator's total SMS traffic being accounted for by the consumer applications like messaging, chat, voice and fax mail notifications, email alerts, ringtones and information services.

SMS continues to be one of the most popular applications around. (IM is mostly a desktop phenomenon, although this is changing rapidly. The wireless version is commonly known as SMS.) It has been readily accepted and is being used by a diverse population: from teenagers to corporate executives, from Japan to Finland. IM allows users the instant gratification of information exchange, hence its popularity. Figure 7.4 shows the popularity of IM amongst various Internet applications in different countries. More recently, enterprise users have embraced IM as well. According to

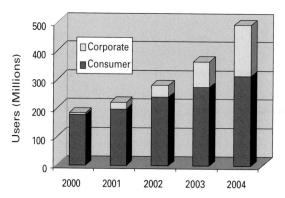

Figure 7.4. Global growth in IM subscribers (source: IDC).

the IDC, there will be close to half a billion users of IM by 2004. That is an astonishing number. With the advent of presence-based technologies like SIP and standards such as SIMPLE, the capabilities and popularity of IM will continue to skyrocket. The end-user devices can vary from desktop to wireless phone to a simple LAN phone. Also, it should be noted that IM is just one of the steps for MMS (Multimedia Messaging Service). An early introduction of MMS is strongly expected to enjoy a wide variety of wireless multimedia services.

To meet the 3GPP standard Multimedia Messaging Service (MMS) is to know complete communication. With virtually no limit to the content that can be transmitted mobile-to-mobile or network-to-mobile, MMS offers total freedom to convey ideas, to supply or exchange information or to express oneself. Beginning with the enormously popular Short Message Service (SMS) for simple text messages, the exciting Enhanced Message Service (EMS) for illustrated text messages with sound has entered the market as the second step. MMS will become the ultimate messaging application, allowing users to create unique messages, using all types of multimedia.

The messaging evolution fosters steady expansion of the marketplace for network operators and service providers. Studies show that users are willing to pay more for multimedia messages than for ordinary text messages. There is virtually no limit to the kind of services that network operators and service providers can offer users.

Subscriptions to the comic strip of the day, weather forecasts in pictures, all some examples of MMS applications. A picture says more than a thousand words and is more fun to look at. Using the Wireless Application Protocol (WAP) as bearer technology and powered by the high-speed transmission technologies EDGE, GPRS and UMTS (WCDMA), Multimedia Messaging allows users to send and receive messages that look like PowerPoint-style presentations. The messages may include any combination of text, graphics, photographic images, speech and music clips or video sequences. MMS will serve as the default mode of messaging on all terminals, making total content exchange second nature.

Although MMS is a direct descendant of SMS, the difference in content is dramatic. The size of an average SMS message is about 140 bytes, while the average size of an MMS message will (in the early stages) be around 30 000 bytes, but is actually unlimited. In the future, the user will be able to store a large number of messages, including those with video clips. The size of these messages will be about 100 000 bytes. That is why the key word to describe MMS content is "rich". Complete with words, sounds and images, MMS content is endowed with the user's ideas, feelings and personality.

The MMS standard, just like SMS, offers store-and-forward transmission (instant delivery) of messages, rather than a mailbox-type model. MMS is a person-to-person communications solution, meaning that the user gets the message directly into the mobile. He or she does not have to call the server to get the message downloaded to the mobile. Unlike SMS, the MMS standard uses WAP as its bearer protocol. MMS will take advantage of the high-speed data transport technologies EDGE and GPRS and support a variety of image, video and audio formats to facilitate a complete communication experience.

Users can easily get MMS into their phone. MMS supports OTA, meaning that the user does not have to configure the settings manually. The configuration is done by the operator. Currently being standardized, MMS is likely to support the following formats:

- image – JPEG and GIF 87, 89a, WBMP,
- video coding – ITU-T H.263, MPEG4 (simple profile),
- audio – MP3, MIDI, AMR/EFR (for speech),
- video – MPEG4, H263.

7.12   Software agents

Mobile agents are software programs that can move through a network and autonomously execute tasks on behalf of users. These devices are useful because of the quantity of information content on the Internet, which is growing at a breathtaking pace. Hundreds of Web sites are created every minute; countless documents are loaded onto intranets and extranets. Furthermore, seemingly useful information resides in disparate databases. Hence we need intelligent information agents that can act like secretaries and help do the leg work that is often required in information mining. Information agents can go even further than mobile agents and can accomplish a sequence of tasks.

We have already begun to see some form of mobile agents being implemented by various online businesses. For example, many online comparison engines utilize agents that go out into cyberspace and bring back pricing and item availability information in real time or on a predetermined frequent basis. Mobile agents are also used for discovering, updating and indexing content for various search engines. These mobile agents are also referred as "bots" or derivatives (such as "shopbots"), a term related to robots. According to a survey conducted by Neilsen NetRatings, shopbots were the second-fastest growing e-commerce market segment during the 1999–2001 holiday shopping season. Agents are also used for personalizing Web content. Software agents can mine the plethora of user- and device-specific databases and quickly generate a dynamic page as the user clicks through the site.

These mobile agents are generally based on fuzzy logic or artificial intelligence algorithms that train the agents to behave in a certain way, depending on the request or input they receive from the user's navigation patterns. In some of the more sophisticated set ups, mobile agents of one system can interact with mobile agents of another system (even systems belonging to different companies) to negotiate for the best rates or deals on products or services. In the future, mobile agents will be able to handle much more complex transactions. So far, agent technology has been limited to the desktop environment, but agent technology is gradually moving in its natural progression to wireless devices.

The mobile agent concept can be further expanded to the execution of multiple tasks and, more interestingly, to coordination and negotiation among the various tasks.

7.13   Middleware/gateway components

Wireless middleware is software that insulates applications from the underlying wireless network, making it easier to develop new wireless applications, as well as to port existing applications to the wireless environment.

Wireless middleware usually consists of client and server software. The client portion resides on the mobile computer and accepts messages from applications on the mobile computer. It reformats these messages and forwards them across the wireless network using application-layer protocols optimized for wireless communications. The messages reach the middleware server, which typically resides on the destination LAN. The middleware server functions as a gateway to other servers and hosts on the LAN, acting as a proxy for the mobile computer.

Middleware performs the following types of function, although specific details will vary depending on the actual middleware:

- isolates the application from connectivity issues such as intermittent connections and varying throughput,
- minimizes the amount of data sent over the wireless connection,
- reduces the number of back-and-forth messages required to complete a transaction,
- queues messages when a connection is not available,
- provides a consistent API regardless of the underlying network.

Some wireless middleware products come as toolkits with which customized wireless applications can be developed. Others work in conjunction with existing applications to make these applications effective both from a cost and performance perspective.

7.14   Service discovery and synchronization

With the proliferation of devices, it is becoming increasingly important (a) to make the services available on a device available to other devices and (b) to synchronize the data between various devices on an as-needed basis.

7.14.1   Service discovery

All devices have certain features that other devices can take advantage of. For example, a printer or a fax machine could be made available not only to a desktop PC but also to handhelds or phones or TVs or other gateways and servers. Similarly, a TV or a monitor could become the display device for all other devices in the room. Traditionally, to make the above happen, someone has to configure and physically connect the two devices to enable communications. However, with a new suite of software that aims at service discovery, devices discover each other and their capabilities by their mere presence in each other's vicinity and thus enable dynamic interaction. Three of the most promising technologies in this area are Jini, Universal Plug and Play (UPnP), and Salutation.

Jini

Jini is the brainchild of Bill Joy, founder and Chief Scientist of Sun Microsystems. It is a distributed computing environment for network plug and play capabilities. The concept is similar to the Service Location Protocol (SLP), which helps to announce device presence to the network by communicating with the other devices within the network. Some of the scenarios are as follows (source: Overview of Jini).

(1) A new fax machine connected to the network can announce its capabilities. A PDA can then use this fax machine without having to be specially configured to do so.

(2) A digital camera is connected to the network. Its user interface could be aware of the presence of a printer in the vicinity and could automatically present that option without the need for user set up and configuration of the printer.

(3) If new capabilities are added to the hardware, these capabilities can be automatically communicated to the other network devices.

(4) Services can also announce changes in their state. A fax machine could tell you when it is running out of paper, or a printer could alert you when its toner is running low.

There are five key concepts of Jini (source: Core Jini).
(1) discovery,
(2) lookup,
(3) leasing,
(4) remote events, and
(5) transactions.

Discovery is the process by which network devices introduce themselves to the network by announcing their presence and capabilities. Jini supports multiple situations with the use of multicast request, multicast announcement, and unicast discovery protocols. Lookup is the ability of devices to find the services and network functionality devices are looking for. In our previous example of a printer and a PDA, the PDA would query the network community for a device that is capable of printing. The network devices together make what is called a Jini community of devices. In case of machine crashes and network failures, it is critical for the community to self-heal and come back to its original configuration so that disruption in services is minimal. To get around this disruption problem, Jini uses a technique called leasing where the access is loaned for a fixed period of time, rather than granting it for an unlimited amount of time. If there is no interest in the services of a given device, the lease expires. This resource allocation expiration helps to avoid stale states of network devices in the case of disruption.

Network devices change in state all the time. A printer could need new toner, a fax machine could run out of paper, a new hard disk could be added to a computing device, and so on. State information is communicated to other devices through remote events. Finally, transactions are a way to allow communications and computations between various devices like printing, faxing, emailing, and so on.

The vision of Jini is to enable any device to be networked smoothly and reliably: from toasters to enterprise servers.

Universal Plug and Play

UPnP is Microsoft's technology for service discovery. It is based on TCP/IP and uses XML to describe and communicate the services and capabilities of a device. Each device must have an IP address, which is assigned by a Dynamic Host Configuration Protocol (DHCP) server. If the DHCP server is absent, the client uses Auto IP to obtain an address, which is a means to obtain an unused address from a range of reserved addresses. The protocol consists of the following five main steps:
(1) discovery,
(2) description,
(3) control,
(4) eventing, and
(5) presentation.

Salutation

Salutation is an industry organization, which defined the Salutation architecture to enable devices to discover and use services provided by other devices in the network. The Salutation manger is the core component of the Salutation architecture. Each device registers with the Salutation manager, which could be local on the same device or remote in the network. The Salutation manager handles discovery of services as well as the communication of the client with the service. The manager itself is network-protocol independent. The transport-dependent parts are encapsulated in a so-called Transport manager. A registry is part of each Salutation manager, which contains information about the locally connected services.

The concept of a "service" is broken down into a collection of functional units, each unit representing some essential feature (e.g., fax, print, scan or even sub-features like rasterize). A service description is then a collection of functional unit descriptions, each having a collection of attribute records (name, value). These records can be queried and matched against during the service discovery process. Certain well-defined comparison functions can be associated with a query that searches for a service. The discovery request is sent to the local Salutation managers, which in turn will be directed to other Salutation managers. Salutation managers talk to one another using Sun's ONC RPC (Remote Procedure Call). Salutation defines APIs for clients to invoke these operations and gather the results.

7.14.2 Synchronization

In the pervasive computing world, one user can carry and use multiple devices: a computer at home and work, PDA, wireless phone, pager, and more. Once information appliances (an appliance specializing in information: knowledge, facts, graphics, audio, video and with the ability to share information) become more pervasive, this list will grow further. One obvious problem is synchronization. Standards like Bluetooth and SyncML were initiated to resolve this problem. If pervasive computing is to be embraced as widely as predicted by the industry pundits, synchronization is absolutely the key to wide acceptance. Simple things like address book, phone numbers, emails, to-do lists and calendars need to be synchronized continuously amongst various devices and with the least amount of inconvenience. Users can not be expected to buy and carry an inordinate amount of connectors and cables to make the synchronization possible.

So far there are mostly proprietary synchronization products on the market like AvantGo, IntelliSync and TrueSync, which allow synchronization only amongst a small subset of available devices. Industry leaders such as Ericsson, IBM, Lotus, Motorola, Nokia and Palm, with the vision of a specification for universal data synchronization format and protocol, launched the SyncML initiative in February 2000. SyncML includes a universal data synchronization format that is defined by an XML DTD. This format is exchanged as SyncML messages be-tween network devices, the messages being an XML document. The strength of SyncML is in the fact that it is independent of the underlying transport layer and can be used in wireless, as well as wired, environments. SyncML frame-work consists of SyncML objects, a conceptual SyncML adapter, and the SyncML interface.

The SyncML adapter is responsible for maintaining a transport connection with the other network device and for marshalling synchronization commands and data into SyncML format. The SyncML interface is an API programmable mechanism for communicating with the SyncML adapter. The Sync agent is responsible for han-dling the SyncML-based data synchronization. The Sync agent provides the SyncML support to generic data synchronization engines.

7.15   Transcoding of content

Transcoding is defined as the transformation and reformatting of the content to match the client (device) needs and user preferences. It is an extremely critical com-ponent of the middleware that talks to the enterprise data and logic on one end and an array of end-user devices on the other. The end-user devices could range from small four-line display phones to a 21-inch desktop, from a two-line display pager to a 52-inch TV, from a rich graphical user interface environment to voice-only input and output situations. It is up to the transcoding engine to take the processed data effec-tively from the server and present it to the user based on device capabilities, network conditions, and user preferences, all in real time. Figure 7.5 shows typical application architecture with a transcoding engine at its core. Initially, the transcoding engine used to be a separate component that needed to work with application server archi-tecture but gradually all server players are incorporating the transcoding engine into the application server architecture. The transcoding engine manipulates the different modalities of content: text, graphics, audio and video, and makes then available to the end-user. Although we have made several strides in real-time transcoding, there is still a lot of room for improvement. End-user devices will continue to have diverse user interface and capabilities, so having robust, future-proof transcoding architec-ture is critical for any pervasive computing application or service, otherwise (a) it will fail to meet users' expectations and (b) maintenance of content will prove to be very expensive and tedious. For applications and services that serve small display devices, user interface defines its success. With the help of a well-tuned transcoding engine, application developers can achieve that goal. We will be discussing the issue of transcoding further in Chapter 10.

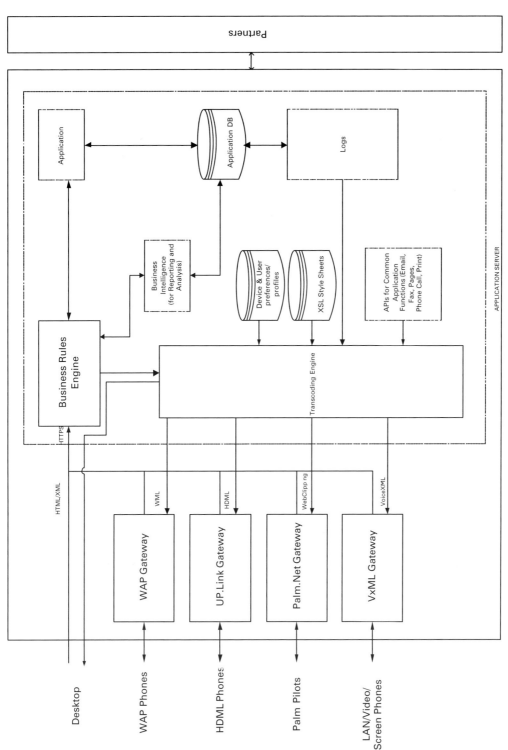

Figure 7.5. Transcoding as a component of the middleware engine.

7.16   Mobile Internet

Internet access service by a cell-phone or a wireless PDA is called a mobile Internet or wireless Internet service. As these names indicate, a feature of mobile Internet is the integration of mobile/wireless and Internet. The launch of the 3G commercial service provides not only an Internet access service, but also increases wireless multi-media services such as video clipping and music distribution. Contents and devices for 3G should have advanced capabilities to handle the multimedia content.

When we compare the functionality of a cell-phone from five years ago or a voice-only phone with that of an Internet access phone like i-mode, we find out that the shapes of the devices do not change very much (although they are becoming smaller and lighter by the day). So, what enables the difference in functionality between the two? It is software inside the device. To achieve mobile Internet connectivity, a wireless device needs two major functions, one is to establish the link in the air (mobile/wireless) and the other is to connect the path to the net (Internet). Rapid progress and development of recent software technology enables us to pack these complicated functions into a small device. From a software point of view, the following issues are essential to support mobile Internet services.

7.16.1   Programming language to describe contents

WAP-WML

Although the industry has been buzzing with talk of mobile data applications and services, it was not until recently, when WAP started to hit the developer and consumer community, that the promise of wireless applications and services really became important. In mid-1997, three of the largest phone manufacturers – Ericsson, Nokia and Motorola – joined forces with Openwave (previously known as Unwired Planet and Phone.com) to form the WAP Forum to work on a common goal of forming a development framework that would expedite the development and implementation of wireless Internet based applications (Figure 7.6). WAP architecture consists of a suite of protocols and a markup language designed for the wireless environment – Wireless Markup Language (WML), which is drawn from Handheld Device Markup Language (HDML) from Phone.com and is based on Extensible Markup Language (XML).

One of the key features of WML is the concept of "Card & Deck". "Card" refers to a set of information on a display of a cell-phone and "Deck" is a unit of download with multiple cards. This technology enables high radio frequency utilization efficiency. WAP architecture (see Figure 7.7) is drawn from Open Systems Interconnection (OSI) introduced by the International Standard Organization (ISO). Figure 7.8 shows i-mode/WAP architecture.

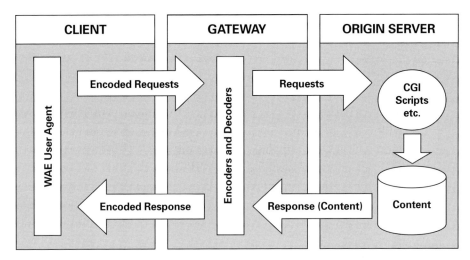

Figure 7.6. WAP programming model.

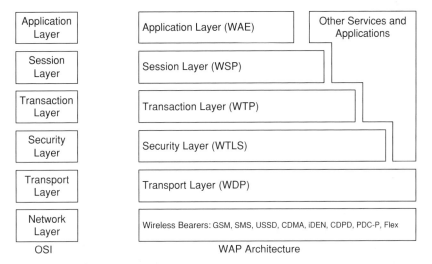

Figure 7.7. WAP architecture.

WAP is made up of the following layers:

(1) Wireless Application Environment (WAE). This top layer consists of the following components.

- Microbrowser specification.
- Wireless Markup Language (WML). WML is based on XML and provides support for text and images (wireless bitmap format, WBMP), user input, user navigation, multiple languages, and state and context management features.
- WMLScript is a scripting language very similar to Javascript and extends WML's functionality.

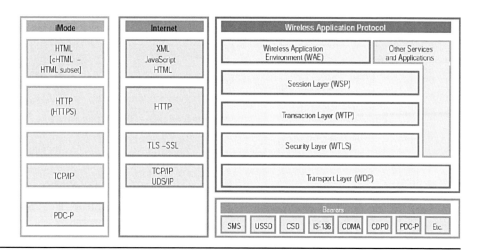

Figure 7.8. Comparison of i-mode/WAP architecture, Source: WAP Forum, W3C, NTT DoCoMo

- Wireless Telephony Applications (WTA). WTA provides the framework for accessing telephony commands using WML and WMLScript.
(2) Wireless Session Protocol (WSP). This is very similar to HTTP and is designed specifically for low-bandwidth, high-latency wireless networks. As discussed previously, WSP facilitates the transfer of content between the WAP client and WAP gateway in a binary format. Additional functionalities include content push and suspension/resumption of connections.
(3) Wireless Transactional Protocol (WTP). WTP provides the functionality similar to TCP/IP in the Internet model. It is a lightweight transactional protocol that allows for reliable request/response transactions and supports unguaranteed and guaranteed push.
(4) Wireless Transport Layer Security (WTLS). WTLS is based on Transport Layer Security (TLS, formerly known as Secure Socket Layer (SSL)). It provides data integrity, privacy, authentication and denial-of-service protection mechanisms. WTLS can be invoked similar to the way that HTTPS is invoked in the Internet world.
(5) Wireless Datagram Protocol (WDP). WDP provides an interface to the bearers of transportation. It supports CDPD, GSM, Integrated Digital Enhanced Network (iDEN), CDMA, TDMA, SMS, Flex protocols.

WML was firstly used in WAP Version 1.0, and was upgraded to Version 1.2.1 in June 2000 and Version 2.0 in August 2001.

Compact_HTML/ i-HTML

Compact_HTML (c-HTML) is a markup language for mobile devices. c-HTML is a subset of HTML 2.0, 3.2, 4.0, which are W3C internet standards. i-HTML is a markup language developed by NTT DoCoMo based on c-HTML. The relationship

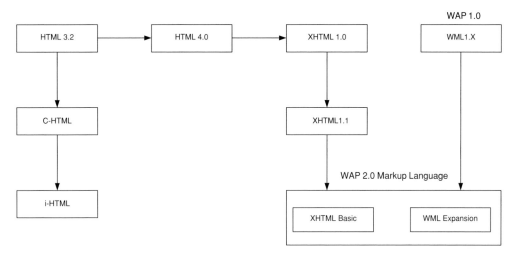

Figure 7.9. Concept of XHTML, i-HTML and WML.

between HTML/c-HTML/i-HTML specifications is shown in Figure 7.9.

Thanks to the compatibility between HTML/c-HTML/i-HTML, existing contents providers for wired Internet business can easily provide contents for i-mode with few modifications. The key features of c-HTML/i-HTML are interactions between browser and communication functions such as "phone to" and "mail to". These functions easily enable users to make and receive calls/emails with one-touch button operation.

XHTML

XHTML (Extensible HTML) is a markup language for next-generation mobile systems. The desire to promote the standardization work of XHTML comes from the following facts:

(a) unsuccessful deployment of WML (WAP 1.0) service owing to its incompatibility between terminal/server vendors,
(b) importance of compatibility with the Internet standard (HTML),
(c) successful deployment of i-mode,
(d) launch of 3G systems with transmission rates 40 times higher than 2G systems.

Considering the above, the WAP Forum has initiated the study of XHTML Basic as a basis of next-generation markup language and WAP 2.0 specs opened in July 2001. XHTML Basic is a group of modules of XHTML and it provides a basis for small devices such as cell-phones and PDAs.

7.16.2 Software environment for wireless (JAVA)

Programming Languages such as WML and c-HTML are suitable for describing "static" content information on a display. However, "dynamic" contents can not be described by these markup languages. Java is considered a solution for this purpose.

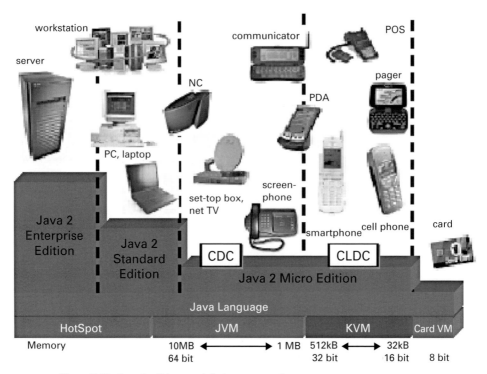

Figure 7.10. Java 2 editions and their target markets.

Java is an OS-free software environment and Java can be executed at any hardware platform. With the advent of PersonalJava™, Embedded-Java™, and other Java technologies, the benefits of the Java platform are being extended to screen phones, set-top boxes, and even deeply embedded devices. In order to provide compelling Java technology solutions for manufacturers building devices across the spectrum from palmtops to desktops, Sun introduced the Java™2 Platform, Micro Edition (J2ME™) software. J2ME is a new, very small Java application environment. It is a framework for the deployment and use of Java technology in the post-PC world. Sun will provide J2ME software in configurations suitable for a variety of market segments. A configuration is composed of a virtual machine, core libraries, classes and APIs. Currently, there are two J2ME configurations: the Connected Limited Device Configuration (CLDC) and the Connected Device Configuration (CDC); see Figure 7.10.

CLDC is designed for devices with constrained CPU and memory resources. Typically, these devices run on either a 16- or 32-bit CPU and have 512 kbytes or less memory available for the Java platform and applications. CDC is designed for next-generation devices with more robust resources. Typically, these devices run on a 32-bit CPU and have 2 Mbytes or more of memory available for the Java platform and applications.

Built into this core platform is the capability to receive not just an application code, but libraries that form part of the Java 2 platform itself. This enables a J2ME environment to be dynamically configured to provide the environment that the consumer needs to run an application, regardless of whether all the Java technology-based libraries necessary to run the application were present on the device when it was shipped. Configuration is performed by server software running on the network. The network architecture and configurability of J2ME software, combined with Java technology's inherent ease of development, simplifies the creation and deployment of intelligent, dynamic Java content. To enhance further the value of the J2ME environment and assure its ability to provide a focused solution to particular device categories and industries, Sun allows industry groups to define Java technology-based profiles specific to their industry. These profiles are specifications that define the Java platform and applications. Industries benefit from the extensive flexibility in defining only what they need for a class of device.

7.16.3   Speech coding technology

Coding technologies for mobile systems have made a tremendous impact on system design and the product development of handsets. Intensive research on coding technology and digital modulation technology such as GMSK has achieved the higher frequency utilization efficiency of 2G than 1G (analog), which made the digitalization of mobile systems in 1980s. Without the improvement of coding technology for mobile system, we would be obliged to use 1G (analog) systems even today.

For 2G systems, several different coding technologies, such as RPE-LTP (Europe) and VSELP (America, Japan), were introduced it in the late 1980s and since then the improvement of coding performance has been intensive. For example, PSI-CELP (Pitch Synchronous Innovation_CELP) can achieve coding rates of 3.45 kbit/s, which represents the highest frequency utilization efficiency. Coding technologies are not only important for mobile use, but coding technologies for audio and video signals are also important for wireless multimedia applications. ISO/IEC standard MPEG-4 (Moving Picture Expert Group) is widely used for Internet and mobile Internet applications.

7.16.4   Human interface functions

The last topic on software in handsets is human interface issues of mobile terminals. This is not purely software, but it has a combination of software and hardware aspects. For end-users, human interface or character input for wireless devices is one important factor of its sales and popularity.

Take a careful look at your cell-phone. How many keys does it have and how do these keys stand in a line? How does it differ from the key-pad of your PC and the remote-control device for your TV? How about comparing the keys of a cell-phone in the USA and in China? Are they all same or not? If you are in the USA, you will find in your cell-phone that "abc" is attached to "2", not "1". But, in Japan, "çた" (this is the first character in the Japanese 84 hiragana characters like "a" in the Latin alphabet) is attached to "1", not "2". Is it not interesting?

The reason for these differences is deeply related to the history of telecommunication and language structure in each country. So, what is a requirement for a user-friendly mobile phone? One can say fewer input numbers, easy to remember, easy to touch, etc. But the response varies quite dramatically depending on personality, age, sex, language and country.

What will be the future input methods for a mobile device? There are so many proposals and prototypes now available in the world. Voice input (recognition) technology is starting to be employed in some cell-phones/cordless phone products and we will see many other interesting examples such as a finger-ring-type keypad.

The conclusion here is that we need to pursue perfect human interface methods to satisfy consumers' needs and continuous research needs to be done to improve human interface issues with future technologies and services.

The discussion of software technologies concludes our discussion on global wireless technologies. We have covered the entire spectrum of wireless technologies: from systems and architectures to network technologies, and from access technologies to key software components. In the next chapter, we turn to discuss the business side of things. Then we will discuss the wireless value chain and the various business models being adopted in more detail. We will also take a deeper look into various business models for different players in the value chain, and also at i-mode, the most successful wireless data service to date.

8 Business models and strategies

8.1 Introduction

We have already reviewed the various segments that make up the wireless value chain, looked at various players in those segments, and discussed why it is so important to have a healthy ecosystem of players and service providers. In this chapter we will review another key element of wireless business – the business models and strategies. With the advent of wireless data applications and services, the old value chain has been disturbed with the emergence of a new breed of players providing valuable services to the industry; for example, content providers and game developers who did not have a place in the chain before constitute an important component of the ecosystem today.

In this chapter, we will also look at the success factors behind i-mode, the service that changed the landscape of wireless Internet applications and services forever. We will also clear up some misconceptions regarding the success factors of i-mode that are common amongst analysts and the press. We will look at various business models, their advantages and disadvantages and how various players open the revenue stream for themselves.

Wireless applications and services are being used extensively in situations that require personalized and/or time-critical information. Location-specific transactions are also becoming possible now, opening up a new range of commercial opportunities. Broadly speaking, wireless data services will embrace infotainment services, lifestyle facilitators and transaction-based services, with financial services and travel and transportation being the most obvious industry segments to go on the wireless data world. However, non-human uses of the wireless network will also open up to wireless operators a vast range of revenue streams that are not available today.

The following situations are likely to be those most suited to a wireless data environment:

- those where consumers wish to respond to personalized and time-sensitive information (e.g. stock market trading and online auctions);

- those where the transaction is simple (consumers are unlikely to undertake complex transactions such as buying a car, or processes such as researching a stock, via a mobile device);
- those where a transaction is location specific (e.g. finding local hotels/restaurants in an unfamiliar city).

Potential wireless data applications for both the consumer and business markets are endless, with the applications divided broadly into communications, infotainment, lifestyle facilitators, and transactions.

In this chapter, we focus on the services and capabilities that have either been introduced already, or that have been announced and can be realistically foreseen within the next few years. In the next section, we will take a detailed look at NTT DoCoMo's i-mode phenomenon that has set an example for the industry to follow.

8.2   The i-mode phenomenon

Since its launch in February 1999, i-mode has amassed over 36.6 million subscribers (as of January 2003), over a quarter of the Japanese population. The service allows subscribers to connect to i-mode and other Internet sites, as well as use i-mode email and online services. There are as many mobile handsets on a Japanese commuter train as on the commuter trains of any other rich economy. Yet in Japan, the telephones are curiously silent. Instead of speaking into sleek silver gadgets, users are busy sending emails or downloading pictures of their favorite pop stars.

Japan's growing band of email "junkies" has driven the phenomenal success of NTT DoCoMo's i-mode, the mobile phone service that offers continuous Internet access. According to DoCoMo, the largest mobile phone operator in Japan, by June 2002, its mobile Internet service boasted 34 millions subscribers – almost one in five Japanese.

European and North American operators look on with envy – and desperation. Wireless application protocol (WAP) based services have failed to make an impact. Meanwhile, operators in Europe have spent billions acquiring third-generation (3G) mobile licenses to support even more advanced services. The more European telecom shares are hit by skepticism about WAP services and worries about the debt taken on to pay for new investment, the more European operators turn to Japan for inspiration and hope. Hence the importance of understanding the reasons for i-mode's success[1], and the worrying possibility that western excitement about wireless Internet services may be based on several i-mode myths. In the next few sections, we hope to dispel these myths and point to some fundamental reasons as to why i-mode has been so popular and wildly successful in such a short amount of time.

---

[1] A good text to read is Mari Matsunga's *The Birth of i-mode* published by Chuang Yi Publishing Pte Ltd.

Western analysts often point out that the success of i-mode is the result of several special characteristics and the environment in Japan, for example:
- Japan is pedestrian-centric country,
- the penetration rate of PCs in Japan is very low, so people have to use a cell-phone instead of a PC for Internet access,
- the Japanese language is different from English and other languages.

All of these reasons are partly true, but mostly false. For example, most Japanese people who live outside of metropolitan areas such as Tokyo or Osaka commute by car just like people in Seattle or Los Angeles. It is only in areas within Tokyo and Osaka that people mainly commute by trains and subways. This kind of misunderstanding stems from the fact that 99% of westerners visit only Tokyo or Osaka and they arrive at their conclusions after a week's stay in these cities. The thinking is analogous to someone generalizing life in the USA after a visit to Manhattan, NY. The lifestyle in Japan is very similar to western countries.

Though PC penetration in Japan is far less than that in the USA, Japan is still among the top five industrialized nation in terms of PC usage. In 2003, consumer penetration was forecast to reach 46%, while business usage would be above 67% (source: Gartner Group).

Now, let us consider the language situation in Japan. No doubt, it is different. Japanese languages are composed of three different character sets and they are always mixed for use. This is what makes it so complicated. Does the success of i-mode have something to do with the Japanese language? Some analysts try to link language with i-mode, but then how do we explain the fact that thousands of foreigners in Japan have also become addicted to i-mode? They are e-chatting or emailing or e-surfing in English or Portuguese (most i-mode phones are bilingual). Consumers can also enjoy English content such as news from CNN, Bloomberg, Dow Jones and of course Disney. Nokia provides three content services in English, namely Tokyo-Q, Tokyo Food Page, and Tokyo Wine News for i-mode. Brazilians can get the soccer and Latin music information in Portuguese with a service called POKEBRAS, which is free of charge and extremely popular. This indicates that the Japanese language has no relationship to i-mode's success.

There is a widespread misconception among western carriers and players that the wireless Internet is all about technology. One view is that once i-mode-style packet switching replaces the circuit switching of traditional telephony, the demand for wireless Internet services should take off. With packet switching, the telephone can be continuously connected to various websites. It also lowers the cost of sending data.

Of course, i-mode does have some technical advantages. For one thing it uses a variant of HTML, the standard script of the Internet, unlike WAP which runs on a completely different protocol/language, hence easier development support (although this advantage goes away with the introduction of XHTML). In addition, packet switching saves users the hassle of calling up a website every time they need to interact with one.

Another common European view is that the faster access speeds of 3G networks will finally make possible the wireless Internet revolution. But at 9.6 kbit/s, i-mode's data transmission rate can hardly be described as fast. In densely populated cities, it would be more practical to download short, 10- or 15-second clips of music and video. Having "always on" connectivity makes a difference, the higher speeds help but are not a prerequisite to good application design. If customers like what they see, they can order a full-length recording of a pop song or sporting event to enjoy at home downloaded over fixed telecommunications networks. That is one of the reasons why DoCoMo was keen to tie up with America Online in developing services that combine the benefits of both fixed and mobile Internet access (more on AOLi service later in the chapter).

Japanese operators are wary of technology for technology's sake. "WAP failed [in Europe] because operators concentrated too much on the technology rather than the content," says Keiichi Enoki, board of DoCoMo and charisma of the i-mode project team. "It is like worrying about the quality of television sets before you have any programmes."

By contrast, content has been central to DoCoMo's success. The secret lies, above all, in the variety of what is on offer. Amid the neon of Tokyo's Shibuya district, Japanese teenagers in knee-high boots dream of downloading the latest cute cartoon characters to decorate their screens. Businessmen try their luck at a virtual fishing game, or check the latest scores of their favorite baseball team. "The package of content that we put together was the killer application that helped i-mode to take off," says Toshiharu Nishioka, former manager of DoCoMo's gateway business department.

This points to yet another myth that has fed the frenzy of European operators: that by putting existing content, such as films and popular songs, on to mobile telephones, the mobile Internet business will provide a quick and easy payback.

"In Europe there is an overly optimistic view that if they just take content off the shelf it will offer them a huge business opportunity," says Kiyoyuki Tsujimura, board of DoCoMo, head of global business department. Content will need to be tailored to the evolving needs of users and the limitations of what 3G can do. Keiji Tachikawa, president and CEO of DoCoMo, noted that the winners in 3G will be the people "who are creative enough to find out what kind of content is most desirable in a mobile [phone] environment".

Japanese youth are far more mobile than, say, American teenagers. The i-mode phenomenon has coincided with changing social behavior, particularly among Japan's urban youth. "Young people have a very different attitude towards personal relationships," says Masahiro Yotsumoto, research director at the Dentsu Institute for Human Studies. "Instead of having one good friend, they prefer to have 200 mobile phone friends."

Sending short messages, such as saying good night to a friend, is one of the most popular uses of i-mode, Mr Nishioka points out. The lack of real privacy in small Japanese houses has also helped to make mobile phones popular. "In Japan, especially for young people, their home is their cell phone," says John Barber, former director of AOL Japan. (Note: in Japan data messages cost less than voice messages, which is not always the case in the USA.)

Even with i-mode's advantages, DoCoMo predicts that it may take four years to make a profit on 3G, and many more years to recover the cost of investing in 3G. Europeans may have to wait even longer.

As we have already discussed in Chapter 3, the Japanese have overwhelmingly embraced wireless Internet services – from teenagers to business executives alike. In a short period of three years, the total number of wireless Internet services users eclipsed 50 million, which is nearly 74% of the total number of cellular/PHS users. The success of i-mode is often a discussion topic of news media, conference debates, and analysts reports, but two other rival services – EZ-web from KDDI and J-Sky from J-phone – have also done well. EZ-web has over 9.3 million users while J-Sky has been able to garner over 9.7 million subscribers in less than three years (both J-sky and EZ-web had over 10 million subscribers as of June 2002). It is important to point out that these two services are using the legacy WAP technologies (WAP 1.X). However, the response amongst the subscribers has been quite different if you compare the feedback from consumers in Europe and the USA where the WAP service has been already been branded "dead". So, why is there such a wide gulf in user behavior? Let us examine some of the aspects in more detail.

Mr Takeshi Natsuno, executive director of i-mode planning division of NTT DoCoMo, nailed it down as follows: "The true mechanism of the great success is that the Operator has made a function of coordination of total value chain."

Let us explain this with some examples. First, NTT DoCoMo introduced the BOBO (Billing On Behalf Of) function (as shown in Figure 8.1) which provided a critical billing mechanism for content providers. Some portion of the income from each subscriber went back to the content provider automatically. This mechanism was a breakthrough for the wireless Internet world, since content providers have no direct way of collecting revenues from consumers without using BOBO functionality. Also, the operators have a high degree of success in collecting money from the cell-phone users (more than 98% in Japan), so the risk of losing money for a content provider is very low. This is a very effective billing mechanism for relatively small/medium size content providers who can not prepare a billing system for themselves owing to financial and technical reasons.

Natsuno further explains that the BOBO mechanism is similar to AOL's strategy where it acts as a platform for Internet businesses and successfully invites a lot of content providers on its portal. In that sense, the business model for i-mode is a global one, and not Japan specific.

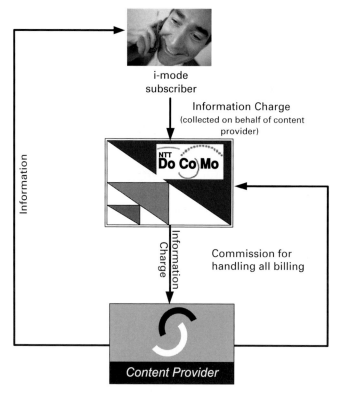

i-mode
subscriber

Information Charge
(collected on behalf of content
provider)

Information

Information
Charge

Commission for
handling all billing

Content Provider

Figure 8.1. BOBO mechanism (source: NTT DoCoMo).

Second, terminals play an important part in the ecosystem. The total value of the service is provided by the combination of useful content and attractive terminals. Black and white display terminals can no longer convey the charms of Disney characters. The interaction between terminal functions and content is indispensable for the success of wireless Internet applications and services. Because of the physical and functional limitations of cell phones, to provide good service and value proposition, its design needs to be optimized. The wireless Internet operator should play an important part to bring content providers and the terminal vendors to the same table by presenting specifications of wireless Internet terminals.

Finally, the interaction with other fixed/cellular telephone operators and ISPs is very important.

Wireless Internet has had a dramatic impact on voice ARPU as well. The *Phone to* function of i-mode allows users to make a voice call easily by just clicking on the number displayed on a Web site or the message screen. These voice calls are often routed to fixed telephone numbers, so the *Phone to* function contributes to the increase in fixed network traffic. Some other examples are the *Mail to* and *Web to* functions. The *Mail to* function facilitates into the display of a mail composition screen

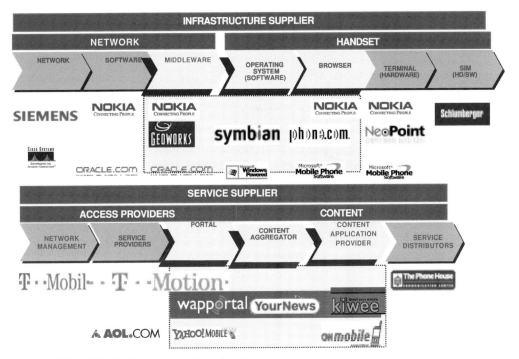

Figure 8.2. Total ecosystem of wireless Internet.

instantly by just clicking on the email address displayed on a Web site or the message screen. Similarly, the *Web to* function allows users to browse with ease, clicking on the URL displayed on a Web site, or the message screen will connect you directly to that Web site. These simple, yet convenient, functions increase the access to existing ISPs.

It is extremely important to understand the wireless Internet ecosystem – locally and globally, the players, the value chain, and an environment where all players in the chain benefit from new and popular applications and services. Mobile data involve the integration of programming environment, content, transport, and end-user terminal. No carrier outside Japan has executed end-to-end to the same extent as DoCoMo. The company has leveraged its channel dominance to corral handset vendors – Sony, Sharp, Kyocera, Matsushita, and even Nokia – to do its bidding. Figure 8.2 shows the wireless Internet value chain from the infrastructure and service supplier angle. With the advent of wireless Internet, the traditional value chain was disrupted and new players (highlighted in the middle) appeared, who need to be considered seriously for new applications and services.

In the next few sections, we will take a look at the business model and strategy scenarios as they affect the players in the value chain.

8.3   Business model for content service providers

Next-generation mobile networks will pave the way for the development of a host of mobile applications. One mobile application category, applications that make use of digital content, will be key for operators to pay back their huge investments in these networks. In Europe alone, mobile operators have spent more than 100 billion euros in purchasing third-generation (3G) licenses and will have to invest at least that much again in deploying a 3G network infrastructure.

To make these huge investments pay off, operators are now faced with the challenge of putting theory into practice. To be successful, they need to ensure that services and applications that are developed for distribution on their infrastructure truly add value for end-users. Most industry observers today acknowledge that, rather than one killer application, an aggregation of different applications (or man-slaughter apps) addressing different lifestyles and uses of mobile communication devices will coexist and generate revenues from different business models.

In Japan, for instance, NTT DoCoMo has found that i-mode data users make up to 15% more voice calls than users that have not subscribed to i-mode services. This is partly the consequence of some applications, such as using a mobile device to search for a restaurant, increasing the likelihood of a subsequent voice call – such as a call to a restaurant to make a reservation. In the UK, a mobile operator survey has found that the most frequent content of premium SMS messages is jokes. The transmission of the jokes creates a cascade of messages as people forward messages on to friends, thereby generating additional revenue for mobile operators. The same is also likely to be true for the delivery of horoscopes, news, pictures and sports information.

In most cases, the provision of content-based applications will result in an increased ability for operators to generate revenues from complementary data or voice applications. This is particularly the case for operators taking a commission on brokerage transactions. "Day traders", for example, will make financial transactions based on information alerts delivered to their mobile devices.

Furthermore, mobile content applications could have additional value for operators in allowing them to reduce costs associated with customer churn. Customers who become accustomed to the mobile content applications made available by an operator in terms of an application's content, format and interface are less likely to switch operators, thus allowing operators to spend less on customer retention.

One key to the development of the mobile content market is to make the economics attractive to all players involved. Lessons can be learned from Japan, but the unique qualities of the European and North American market mean that it must define a new approach to revenue sharing.

There are various revenue streams in wireless data. Players have the opportunity to derive revenues from:

- service fees,
- commissions on monetary transactions,
- content, and
- advertising.

Let us evaluate each of these major models in more detail.

8.3.1 Service fees

Several wireless data services may be available by paying a fee, which may take different forms: fixed subscription, per-usage fee (and premium per-usage fee), or premium on airtime price. Service fee revenues are likely to benefit content providers, portals and mobile operators. However, it is expected that increasing numbers of services will be provided for free or bundled with a bucket plan in the future.

Fixed subscription

Customers pay a fixed monthly charge to have access to the services. A couple of examples of this are Telia Mobile and AT&T Wireless, who are offering an Internet WAP portal access for a fixed monthly rate.

Per-usage fee

Each time a service is used, a fee is charged and included in the mobile phone bill. An example is Sonera, which charges around $0.50 for each access to CNN news or stock quotes.

Premium on airtime

The mobile operator controls the price of access to wireless data services. Some operators charge airtime (like Sprint PCS) at a premium rate for access to their wireless data platform. This type of charging model will disappear once GPRS networks are implemented, because of packet switching.

Airtime

It is inevitable that network usage will increase as a result of wireless data. However, what is not clear is whether network usage revenue will increase for operators in the long run. The current per-minute charging system used by network operators will lead to increased revenue. Increased network usage revenue will also be directed towards content providers, ISPs and portals. An example of this is Reuters, which has developed its own portal offering. Vodafone pays Reuters 20% of access revenues for exclusive access to this content. In the long term, decreasing ARPU coupled with the idea of free Internet access will mean that network usage becomes a low-margin business. As explained elsewhere, this development has major implications for operators, who must reposition themselves as service and portal providers in order to avoid becoming mere bit pipes. Some of the advantages and disadvantages of this model are as follows.

Advantages
- Simple pricing per minute model that customer can understand.
- If the content, applications, and services are good, traffic will be increased and maintained.
- The approach is consistent with the Internet today.
- Systems implementation issues are simplified: key issues are securing content delivery and billing access time.

Disadvantages
- The model gets more complicated in the GPRS and 3G world, where the access-per-minute model will need to be replaced by a subscription-based content pricing model.
- This potentially sets the price of content below what at least some of the market would be willing to pay.
- It is difficult to change the consumer mindset once content is given away free.
- Certain content providers will be unwilling to provide content on this basis.

It should be noted that one of the flaws with WAP has been circuit-switched billing for certain implementations. It is less efficient and more expensive to end-users and hence less popular.

8.3.2   Advertising

There are two main methods for obtaining revenue from advertising: content and brand. Advertising will probably become a significant revenue stream for content providers and portals.

Advertising content

The implications of advertising content on mobile devices are only just being investigated. However, the present authors believe that direct advertising revenues are unlikely to be generated in the short term. Instead, partnerships between operators and content providers are likely to share call traffic or sponsorship revenues. Push advertising will only work with the user's permission. That is, users will accept advertising in return for some form of benefit, such as lower call charges. Hybrid approaches of sponsoring content are also being explored by operators.

Advertising brand

In a rapidly evolving and highly uncertain marketplace, trust and brand image are vital concerns. All players will look to strengthen brand image through advertising.

This will be of particular concern for partnership members, who must guard against the effects of brand dilution. We will discuss more about advertising and advertising strategies later in this chapter.

8.3.3   Commission on monetary transactions

Another way to collect revenues is by charging commission fees on transactions carried out over wireless devices. This offers real benefits to content and service providers alike. On even small transactions, given the future growth expected in mCommerce, the return could prove to be huge. There are two ways of charging commissions on monetary transactions: charging a percentage of the transaction value, or charging a flat rate. For example, a content provider selling theatre tickets pays a fixed monthly rate to appear as first choice on a mobile portal.

The transaction fee might be charged to the customer. However, a likely model in the future would be to charge the transaction fee to the goods/service provider. The companies that will benefit from those revenues are the players that succeed in becoming the mCommerce distribution channels.

8.3.4   Content management

Content aggregation provides a key source of revenue owing to the value added in bringing related content together in one place. Portals are very keen to market themselves as the content aggregator of choice, with established players such as Yahoo! leveraging their Internet presence.

8.3.5   Free content is not a business model

There is general agreement that the free-content model pioneered on the fixed-line Internet is dead. For the most part, content providers are not eager to repeat their unfortunate free-content, advertising-based business-model experiences. And who can blame them, when recent surveys show that, even today, only a small number of content-based fixed-line Internet sites have a positive operating margin? Instead, content providers and aggregators alike want, and need, their mobile distribution channels to be profitable, even when considered on a stand-alone basis. Fortunately, end-users seem willing to pay for certain types of content delivered on mobile devices. The experience in Japan with i-mode, the explosion of premium short message service (SMS) messaging, and the current market for downloading ring tones and icons indicate that users are already prepared to pay for data services that they perceive as value added.

Sources of revenue for mobile content applications

Players in the mobile content market will continue to find sources of revenue beyond basic service fees. For example, end-users are willing to pay for network data traffic created by the delivery of content-based services. Although likely to be substantially less than on the fixed-line Internet, advertising and sponsorships will also provide an alternative revenue source. Additionally, early SMS sponsorship experiences are quite promising. It is worth noting, however, that such enthusiasm also appeared in the first days of the fixed-line Internet, ultimately ending in disappointment.

According to a financial model recently developed by Andersen to evaluate the size of the mobile content market, the split by revenue source for mobile content applications in Europe shows that, in 2005, approximately 46% of market revenues could be generated through end-user service fees, 45% through the payment by end-users for network usage, and 9% by advertisers or sponsors.

Generating revenues presents the obvious first formidable task. Defining how these revenues will be shared, however, poses an increasingly important challenge as it will be a decisive factor as to whether different players will enter the mobile content market, which, in turn, will determine how, and how quickly, the market will develop in Europe. In the next section, we will cover the revenue flows amongst the major segments.

8.3.6   Flow of revenues

In this section, we will discuss briefly how the revenues flow amongst different players in the value chain. There are four main models in existence:
- operator-led,
- content or portal-led,
- financial service provider-led, and
- cooperative.

Operator-led

Operators have the opportunity to lock in customers through ownership of the initial login stage or portal. They also have access to subscriber data and the location of the customer. However, they lack experience in the content business. This means it is likely that operators will look to form partnerships with content providers or alternatively use their current valuations to acquire the missing skills, such as obtaining a banking license or portal provider.

Players such as Vodafone UK have been planning and launching their own portals but lack experience in this area. This had led to partnerships by Vodafone with content providers such as Charles Schwab and Infospace. An example of a successful operator-led partnership is NTT DoCoMo in Japan, whose partners include banks,

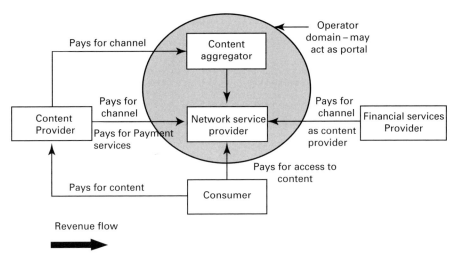

Figure 8.3. Operator-led revenue flow scheme (source: Ovum).

credit card companies, ticket providers and news providers. Figure 8.3 shows a typical operator-led revenue flow scheme.

Portal-led

The portal provider holds a powerful position in the value chain through acting as the main point of contact for the customer. As such, the mobile portal arena is set to become one of the most difficult areas in which to stake a claim as most players look to set up a portal offering. Existing Internet portals such as Yahoo! Mobile and AOL Europe have already set up mobile portals and pose a threat to operators and service providers. Figure 8.4 shows a typical portal-led revenue flow scheme.

Financial service provider-led

Financial service providers such as banks and brokerages have the potential to expand rapidly into the wireless data arena through established customer relationships and strong brand images. The information they offer is also suited to mobile access since it is often time sensitive and content light. In most markets, the strength of network operators makes it unfeasible for banks to dominate a partnership. One exception is in Italy where Smart.com has begun distributing mobile phones to its premium customers to conduct secure stock transactions. Figure 8.5 shows a typical financial service provider-led revenue flow scheme

Cooperative model

In the cooperative model, network operators, service providers and content/portal providers act together for mutual benefit. No one player takes the lead or accepts

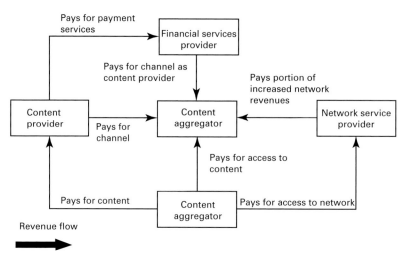

Figure 8.4. Portal-led revenue flow scheme (source: Ovum).

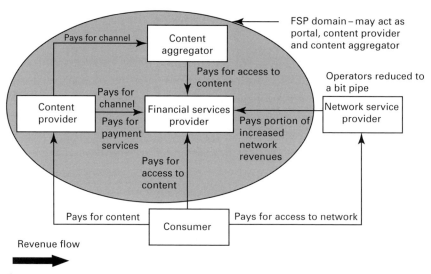

Figure 8.5. Financial service provider-led revenue flow scheme (source: Ovum).

the entire venture risk. Given the high value associated with occupying the portal space and thereby becoming the customer's choice of Web entry, it is unlikely that players will cooperate fully in the early stages of mCommerce. The question is not so much whether network operators and service providers reinvent themselves as portal providers, but rather whether they can afford not to. Figure 8.6 shows a typical cooperative revenue flow scheme.

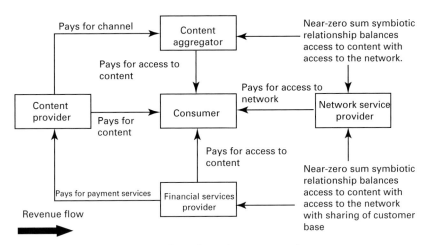

Figure 8.6. Cooperative revenue flow scheme (source: Ovum).

8.4   Business model of advertising and publishing services

Wireless Internet access with a cellular phone is one of the access mediums that content providers can use as a tool connecting end-users to other access mediums like TVs, radios, magazines and newspapers. The strength of a cellular phone as a medium is different from other channels in terms of longer hours for use per day and the greater number of end-users.

Let us consider the total amount of time a consumer spends being in touch with each of the main communication mediums.

Cellular phone	10–15 h
Television	2–4 h
Newspaper	0.5–1 h
Computer	5–10 h

Since the cellular phone is such a personal device, one can easily imagine the potential of using it as an advertising tool. Leveraging wireless Internet for advertising is very promising because very high click-rates on a banner advertisement can be expected. Natsuno commented in his book that the highest average click-rates on a banner advertisement in i-mode reached over 22%.

In addition to advertisements, contents providers can get customer segment information through services based on locality, gender, age group, etc. Armed with this information, content providers can promote pier-to-pier marketing over cellular phones.

In 2002, a service to download e-books began in Japan. Using a modern, colorful and large display terminal, the Japanese did not find it hard to read a book on wireless

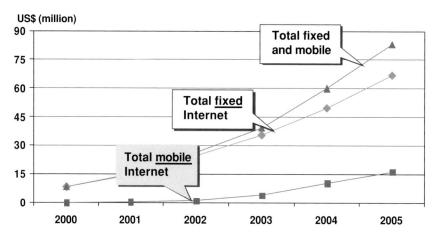

US$ (million)

Figure 8.7. Worldwide advertising market forecast (source: Ovum report).

phone. The service is mainly targeted at young office women who are otherwise reluctant to have full handbags during their normal commute.

All content providers can utilize simple forms of advertising on their sites. For most companies the advertising is a side business and the main source of income is from elsewhere. However, for some unofficial sites, advertising can be the only revenue stream. There are content providers whose main business is advertising. They make all their money from offering content, which has the duplicate role of being information for the users and a way of presenting themselves for the advertising company. The content providers can be established businesses as well as new businesses. Figure 8.7 shows the worldwide forecast of the advertising market and compares mobile and fixed wireless advertising revenues potential.

8.4.1   Advertising strategies

The mobile Internet brings a range of new possibilities in the way of advertising. Companies are beginning to see the possibilities and how they can benefit from other companies' needs to advertise. New methods and business models are appearing in the market and the advertising is becoming more and more creative.

Banners are one of the most basic forms of advertisement that a content provider can have on their site. So far banners have been more successful on the mobile Internet than on the fixed Internet. Click-through-rates (CTR) measure how many people actually click on a banner that they are exposed to and present the result as a percentage. When the advertising company ValueClick started to test their banner provision system in February 2000 the CTR was more than 6%, compared with the Internet figure of less than 0.5%. The high rate at the launch can easily be explained as people's initial curiosity and since then the rate on the mobile Internet has stabilized at 1–1.5%. This is still higher than on the fixed Internet, and as a consequence the advertising companies are prepared to pay more for advertisements on the mobile Internet.

A site called Keitai Net and the Japanese Postal Savings Account System have together created an innovative service with regards to banners and CTR. After users have registered on the site they get paid to view advertisements. Currently the users receive 15 yen for each advertisement that they click on, money that is deposited in a postal account. The user can use the money to purchase products featured on the site, or they can download the cash from the account.

Another obstacle that has prevented banners from becoming widely used is the fact that the users themselves have to pay to have data downloaded to their phones. So far the user has been paying the traffic cost for both downloading the banner advertisements as well as the cost of clicking on them. This is due to change, however. One operator has now introduced a system for charging the advertising company instead of the user for the traffic that is generated when a banner is clicked. Content providers will thereby be more open to having advertisements on their site since it does not imply any additional costs for the user.

Charge for loading content: a number of content providers offer sites with the main purpose of informing users about other companies' products and/or services. The latter pay the content provider for making this service available to them. Examples of this are townguides, where restaurants and shops pay the content provider a fee to be listed on the site. As an example, restaurants pay a fixed cost of 100 000 yen per year to be listed on a guide by the ticket company PIA. For the users the service is free of charge.

A variation on this theme is so-called promotional sites, in Japan called Kenshou (= prize). Companies that want to promote themselves and attract traffic to their own mobile Internet site can have them listed on a Kenshou site. The content provider of the Kenshou site charges a fee for this. Users are drawn to the site with promises of prizes and of being entered in various lotteries if they register and enter their email address. They can then choose to receive, for example, targeted emails about categories of products that they have registered beforehand as interesting. They can also enter an advertising company's site, and perform certain actions such as finding specific information. As a result the user is entered in another lottery and the company has exposed its brand to a potential buyer.

Sponsorship: another form of advertisement is when companies sponsor different parts of the content. The advertising company's name or products would be exposed in some way when the user accesses the content. An example of a sponsorship model could be that a soft drink manufacturer sponsors an activity on a content provider's site. If the user buys a drink of the manufacturer's brand, he or she gets a number, which is printed on the can. When entering the number at the content site the user gets access to, for example, a game. The advertising company pays a fee or commission to the content provider for this service.

The form of advertisement that will be most successful on the mobile Internet is the kind that is asked for by the users. Charge for loading content, a model mentioned above, is an example of this. In this case the users view the content as information, while at the same time it is an advertisement for the advertising companies. To compensate users for receiving an advertisement is another method to make them want it. One way could be to give them credit points or cash, as mentioned in the example of Keitai Net above.

Sponsorships also belong to this category: users accept marketing in order to get free access to content they otherwise would have to pay for. As mentioned earlier, the mobile Internet advertising business is still rather immature and many new solutions will most likely see the light of day before long.

In order for content providers to be able to obtain revenues of a significant size, it is essential for them to have a large user base. If not, no company will be willing to pay for having their advertisements on the site, or if they are they will surely not expect to pay much.

8.5   Business model for device manufacturers

The competition amongst device manufacturers is intense. With an increased pressure on churning out new models on regular six month intervals, even larger players like Ericsson and Nokia are having a tough time competing. In addition, handheld and PDA players like RIM, Palm and Handspring are getting into the market by providing a combination of PDA and phone in one device. As such, handset manufacturers must worry not only about launching new products but must also think about diversifying their approach to the market. Recently, Ericsson and Sony – arch-rivals in the handset business – teamed together to do joint research and development of future-generation phones.

Let us consider RIM's approach to the wireless market. Blackberry is a handheld device with a built-in wireless modem introduced by Research in Motion (RIM), a Canadian company. Blackberry offers functions such as secure, remote corporate email access and also PIM tools. By the end of 2001, RIM had over 289 000 subscribers with 13 200 companies, and announced that 82% of its revenues is generated by the product. RIM's Blackberry has been extremely successful in corporate North America, similar to the popularity of SMS in Europe and i-mode in Japan amongst the consumers. The key to their success has been a secure and efficient way of accessing corporate email, which mobile professionals consider the most important requirement for wireless access. Also, Blackberry is designed to be "always connected" to the packet network, which gives a similar end-user impression to the i-mode service with packet transmission. The single most important reason why Blackberry is so popular in North America is the fact that RIM was able to package application, network, and a device into one compact unit that is very easy to use. The business model for Blackberry can be classified into the following categories.

The first model is operation on two-way packet-data networks such as 800 MHz DataTAC and 900 MHz Mobitex. As paging is still popular in North America, the Blackberry paging service is well accepted in North Americas in spite of the non-voice call function of the device. The killer application at this stage is mostly real-time access to corporate email. RIM enjoyed the direct sales of the Blackberry devices to end-users as data-traffic generated income rapidly.

The second model is operation on WAN such as GPRS and iDEN. As GPRS networks have been widely rolled out in Europe and Asia and also parts of the Americas, Blackberry on GPRS can create a new revenue model on a global scale. Also, the new Blackberry device (5820 in Europe, 5810 in North America) added voice functionality and this function will surely increase the voice and data traffic. The killer application at this stage is the total business solution for mobile professionals, which may dramatically increase productivity with the combination of email and phone functionality. For example, AT&T Wireless's Blackberry access plans are $39.99 monthly including 3 Mb data and Enterprise server software with 20 client access licenses is available at $2999 (as of Q1 2003).

As discussed above, the business area for RIM has enlarged dramatically from just a domestic-email device maker into a global wireless solution provider.

The third model is to license RIM's technology and intellectual property (IP) to other device venders. RIM announced in April 2002 that the company will offer its Java-based technology (J2ME) as a platform to other competing vendors. Because of this, RIM's platform could be used for a variety of devices and this strategy will increase the future sales and income from the licensing fee for the company.

The last strategic resort is a global market. RIM's business area is not limited to North America; it is expanding into Asia and Europe. The company has announced alliances with UK and Hong Kong wireless carriers. However, similar to the experiences of Vodafone and DoCoMo, RIM will face the differences of habits, mentality and culture on each continent. Success in the North American market does not automatically mean success in Asia and Europe. The paging service has almost passed away in Europe and Japan (like telex), but it is still alive in North America!

It is important to discuss the differences in the way carriers work with handset manufacturers worldwide. DoCoMo's overwhelming technical leadership amongst the PDC handset vendors goes a long way toward explaining its domestic success over global players like Nokia, Ericsson and Motorola, whose advantages in other major wireless markets – economics on the scale of international mass production, experience with multiple wireless technologies, and global brands in the cell-phone business – do not mean much in Japan. Nokia, far and away the world's wireless handset leader, with over 37% of the global market, has only a small share in the PDC market.

In particular, Japan uses a wireless protocol called personal digital cellular (PDC) that is used nowhere else in the world. (By contrast, the GSM standard is used throughout Europe and in parts of the United States and Asia. Global players like Nokia, therefore, have made massive investments in GSM technology but have been more reluctant to do so for PDC phones.)

Beyond the DoCoMo factor, there are challenging market differences outside of Japan. Some of these derive directly from Japan's lead in wireless Internet access but its equally pronounced lag in fixed-line Internet. The already-proven Japanese handsets certainly might appeal to network operators in the USA, where wireless data services are still in their infancy, and in Europe, where manufacturing difficulties have plagued the phones built to use wireless application protocol (WAP), and the dominant mobile Internet technology. Japan's lead in the wireless market also presents a challenge to the handset makers who want to sell where markets are less developed. Even Sony, one of the strongest brand names in the world, pulled out of the US mobile handset market in the late 1990s after disappointing results and says it will not try again until broadband wireless access becomes widespread. Meanwhile, Nokia, Ericsson and Motorola have entrenched themselves further in the USA. As we have been discussing throughout this book, to create a global market requires a player to deal with both technical and non-technical challenges in a time-efficient and economic manner.

8.6   Business model for operators

Network operators have an extremely important role to play in the wireless Internet ecosystem. It is estimated that the ARPU for operators will decrease and the churn rates will increase as penetration rates of cellular subscribers increase. This stems from the fact that higher penetration rates accelerate the user segment changes and there is a subscriber shift from business executives to younger generations who are always looking for "cooler" devices and affordable prices. Additionally, the cellular operators are also competing with each other to get new subscribers by reducing the entry barriers. Most of the operators are still struggling with this contrasting issue, but we can find some good examples from leading-edge wireless data services such as i-mode on PDC and FOMA. DoCoMo has kept high ARPU and very low churn rates even after they secured millions of subscribers. The churn rates have decreased from 1.97% (in 1997) to 1.61% (in 1999) and 1.18% (in January 2003) in proportion to the increase in i-mode subscriber rates. Total ARPU (voice + data) remained almost the same in March 2000 and March 2001 in spite of the voice ARPU decrease of 10%

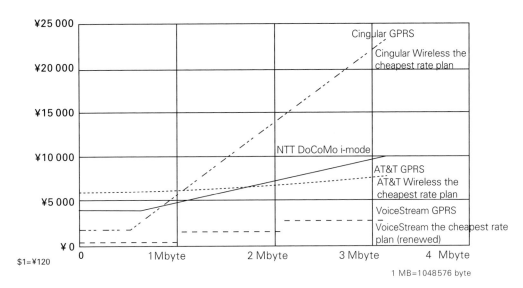

Figure 8.8. Wireless data tariff comparison (source: NTT DoCoMo).

in this one-year time interval. An increase in i-mode ARPU (from 120 yen in March 2000 to 850 yen in March 2001) has helped in recovering from this reduction (as of March 2002 the ARPU was 8480 yen). Now let us consider the following.

- Packet transmission fee,
- information fee (optional).

First, additional monthly charge for wireless data differs by operator. This charge is to recover the implementation cost for wireless-data specific equipment such as a gateway server. In Japan, it ranges from 100 yen to 300 yen. Charges appear in the same monthly bill provided by operators.

Second, the packet transmission fee differs not only by operator, but also by services: for example, 0.3 yen per packet (1 packet = 128 bytes) for i-mode and 0.2 yen per packet for FOMA. Considering the difference in data rates of i-mode (9.6 kbps) and FOMA (64–384 kbps), normalized tariffs per data rates (yen/packet, kbps) are as follows:

i-mode (9.6 kbps)   0.031 25 yen/kbps (1)
FOMA (64 kbps)   0.003 125 yen/kbps (0.1)
FOMA (384 kbps)   0.000 52 yen/kbps (0.017).

In order to limit the total payment for a user, the normalized tariff is set low according to the increase in data rates. In other words, the wireless data user can enjoy a more economical tariff plan when the system upgrades from 2 (2.5) G to 3G. Figure 8.8 compares the wireless data tariff of US and Japanese operators.

Third, an information fee is optional and it is only generated for the access to pre-registered fee-based content sites. When a user accesses only charge-free sites,

i-mode services - traffic distribution

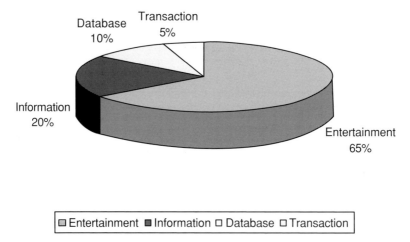

Figure 8.9. i-mode services by traffic distribution (source: NTT DoCoMo).

this charge is not generated. The operator handles the billing on behalf of (BOBO) mechanism for the information/content providers. Charges appear on the same monthly bill provided by operators and some portions of these charges are later channeled back to the information providers. So, the information providers need no billing mechanism, no customer care and no customer centers. They can fully devote themselves to what they do best – providing value-added content.

With the help of the BOBO mechanism, content providers can make a profit in proportion to the number of subscribers visiting their sites. Natsuno reported in his book *i-mode strategy* that 48% of total i-mode users had subscribed to charged contents service by October 2000. At that point, the total number of i-mode subscribers was about 14 million, so nearly 7 million users enjoyed at least one content service for which they paid. This is a very high percentage. He also mentioned that an i-mode user subscribes to 2.2 paid content services on average. The content providers who do not charge for their content can increase the number of hits to their wired sites as well and increase brand awareness across multiple access channels and leverage the opportunity to do more consumer marketing.

Using the wireless Internet, content providers can provide unlimited content and entertainment applications such as: ringing tones, karaoke, maps, games, images, traffic, travel, rental cars, fortune-telling, lottery, horse racing, sports, shopping, coupons, jobs, banks, food, recipes, fashion, music, movie, real estate, school, health, security, celebrities, town, politics and much more. Figure 8.9 shows i-mode traffic distribution by services.

The download service is similar to satellite broadcasting services, providing point to multi-point broadcasting like a shower; it is a very efficient mechanism for content service providers. Another aspect that operators are focusing on is the replacement of wireline phones with wireless phones. This is already happening in Scandinavian countries where wireless is becoming the sole communication mode (wireless replacing wireline) and in some developing countries where, instead of focusing on the wireline infrastructure, wireless WAN and LAN infrastructure is being put into place.

8.6.1   Conundrum for operators?

Although DoCoMo's business model has been shown to be one of the key factors in the success of i-mode, there are issues specific to the current marketplace that must be considered when trying to replicate it in Europe. First, European mobile operators are under pressure to recoup heavy investments in 3G licenses and the deployment of the network infrastructure. In light of this, financial markets are pushing mobile operators to take a significant part of the value created in the mobile content market. As a result, European mobile operators are much more demanding in negotiations with content providers on revenue sharing, and several operators have either acquired application developers or announced deals by which they will receive a significant part of the revenues that mobile content applications could potentially generate.

Likewise, certain content providers will be more imperious negotiators than their counterparts in Japan. A Dutch-language newspaper, for example, has a small addressable market, largely limited to a specific region of the Netherlands. The cost to produce high-quality content is, therefore, inversely proportionate to the size of the audience. In fragmented markets such as Europe, where addressable markets are small, content such as news must be priced higher in order for providers to be as profitable as they are in more homogenous countries, such as Japan and the USA. As a result, content providers must negotiate to recover the higher costs of producing content. Additionally, owing to a lack of financial resources for technological upgrades, many European mobile operators currently are unable to bill end-users for service fees on content applications. In the absence of upgraded billing systems, content providers cannot bill the end-user for services. The business models, therefore, must address the sharing of revenues generated from network usage. Given the present context, most mobile operators will consider such a request unrealistic.

Finally, when DoCoMo launched its i-mode service, the Japanese digital content market was nascent. DoCoMo played an important role in putting in place the technological enablers and business models to allow the development of the mobile content market. Content providers at the time had a relatively poor view of the potential of the mobile content market and DoCoMo persuaded them to join the business chain with sweat and tears. However, in Europe, content providers have learned lessons from their painful experiences on the fixed-line Internet.

They have seen that content-based applications can potentially generate significant network traffic revenues for operators and, in light of this, are likely to demand a larger piece of the pie. By imposing unbalanced revenue-sharing models, mobile operators run the risk not only of reducing the size of the overall mobile content market, but also the revenues that they themselves could generate from mobile content applications. According to some financial models, if mobile operators were to adopt a revenue-sharing model whereby only 50% of service revenues were returned to content providers, the overall market size could decrease by more than 50% (source: Andersen). The content providers with the most significant bargaining power (e.g. record labels for mobile music applications) will suffer the most from an unbalanced revenue-sharing model.

The importance of revenue sharing in the development of the mobile content market cannot be underestimated. Although the business model in Japan has proven to be one of the key factors for the success of i-mode, there are specific issues that must be considered when trying to replicate it in Europe or elsewhere. Content providers' and mobile operators' ability to cope with each other's bargaining power will be the key in implementing successful revenue-sharing deals. Furthermore, inadequate revenue-sharing deals might have a harmful impact on the revenues of each value-chain player and, more importantly, on the overall development of the mobile content market. All players have much to gain, and much to lose, depending on their ability to see the larger picture.

The key lesson from Japan is the revenue-sharing model. Carriers outside Japan are finally getting it. AT&T Wireless launched the mMode service based on the BOBO functionality on its GPRS network in April 2002. Similarly, service providers in China are embracing DoCoMo's i-mode business model. If a customer uses a Sohu (Chinese portal) download, for example, the charge shows up on the monthly bill received from their mobile carrier. The phone company keeps between 12 and 15% of the revenue, and Sohu gets the rest. Downloads of mobile phone ring tones and screen designs, as well as news reports and jokes – which cost from 0.1 to 2.0 yuan (1 to 25 cents) – have enticed mobile-phone-owning Web users in China to open their wallets. While the transaction amounts are small, the potential is encouraging given that China's 161.5 million cell-phone customers outnumber Web surfers by about three-to-one.

8.7   Business model for WLAN operators

As has been the case in the past, it appears that technology is seen to be the driving factor in "restructuring" the marketplace, whether we are speaking of 3G or WLANs. Technology indeed can restructure things. However, technology is not sufficient for providing a base from which "business model architectures" are formed.

Since 2001, the WLAN market seems to have been on fire. Although there have been solution providers like Metricom offering high-speed connectivity (128 kbps) and the IEEE 802.11b technology has been around for years, it was only in 2002 that the market was ready to embrace WLAN solutions. Millions of dollars have been poured in by the venture community even though only a few players can survive in the long term. WLAN poses an interesting conundrum to the market, to the operators, and to aspiring WLAN operators who would want to disrupt the value chain. Fortunately, we can learn from the history of companies like Metricom and Mobilestar (also see Chapter 9 for lessons learned from the failures of Metricom and Mobilestar). One of the fundamental issues faced by operators and service providers, especially new ones, is that of infrastructure investment and business model that justifies the same. Some operators have been trying to force the WAN models (business and pricing) on to their WLAN offerings and it will not work either in the short- or long-term. What is really required is to look at the service from an end-user point of view, segment the user base and design services stemming from those findings. The wireless Internet service provider (WISP) retail business model for WLAN deployment is not sustainable. The net present value (NPV) of the flat-rate access model comes up consistently negative owing to high infrastructure and operating costs. The recovery rate is in years and not in months[2]. Hence a business based just on making hotspots available can not survive on a standalone basis. The WLAN wholesale business model shows more promise. Here, the WLAN network operator connects large Internet, telecommunications, and media companies to network access points, whereby service is passed on to end-users. The NPV ends up positive in majority of instances. Another issue that is hotly debated in the industry is that of WLAN cannibalizing 3G or WWAN revenues. WLANs need not be threats to mobile networks, granted that operators recognize the complementary nature of both networks and the revenue opportunities available from offering integrated WLAN and wide-area network (WAN) access (also see Chapter 9 for more discussion on this topic).

The most prevalent business models in the industry are as follows:

- integrated service provider (operator),
- pure Play wireless Internet service provider (WISP),
- non-facility based aggregator, and
- wholesale service provider.

Table 8.1 lists some of the advantages and disadvantages of these business models. Figure 8.10 shows the variance in pricing models.

In Chapter 9, we discuss some of the challenges facing the WLAN industry and, in Chapter 13, we discuss the sister technology mesh networks that will have a big impact on acceleration of WLAN deployments.

---

[2] http://www.outlook4mobility.com/commentaries/April2202.htm

Table 8.1. Advantages and disadvantages of various WLAN models

Business model	Companies	Description	Advantages	Disadvantages
Integrated service provider (operator)	AT&T, BT, NTT DoCoMo, Nextel, T-Mobile	Established operator adding WLAN access capability to their customers	Strong synergy with their WAN offering. Have the financial resources for infrastructure investment, and marketing. Experience with service – billing, customer service, etc., are in place. WLANs will bring in additional revenues. The billing relationship with customers can be exploited. GPRS and 3G do not yet offer high bandwidth for data access. 802.11 base stations are cheap to install. WLAN may address a segment of demand that could otherwise be captured by WISP competitors. The complexity of the service escapes most of the emerging WISP providers	If not careful, potential competition to high-speed WAN offerings. Lack of experience with WLAN. Need to negotiate rental contracts with local real-estate owners. WLAN data revenues will cannibalize, to some extent, GPRS/UMTS revenues. New pricing schemes may be necessary to spur demand. Initial investment required. Value chain not yet understood. Need to establish roaming agreements
Pure Play WISP	Wayport, WiFi Metro	WLAN focused service provider	First mover advantage. Create brand recognition with early adopters	Infrastructure costs are huge. Difficulty in scaling. No experience in customer acquisition and retention
Non-facility based aggregators	HP, IBM, hereUare	Providers of branded retail WLAN services, but no investment in infrastructure	Low risk due to limited infrastructure investment. Can be powerful if leverages existing strong brand	Overtime can create conflict of interest with facility partners
Wholesale Service Providers	Concourse, Skypilot	Lease WLAN access	Does not incur cost/complexity of broad retail operations	Does not own end-user. To reach scale may need substantial investment

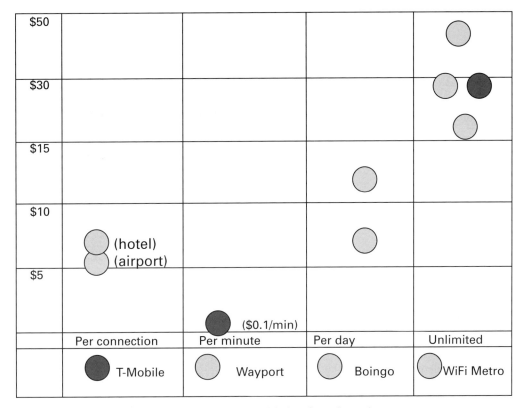

Figure 8.10. Variance in WLAN pricing models (as of March 2003).

8.8   Business model for platform services

In order to increase the business opportunity more widely, it is important that a wireless Internet on a cellular network platform can cooperate with other information or communication platforms such as telematics/car navigation systems, entertainment tools, broadcasting service and traditional vending machines. The combination with other platforms will generate and accelerate the different usage scenes of wireless Internet.

Combination with telematics/car navigation systems is a great tool during driving and it will increase the in-car usage of wireless Internet access. A user in a car can send/receive emails and look up useful information in real time to find out where he or she is and what is in the area via his/her cellular phone and telematics/car navigation system. The Strategis Group, a research firm, predicts that 84% of new cars sold in 2004 will feature some form of telematics equipment. And Forrester Research estimates that telematics will become a $20 billion industry worldwide

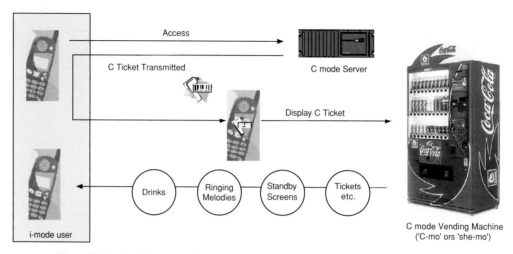

Figure 8.11. Cmode services (source: NTT DoCoMo).

by 2006. The auto industry hopes these developments will invigorate new markets, increase margins, and ultimately grow revenue. Silicon Valley startups are jumping into the fray as well, contributing software and electronic wizardry to this new model for the information age – nothing less than a big computer on wheels.

Telematics services can be as simple as automatically unlocking a car door if a customer gets locked out, or as complex as using a global positioning system (GPS) to help customers find their way. The more advanced telematics features include Internet and email access, both of which will be controlled by voice commands.

Another example is the combination with an entertainment platform such as the PlayStation by Sony. Connecting a cellular phone to a PlayStation enables games to be enjoyed both at home and outdoors. This capability adds the wireless network function to a game device and a user can enjoy online play and download new data.

Cmode links i-mode with Coca-Cola vending machines (Figure 8.11). By combining Cmode vending machines with i-mode, consumers have access to a wide range of consumer services such as cashless shopping and standby screen and ringing melody downloads. This is an approach to creating man to machine communication.

8.9   Business model of fixed wireless integration

The convergence of fixed Internet and wireless Internet may provide the ultimate solution for heavy Internet users. The current fixed Internet service needs a ubiquitous nature (at any time and any location) and the current wireless Internet service can provide ubiquity.

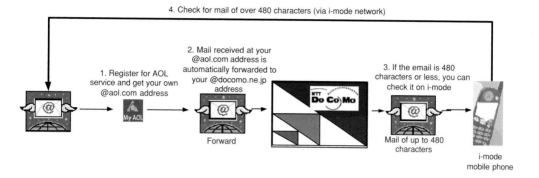

AOLi Service (for received mail)

Figure 8.12. Concept of FWI (AOLi service) (source: NTT DoCoMo).

Readers might remember the concept of UPT (Universal Personal Telecommunication) that was discussed extensively in ITU in early 1990s. UPT, which aimed at seamless service by number portability, is a one-number service and it aims for number portability across a variety of networks for voice and data. Unfortunately, the UPT service was not successful globally since the recent explosion of 2G cellular service overwhelmed it.

FWI is a very general concept but it aims to achieve seamless service capability on fixed and wireless Internet networks. AOLi service by DoCoMo AOL is one step towards FWI (Figure 8.12).

The AOLi service was launched in Japan in 2000 to combine i-mode and AOL email. The service aims to offer users enhanced convenience with one common email address since it allows them to check emails anywhere and it also enables i-mode users to view long messages (max. 2000 transmitting and 25 000 receiving).

The purpose the AOLi service from the marketing point of view is to increase the subscribers of both services by service integration. For AOL Japan (now called DoCoMo AOL) with less than one million subscribers, the huge number of i-mode subscribers is very attractive.

8.10 Business model for 3G services; differentiation from 2G systems

Now, let us consider the business models for 3G systems such as WCDMA and CDMA2000. From a user's point of view, there is not much difference between 2/2.5G systems and the 3G system apart from an increase in data transmission speeds. Most of the business models described above can be inherited from 2G/2.5G systems to 3G systems. The key factors for the success of 3G business totally depends on how the industry has cultivated and educated the market. For Europe and other areas where

Table 8.2. 3G vs fixed network-based video phones

Factors	3G-based mobile video phone	Fixed network-based video phone
Terminal costs	Cheap ($400–500)	Expensive(>$500)
Activation	Very easy	Needs installed fixed line
Setting	Very easy	Needs software
Portability	Yes	No
TV-conference	Not suitable	Suitable
Space	Indoor, outdoor	Mainly indoor

migration from 2G/2.5G systems (GSM) to 3G (WCDMA) is to be introduced in the near future, the success of the wireless data service on existing 2G/2.5G platforms is indispensable. However, the high data speed (64–384 kbps) access technology of 3G can create new services and business opportunities that low data speed (10–64 kbps) 2G/2.5G systems can not provide, such as wireless video mail and image distribution services.

Let us consider the potential capability of such services as a case study. In spite of the incompatibility of radio access technology for 3G (WCDMA and CDMA2000), the mobile video phone standard for 3G systems is expected to be unified as 3G-324M for both WCDMA and CDMA2000. As of February 2002, 3GPP has specified 3G-324M as the mobile video phone standard. 3G-324M selected H.263 and MPEG-4 for the video coding and AMR (Adaptive Multi Rate) for the voice coding scheme (MPEG-4 is not mandatory, but recommended).

Now 3G-PP2 for CDMA2000 is studying its mobile video phone standard and is expected to select H.263 and MPEG-4 as mandatory specs for video coding. Hence, there is the possibility of having compatibility between both 3G access technologies. As a result, 3G operators may need no gateway servers to connect 3G video phones. If we compare a 3G-based mobile video phone and a fixed network-based video phone, both services have their pros and cons as shown in Table 8.2.

What will be the most important applications for mobile video phones? Only time will determine that but let us now consider a few likely candidates.

8.10.1 Video mail service (email with short video message)

Although the fixed-video phone concept has been around for a long time, the phenomenon never really caught on as people were reluctant to show their face to the caller on the other side. However, video mail is just another form of email. The sender can take a short video message using a wireless video phone with a small camera and attach it to the email. Considering the fact that email continues to be the most popular wireless Internet application, people are most likely going to adopt this enhanced

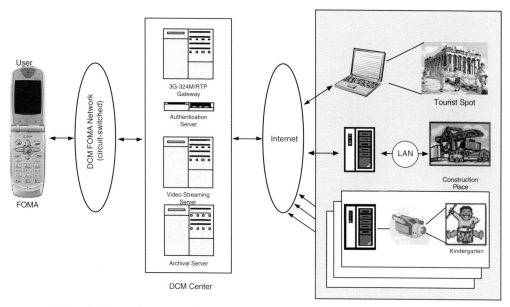

Figure 8.13. Configuration of a remote live monitoring service (source: NTT DoCoMo).

communications style. Another example is J-Phone's SYA-mail service. This is an email service where you can attach still pictures taken using a cellular phone with a small camera. The SYA-mail service has attracted three million users in less than a year. Video mail service may reduce the user's mental barriers and might become the first killer application unique to 3G.

8.10.2   Remote monitoring service

Another promising application is a remote monitoring service. By using mobile video phones, remote monitoring with live images can be achieved easily and economically. Figure 8.13 shows an example of such a system configuration. With this configuration, video phone users can get a remote monitoring service such as event information, kindergarten watching and security control, etc. Some of the enterprise and customer applications already in service are listed in Tables 8.3 and 8.4.

8.10.3   Video clip download service

Many cable and news service companies like CNN have been providing a download service of short video clippings on the Internet and it has become quite popular. Similarly, downloading of short video clipping on the 3G system is a very promising service. DoCoMo started a service called *i-motion* in December 2001 using the FOMA

Table 8.3. Live image applications (source: nG Japan)

Purpose \ Industry	Finance/Insurance	Construction/Real Estate	Manufacturer	Courier Service	Service	Public Sector/Others
Sales/Sales Support	Remote evaluation of insurance assessment **[TV Phone]** / Providing stock information occasionally. **[Multiaccess]**	Construction-procedure management using pictures **[Connecting to private line]** / Reporting real estate information with pictures **[Picture attached to e-mail]**	Assisting sales presentation with moving pictures **[Picture information]** / Utilizing pictures for business trip report for internal use. **[Picture attached to e-mail]**	Retailer training using pictures **[Picture attached to e-mail]** / Developing estimate for moving according to pictures **[TV Phone]**	Effective management of printing-design corrections **[Remote access]**	Remote database access **[Remote access]** / User market research using picture screen **[i-mode]**
Maintenance Support		Confirming construction-design changes on real-time basis. **[Connecting to private line]**	Supporting maintenance personnel for technical matters **[Multiaccess]**	Sending baggage-picture data with pictures upon request **[TV Phone]**	Obtaining updated manual (only revised part). **[Connecting to private line]**	Business-procedure direction using pictures **[Picture information]**
Operation Management/Tele-metering				Sending emergency call with pictures. **[Connecting to private line]**	Controlling vending machines remotely. **[Connecting to private line]**	
Others	• Using group ware by accessing intranet remotely • Managing daily reports using higher speed i-mode service					

Table 8.4. Live image transmission applications (source: nG Japan)

Type of business	Company name	Foreseeable application
Security and maintenance services	Sogo Keibi Hosho	Surveillance of vacant premises and of illegal disposal of refuse
Energy	Tokyo Gas	Surveillance and maintenance of gas appliances
Trading companies	Mitsubishi Corp.	Image transmission
Day-nursery services	Alpha Omega Soft, Life Little,	Virtual visit to day nursery
Travel agencies	JTB	Virtual travel
Retail	AmPm	Monitoring of stores
Manufacturing	Yoshida Original, IBM Japan, Nippon Information and Communication, MegaChips, Mega Fusion	Video catalogs, internal transmission of information, live image transmission
Broadcasting	Asahi National Broadcasting	Live relays
Advertising	Dentsu	Event-information transmission
Aviation	Northwest	Airport surveillance
Medical	Obihiro Univ of Agriculture and Veterinary Medicine	Surveillance of large animals
Welfare	Clinic Ootsuka	Remote at-home medical care
Education	Linguaphone Academy, AEON, Keio Univ Business School, Koriyama Women's University	Virtual English-language conversation practice, adult distance learning, remote lifelong learning
Events	IDG Japan	Even-information transaction
Consulting firms	PwC, Mitsubishi Research Institute	Proposals for corporate systems
Information services	NTT Phoenix Communications, NTT-ME, NTT-X	Transmission of images for companies
Content series	Mobile Television	Live-image transmission
Content production	Digicon	Live-image transmission

platform. The i-motion service employs MPEG4 video coding and provides 10–30s of video clipping for a variety of content; i-motion content can be downloaded to the terminal once and can be replayed any number of times. Also, downloaded content can be stored on the terminal. A FOMA phone can accommodate a maximum of 100 kb of content. The typical content available on i-motion includes movie previews, promotional videos of music and movies, press conferences, highlight scenes of sports, weather forecasts and so much more.

8.11   Conclusion

In this chapter, we have reviewed various business models and strategies for players along the value chain. We also reviewed i-mode success factors and the lessons drawn from the DoCoMo experiment. The wireless data market has a huge global potential; we are just starting to scratch the surface. The key element for all business players will be sustainable business models that can thrive in the local and global ecosystems. There are still no globally dominant players and the playing field is fertile. In the next two chapters, we will take a look at the different technology and business challenges that the wireless industry is facing and how some of them are being addressed.

9 Business issues and challenges

In the previous chapter, we reviewed the value chain and business models, both key components to any industry analysis. In this section we will look at some of the challenges being faced by the wireless industry – in both the business and technology realms. This chapter focuses on the business issues and the next chapter will address some of the challenges in the technology arena. Although many industries have some common idiosyncrasies, the wireless industry is unique and sometimes issues and problems are specific to a geography or nation. For instance, spectrum (which we will discuss later) is squarely a US issue while their counterparts in Japan do not encounter such problems and 3G spectrum auctions are driving many carriers out of business in Europe. While Europe and Japanese markets are approaching saturation in the consumer market, the US market is largely untapped. On the other hand, China (which is the biggest wireless market in the world) has its own challenges as it tries to define itself amidst government control. The business issues and challenges we are going to be discussing in this chapter are as follows:

1. hyping
2. transition from flat rate to à la carte billing models
3. mobile SPAM
4. WLAN business
5. interoperability
6. churn and increasing the ARPU
7. 3G auctions and spectrum
8. consolidation
9. market saturation and search for new markets
10. privacy
11. security
12. position location rollout and privacy
13. mobile fraud.

Some of these issues are geography-specific while others apply to the industry as a whole. We will try to give examples wherever possible to illustrate our points.

9.1   Hyping: a useful business strategy?

Over the past decade, hype has been at the heart of technology like a malfunctioning pacemaker. By racing out of all control it has left the marketplace strewn with casualties, with the dot.com mania and subsequent crash being the most apparent example. History is full of examples of spectacular rises and failures, yet, with every new technology or "buzz", the lessons are soon forgotten and marketing machines get into the action of promoting with inflating claims and misrepresentation and in the process setting themselves up for failures. A perfect illustration of hype running riot is provided by the story of WAP. The meteoric rise and fall of a technology that promised the marriage of the Web with mobile telephony and the birth of a mobile Internet, from the slogan-laden grandiose welcome given to WAP at the Telecoms99 exhibition to the drab reality that emerged a year later.

From ads such as the British Telecom one featuring a cyborg-like animation surfing the dynamic and color-rich waves of the wireless infosphere, to those early adopters standing round the display of a Nokia 7110 (the first commercial WAP phone) trying to read six lines of black text against a greenish background downloaded at unbearably slow speed, it is not surprising that WAP became a flop. Beyond the slow connection speed, the bugs and configuration problems, the fall of WAP (by then known as Wait And Pay and Where Are the Phones?) was also the result of poor content as well as complex user interactions. In turn, the lack of engaging and application-tailored content is seen as deriving from greed, with operators unwilling to share revenue with content providers, thus driving many of the latter out of business. Talk to any wireless industry analyst and you will probably be told that mobile commerce or wireless LAN or UWB will be the next big thing. UK-based Ovum predicts that mobile advertising revenues will reach 16 billion euros in 2005, and Strategy Analytics, USA, claims that 325 million people will use mobile commerce technology in 2006, generating $230 billion in revenue.

As shown in Figure 9.1, the classic hype–reality curve (technologies generally go through a seven-year adoption cycle) starts when the industry or company is doing experiments with new technology such as WAP. Then the expectations of the end-users are pumped up by the parties who have a vested interest in its success; expectations are often inflated several times over like for instance, "WAP is like surfing the Internet on the phone". When consumers get a chance to interact with the technology, there is often a series of issues – from value-chain disruptions to false expectations to poor customer service. After the transition period of absorbing the market reaction, the reality sets in and the industry learns the "real" uses of the technology or applications and then the technology progresses on its "real" merits, not hyperboles.

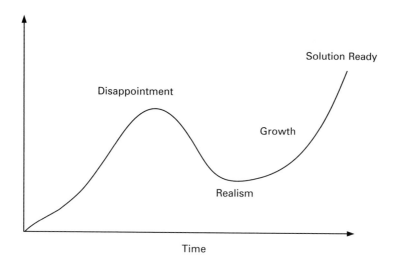

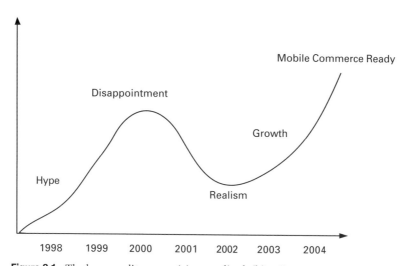

Figure 9.1. The hype–reality curve: (a) generalized, (b) mCommerce.

Government agencies have been helping by clamping down on misinformation. Microsoft and Hewlett-Packard began promoting the Pocket PC in April 2000 in a host of national publications. According to the FTC (Federal Trade Commission), several of the ads touting Pocket PCs pictured the device accessing email or Internet sites. The ads claimed the Pocket PC could access the Internet and e-mail "anytime". The FTC complaint said the ads implied that the Pocket PCs contain everything consumers need to access the Internet at any time from anywhere. But, said the FTC (and consumers), this was not so. To connect to the Internet and receive email, Pocket PC users need additional equipment – such as a modem. A separate landline modem can cost about US$130, and wireless modems can cost up to $350, the FTC said – about as much as the Pocket PC itself.

Under the terms of the settlement deal, Microsoft and Hewlett-Packard are barred from engaging in similar advertising practices in the future. The agreement applies to PDAs and other handheld devices that do not come with built-in wireless Internet and email access. When making claims about the Internet or email access for PDAs, Microsoft and Hewlett-Packard have to disclose clearly and conspicuously the need for any additional products or the need to subscribe to a special Internet or email access service.

In Sprint PCS's original mobile data services campaign, an X-Files-type guy in an overcoat stood outside buildings saying, "You can come out now". The implication that people could complete the same tasks on wireless devices as on their desktop computers was misleading because it implied that consumers could expect a true Web experience through Sprint's data services, but since then, Sprint has done a good job of mending its ways and showing what you can really do with the mobile data services that they offer.

Although US wireless carriers have toned down their hyperbole about the "wireless Web" and their far-out depictions of how wireless data services work and substituted more realistic advertising, there is still work to be done to enthuse consumers about mobile data services. Carriers need to highlight "what things you can do that are real to people and what you can enjoy on a tiny phone". This is, and has been, a basic concept of the i-mode campaign in Japan.

We are seeing similar infatuations with the promotion of 2.5G services in the USA. Even though some carriers are installing only the 2.5G network, they are promoting it to their customers as 3G. Even though the 2.5G network might be available in only select cities, it is promoted as a nationwide service. The fact that the 2.5G phone will not work in some of the major cities does not even come up. Some carriers are positioning 2.5/3G as DSL to a phone even though (a) speeds are substantially lower in initial rollouts and (b) there are no applications available on the phones that will use 3G theoretical bandwidths.

Geoff Varrall, an eminent industry analyst of RTT Systems, a consulting company in the UK, divides the progression of wireless technology into three distinct phases: pain, pleasure, and perfection. It takes on an average five years to go from one phase to another in a 15-year maturation cycle. So far, his assessment is impeccable. Table 9.1 depicts his analysis.

As shown in Table 9.1, any given wireless technology has, on average, a 15-year maturity life cycle, divided distinctly into three phases: pain (coverage is poor, interference is high, customers are unhappy, devices are expensive and scarce), pleasure (coverage is the best it can be, interference is low, customers are happy, devices abound), and perfection (everything works as hoped for in the first phase, so it is time to move on to something new).

Table 9.1. Timeline of major global wireless technologies, 1979–2020

Japan	1979–84	1984–89	1989–94	1994–99	1999–04	2004–09	2009–14
1G: ANALOG	*Pain*	*Pleasure*	*Perfection*				
2G: PDC			*Pain*	*Pleasure*	*Perfection*		
3G:					*Pain*	*Pleasure*	*Perfection*
Europe	1982–87	1987–92	1992–97	1997–02	2002–07	2007–12	2012–17
1G: TACS	*Pain*	*Pleasure*	*Perfection*				
2G: GSM			*Pain*	*Pleasure*	*Perfection*		
3G: UMTS					*Pain*	*Pleasure*	*Perfection*
United States	1982–87	1987–92	1992–97	1997–02	2002–07	2007–12	2012–17
1G: AMPS	*Pain*	*Pleasure*	*Perfection*				
2G: TDMA			*Pain*	*Pleasure*	*Perfection*		
			1995–00	2000–05	2005–10	2010–15	2015–20
2G: CDMA			*Pain*	*Pleasure*	*Perfection*		
					2002–07	2007–12	2012–17
3G:					*Pain*	*Pleasure*	*Perfection*

Key.
Pain = coverage is poor, interference is high, customers are unhappy, devices are expensive and scarce.
Pleasure = coverage is the best it can be, interference is low, customers are happy, devices abound.
Perfection = everything works as hoped for in the first phase, so it is time to move on to something new.
(Copyright © 1998 by RTT Systems, reproduced with permission)

So, all players in the value chain should pay attention to the "state" the technology is in at any given moment to truly capitalize on opportunities. Also, it is very important to be absolutely honest with your customers about the capabilities of your applications and services and provide a future roadmap of your technology rollout. It will help in keeping your customers loyal.

9.2   Transition from flat rate to à la carte billing models

Billing models have been a key component of any carrier data strategy. European and Japanese carriers have successfully used à la carte, per-transaction billing while in the USA, where consumers are more accustomed to an "all you can eat" model[1], consumers are used to paying a flat fee for services. With the advent of 2.5/3G services, European and Japanese carriers will thus not have a challenge to set user expectations as to how they will be charged for use of certain applications or services, while in the

---

[1] Most of the US carriers grew by acquisitions and mergers, hence there were disparate billing systems that precluded à la carte billing models.

USA it will be a monumental task. Historically, carriers have been slow at upgrading their legacy billing systems which did not allow them to bill any other way but a flat fee. But with the transition to packet-based networks and packet-based billing, carriers have an opportunity to bill more granularly but face consumer dissent. So, carriers have decided to adopt a hybrid approach of bucketing usage. Unless carriers use the packet-based billing approach, they must also to strike reasonable agreements with application developers as the measurement of revenue flows become murkier.

9.2.1   What are customers willing to pay for?

The renewed optimism in this whole space is a result of traditional wireless carriers deploying packet-based networks, so-called 2.5G networks. These networks, like the GPRS and CDMA1x, are packet-based overlays on top of the existing circuit-switched networks. The 2.5G networks enable carriers efficiently to provide exciting new data services including Internet connectivity from a notebook or PDA. The current methods that use "circuit-switched" data are very inefficient and expensive. Metricom's packet-based data service, despite being technically superior, still failed. The people who use it are very happy with the performance and throughput. It failed because the "economics of data" did not work out, and it can be argued that the deployment of 2.5G networks by wireless carriers does not change this situation, at least in the next two to three years. Why? Data competes with voice for wireless spectrum, and spectrum is scarce and expensive for wireless carriers. So, at least in the initial two to three years, carriers will price the data high and focus on low-bandwidth and high-ARPU data applications and devices instead. Examples are IM, MMS, mobile email, mobile Internet access from cell-phones and pagers (like RIM). NTT Docomo of Japan has priced its data service at approximately $23/Mb (based on $0.003 per packet and 128 bytes per packet) and still has over 30 million customers and an average data ARPU of $14 per month per user. "Circuit-switched data" provides a throughput of 14 kbps (note that throughput is different from peak rate). In one minute, you can transfer roughly 100 kB (14 kb×60/8). If carrier price per minute of usage to be $0.20, it would cost a user $0.20/100 kb or roughly $2/Mb for data. A three-minute MP3 file, which is roughly 3 Mb, will cost the user $6. Also, in 1999 a typical wireline user consumed[2] an average of 200 Mb/month. If this wireline user decides to use wireless for their Internet needs, it will cost $400/month (200×$2/Mb, not including $300–$400 upfront cash outlay for a wireless PC card) instead of $40/month that they might be paying currently. The point of this discussion is that widespread adoption of WWAN wireless internet will not happen till carriers price data more economically for users ($0.20/Mb). However, a carrier who can make $0.20/minute on voice will have no motivation to provide data at $0.20/Mb unless

---

[2] Source: Qualcomm white paper, http://www.qualcomm.com/main/whitepapers/WirelessMobileData.pdf.

it can provide data at average throughput of 1 Mb/minute or 144 kbps. This does not even include the cost of new data infrastructure (in an ideal world, if the spectrum were not an issue and extra capacity were available, carriers could price data aggressively). The current 2.5G technologies do not permit the efficient data pricing that is needed for widespread adoption of wireless Internet access by notebooks. It could be two to three years before the existing technologies mature. Alternate technologies by startups like Flarion and Arraycomm are trying to address these issues, but these are proprietary techniques and adoption is very uncertain.

9.3 Mobile SPAM

In Japan, spam mail using wireless devices has become one of the most serious social problems recently. Since the number of wireless Internet users exceeds 60 million, email to cellular phones is considered an efficient and direct communication tool for marketing and PR purposes.

Several companies have developed the spam mail sender system for advertisement purposes such as solicitation, direct sales, pornography and matchmaking. The system misuses the email address of wireless carriers using a subscriber's telephone number. For example, in the case of i-mode, 090XXXXXXXX@docomo.ne.jp is a primary setting of an i-mode mail address. The spam mail systems randomly create these email addresses and send millions of emails per hour addressed to these randomly created addresses.

Most of these spam mails are sent to non-existant mail addresses but some portion of huge spam mails reaches valid end-users. These spam mails are not only troublesome for end-users, but are also a big nuisance for wireless carriers as network and traffic loads can climb instantaneously. Especially, the gateway server has to handle an enormous amount of spam mail and as a result throughput of normal mails is decreased. Wireless carriers and the Japanese Government are taking this problem seriously and have taken several measures to meet the situation. The easiest and fastest way to prevent spam mail is to change the address to something peculiar of one's own. In addition, some operators have changed the primary setting of address to alphabets such as abcdefgh@docomo.ne.jp, and it is more difficult to create these email addresses randomly. Also, operators are developing systems to block spam mail and are taking a look at legal action that could be taken against spammers.

In July 2002, it was announced that NTT DoCoMo, Inc., and its eight regional subsidiaries would offer a new feature enabling users to block spam mail (unsolicited email advertisements).

New anti-spam legislation, the LAW ON TOPICS INCLUDING THE APPRO-PRIATENESS OF SENDING SPECIFIED E-MAIL MESSAGES (Law No. 26 of 2002) and the LAW FOR PARTIAL AMENDMENT TO THE SPECIFIC COMMERCIAL

TRANSACTION LAW (Law No. 28 of 2002) took effect on 1st July 2002. The new laws require, with some exceptions, that sender's of advertisements begin the subject line of each mail message with the Japanese kanji characters (transliterated as mishodaku kokoku, which means unsolicited advertisement) plus the mark. The message must also include the sender's name, postal address, email address, and the like, so that a recipient can send back an "opt-out" request. Once a user has thus responded, the sender must not send any further email to that address.

DoCoMo will maximize the benefits to its customers of this new legislation by enabling users automatically to block all emails with mishodaku kokoku at the head of the subject line. Users will be able to select this new anti-spam option by accessing a designated website link from the official i-mode portal site, i Menu. The new option was available in October 2002; the transmission fee for selecting it will be free of charge. It is quite likely that such legislation will be copied and implemented in other countries as well.

9.4   WLAN business

As we have already noted, WLAN business models are still being worked out with mobile operators having a clear advantage of leading the charge with deployment and extension of their offerings to the WLAN segment. As more and more WLANs get deployed around the world, the challenges facing the service providers are mainly business related rather than technical. Pervasiveness of connectivity and reliability are two important concerns of users. There are slight differences in terms of what business users vs consumers desire the most; for example, cost is the number one priority for consumers, whereas business users want wide availability and security of their data. The primary challenges faced by the WLAN industry are as follows:
- availability
- roaming and seamless handoffs
- integration with WAN technologies
- billing and pricing (single pricing relationship)
- security and authentication (single sign-on working in concert with roaming)
- consolidation pressures
- customer acquisition, and retention
- branding and marketing
- customer support and service
- spectrum overcrowding
- real-estate ownership.

WLAN roaming is fundamentally different from wireless WAN roaming, as coverage will be limited to hotspots, and spectrum constrains the co-located WLAN networks to three for 802.11b. Even large service providers will have a tough time covering a sizable footprint to include all desirable hotspots. The development of roaming agreements is essential for a viable WLAN value proposition. Despite the rapid rise in the number of hotspots around the world, the number of paying users has been low. One of the major reasons as outlined above is marketing. People are willing to pay for access if they know it exists. WISPs have been concentrating their resources on infrastructure, but without marketing nobody will know about the installation and service availability. This is another reason as to why operators hold an advantage as they can lever their marketing clout and customer-service capabilities. Another aspect that needs to be sorted out by the industry is the billing relationships and some convergence and symmetry. There are widely varying pricing models available and they do not necessarily account for roaming issues.

The tariff level is the most critical parameter for the economic criteria NPV (net present value) and IRR (internal rate of return) – the key profitability factors. Since revenues are linked to usage levels, which mean that a 50% increase in revenue corresponds to a 50% increase in usage, it would be expected that network costs would increase accordingly. However, network costs are essentially dictated by coverage constraints and not by capacity constraints. Hence, an increase in usage translates only as greater revenues, while the corresponding increase in costs is minimal, and relates to core network elements.

The lack of an integrated authentication and billing system is also a hindrance to adoption. Password-based authentication has so far been dominant, but this does not integrate well with mobile operators' billing systems. Hardware-based solutions that allow SIM-based authentication are coming to the market, so that mobile operators can authenticate the user and have access to the user billing record via HLR.

Although there are some key business and technical challenges to overcome, the future is bright for WLAN. WLAN technology is going to be rapidly adopted by consumers not only in the enterprise environment but also in home environments. In addition, mesh networking technology (see Chapter 13) holds significant promise for *ad hoc* networking to provide expanded coverage. This, combined with improvements in VoIP technology, will definitely lead to some disruptions in the models of wireless applications and services.

9.4.1   Economies of scale – lessons from Metricom and Mobilestar failures

A firm gains economies of scale if its (average) cost of production falls when it increases its scale of operations. Think of car manufacturing. Would you expect a car manufacturer that turned out ten cars a year to produce a car as cheaply as another manufacturer that turned out 100 000 cars a year? One reason it would be difficult is that there are technologies (including specialization from the division of labor), machines (including robots) and specialized services (e.g. research and development, engineering skills, managerial expertise and advertising services) that can greatly reduce the average cost of building and selling a car. However, they are costly and their use would make economic sense only if their cost could be spread over a large number of cars. The PC industry typifies the economies of scale rule. The quantities in which most PC components are produced are so large that the final price to the user drops far below the off-the-shelf price of the individual chips, connectors and other hardware.

The large number of units produced means that the Research and Development cost amortization is very low – an extra thousand dollars spent on design costs is not terribly important in the greater scheme of things, indeed if it can save a few cents on production costs it is well worthwhile.

The analogy can be extended to the telecommunications businesses as well, especially for new players with a unique technology or business concept. It is often a chicken-and-egg problem for network or service operators. You will not have enough customers if the service is not available across wide geographies, and you will not have enough funding resources unless you can show that the customers are clamoring for the service. Several players, including some big ones, have been burnt by disregarding the economies of scale equation, so it is important to learn the lessons from the failures of Metricom and Mobilestar, two players who were first to offer high-speed wireless data services but just could not be successful in spite of having good financial backing (Metricom).

Networks and technologies are emerging which are optimized to support users who need better connectivity to the Internet than can be provided by today's mobility networks. These will serve users of laptops, digital cameras, handheld games, audio players, and other devices yet to be introduced, using Internet protocols and standards rather than wireless industry standards. Most importantly they will do it at lower cost, with lower latency and better bandwidth availability than can be offered by 3G.

According to the Shosteck Group, Metricom, a high-profile public venture funded primarily and liberally by WorldCom and Vulcan Ventures entered this market in 1996 but failed in mid-2001 because its technology could not be built out affordably on a nationwide basis. The underlying cause of this was its inability to garner subscribers, which would have provided ongoing operational revenues. It ultimately achieved less than 1% of its perceived target market. A major reason for this was a gross overestimation of the number of people qualified and motivated to subscribe to the network. The Metricom failure was exacerbated by its limited network coverage, proprietary technology, poor marketing, high end-user cost, and poor scalability of its "micro-cellular" technology.

MobileStar entered the market in 1997 as a Wireless LAN "hotspot" network. It failed in late 2001 because it also could not support network build out on a nationwide basis, despite a markedly lower cost for infrastructure. As with Metricom, it was unable to achieve subscribers owing to poor marketing, limited network coverage, and poor scalability, but the key cause may have been a much smaller market than was originally perceived. An ill-advised agreement to provide the service in Starbucks coffee shops accelerated MobileStar's failure. Other WLAN hotspot networks are still in business because they have effectively shifted cost and risk for network build out to others.

Both failures point to a need for future portable Internet access networks to serve a wider group of users and devices (meaning mass-market appeal) and/or be substantially less expensive to deploy on a nationwide level. "Better" technology will not be sufficient, and business users cannot be relied upon to subsidize an overly expensive network build out.

Network operators are considering building or buying WLAN hotspot networks to augment their 3G offerings. For the portable Internet access market, cost and ubiquity will be paramount, followed by technology capabilities that can closely reproduce the landline Internet experience.

Beyond the direct lessons of MobileStar and Metricom are theories about the nature of successful portable wireless Internet access networks. For example, Metricom and MobileStar demonstrated that defining the market by laptops alone creates a tight economic restriction on the cost of a technology approach. A price point which appeals only to business users exacerbates the situation as economies of scale will be extremely difficult to achieve. Even carriers who are planning to roll out WLAN services need to strategize and implement plans to attract "average" customers rather than just focus on business users.

For players considering rolling out new WLAN and WWAN technologies, economies of scale are probably the most important factors to consider or else even the most versatile and unique technology will falter as nobody is going to be willing to invest in the infrastructure. Similarly, players in the handheld and phone manufacturing business need to assess "economies of scale" issues as they prepare to roll out new device models around the world.

9.5   Interoperability

Interoperability is one of the biggest, if not *the* biggest issues, facing the wireless industry. There are serious interoperability issues amongst vendors who sometimes do not have the incentive or the time to work with their rivals in winning the lucrative infrastructure contracts. As such, it is not uncommon for wireless carriers to fight the battle between switch vendors and handset providers if they do not happen to be the same. The time and energy it takes to get them to work together is a drain on resources and an unfortunate reality. Interoperability issues were not even addressed by the WAP forum until too late into the process of conforming to standards and until application developers complained vociferously about the difficulty of building WAP applications and services.

Similarly, Bluetooth was plagued by interoperability issues from the beginning. The resolution of interoperability cycles have some parallels to hype–reality cycles: they tend to follow the same pattern. Initially, vendors are in a mad rush to roll out their devices, applications, and services to score marketing points. But, gradually, players in the ecosystem realize, interoperability needs to be really resolved if industry or the segment they are serving can move forward.

Interoperability is also used by some large vendors as a tactical ploy to keep competition, especially from smaller players, out of the mix. One does not have to look beyond AOL and Microsoft to see how quickly interoperability becomes a key component of product and marketing strategy. Even though both players have acknowledged that consumers would benefit from interoperability between their IM services, it has not happened yet and it is likely to stay that way for some time to come.

Overall, although interoperability continues to be a big issue, with the gradual embracing of the standards by players like IBM, which used evangelism for standards to its advantage (strategy and marketing), the cycles to interoperability have accelerated. Instead of several years, it is now taking months to reach agreements amongst vendors.

9.6 Decreasing churn and increasing ARPU

Morgan Stanley recently concluded that the industry churn should remain in the 2.5–2.3% range for the years 2003–2005. This they derived from the fact that the factors working to increase and reduce the churn will offset each other. Factors that could cause an increase in churn are number portability, intense competition, more pre-paid subscribers, and convergence of handset technology. On the other hand, the forces that are forcing reduction in churn are: industry consolidation, improved voice and QoS, and increased focus on retention as penetration rises. Figure 9.2 shows relative cost per gross add (CPGA) for carriers around the world. The average subscriber acquisition costs remain very high in Japan and the USA. If CPGA is assessed as a proportion of lifetime revenues per subscriber, the CPGA level in the USA is still high relative to other regions.

One important issue to consider to understand Figure 9.2 correctly is the difference in pre-paid service penetration in each region. In Europe and Latin America, the penetration ratio of the pre-paid service is relatively very high compared with Japan and the USA. As the CPGA for the pre-paid service is lower than for the post-paid system, total CPGAs in Europe and Latin America become lower than other areas.

One important metric with which to measure the health of a wireless carrier's subscriber base is the net present value (NPV) of lifetime value of a steady-state subscriber – or, in short, lifetime value. It is the discounted cash flow of the average revenue per user minus the cost of providing service minus customer acquisition

Table 9.2. Various performance metrics for US carriers

	Verizon	AT & T Wireless	Cingular	Sprint PCS	Nextel	VoiceStream
CPGA	$261	$333	$326	$323	$435	$365
Cost of service/month	$23	$32.36	$24	$33	$28.29	$32.11
ARPU	$49.26	$63.6	$54.11	$64.03	$71.05	$50.53
Gross profit/month	$26.26	$31.24	$30.11	$31.03	$42.76	$18.42
Monthly churn	2.2%	3.1%	3.2%	2.6%	2.1%	5.0%
Average length of service (months)	45.5	32.3	31.3	38.5	47.6	20.0
NPV of lifetime value	$662.78	$507.16	$462.93	$637.79	$1121.74	−$35.89

Source: Yankee Group, 2002

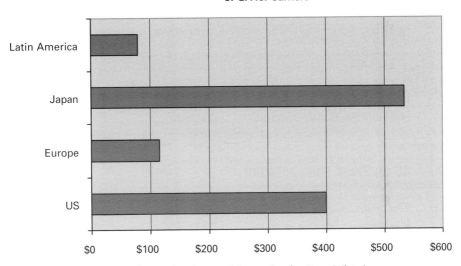

Figure 9.2. Average CPGA for carriers (source: Morgan Stanley Dean Witter).

costs for the average duration of the length of service excluding capital investment. In basic terms, it is what the carrier can expect to receive from the subscriber over the lifetime of the subscription. Table 9.2 shows the net present value of subscribers for the Big Six carriers in the USA for the third quarter of 2001.

There is an increased pressure on carriers to generate more revenue from existing customers and offer services that increase the average revenue per user (ARPU). With voice becoming a commodity in many markets, carriers need to increase their total ARPU with wireless data traffic including machine to machine (M to M) and machine to human (M to H) communications. Also, data driving data/voice scenarios are being implemented by operators and application aggregators like InfoSpace (email and click to call).

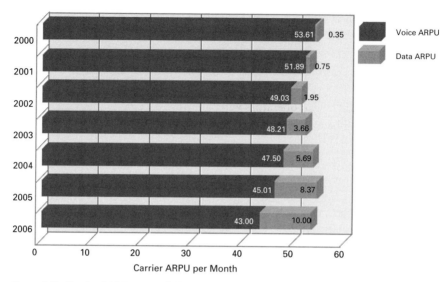

Figure 9.3. Carriers' ARPU/month (in US $) (source: Yankee Group).

By 2006, the Yankee Group believes that wireless data will add $10's worth of incremental revenue per user per month for US carriers (see Figure 9.3). Platform manufacturers, publishers, and content developers are working with carriers to create and implement data applications and services.

9.7   3G auctions and spectrum

The prime example of greed in action is to be found in the sale of third-generation (3G) mobile licenses in Europe. In 2000, while the use of 3G spectrum was given away for free in Finland and Japan, the United Kingdom decided to auction its five licenses. That year every analyst, operator and manufacturer published one of these two statements: "by 2002 there will be over a billion users of data-enabled wireless devices" and "by 2003 more people will access the Internet though their mobile phones than by using Personal Computers". The conditions were right to rake in the money. The British auction ended up bringing in a staggering 22.47 billion euros for the government, setting an example for the rest of Europe to follow. Germany made 50.4 billion euros, while other countries, such as Spain, backtracked on previous agreements to join in the windfall. Behind this mad scramble to grab a 3G license at whatever cost, there was little tangible evidence that this was a smart move.

In the USA, as the costs and benefits of rolling out 3G infrastructure (WCDMA) become apparent (following the trends in Japan and Europe), the issue of spectrum availability will become extremely important for the US operators who plan to implement WCDMA as their 3G alternative. Unlike Europe, where separate frequencies have been auctioned for 3G licenses, US operators will most likely have to implement WCDMA within the spectrum that is already occupied by them. WCDMA needs at least two (up and down link) channels to operate and may require sufficient guard bandwidth for coexistence with 2/2.5G technologies such as GSM and GPRS.

By using globally standardized IMT-2000 frequencies specifically assigned for 3G, European and Asian operators will avoid any frequency issues. This global harmonization will be extremely important to implement the 3G global roaming mechanism in the near future. Customers who do not want to migrate will have an option. Clearly, resolving the spectrum issue in a timely fashion is the single most-important issue for carriers.

9.8   Consolidation

One of the aftereffects of the 2000 economic downturn was a cleansing effect on the wireless industry that is accelerating the carrier consolidation issue around the world. Although all players in the value chain have to somehow contend with it, it is the wireless carriers with unbelievable debts who are being forced to reassess their futures. This is especially true in the highly competitive US market. The government allowed the number of large players who could compete at a national level to rise to six – AT&T Wireless, Sprint PCS, Verizon, Cingular, VoiceStream (now part of T-Mobile) and Nextel. In addition there are several smaller regional and rural players. This makes for the most competitive market in the world, with the lowest margins. Six different players meant different strategies, visions, and technologies in the market that continue to confound the consumers and analysts alike. The USA is the only major market that has a myriad of technology installations and evolution paths – Analog, iDEN, TDMA, CDMA, GSM, CDMA2000, GPRS, EDGE and WCDMA (forthcoming) installations. In such an environment, carriers are forced to roll out dual- or tri-band phones and interoperability is a big problem.

In addition, the US wireless industry has been bedeviled by a shortage of spectrum since its inception, and lifting the spectrum cap as the FCC did recently does not increase the overall amount of airwaves available; it simply shifts them around. Wireless spectrum is a finite natural resource and one that the wireless industry must start learning to use more efficiently. From 2003, when the cap has been eliminated entirely, the industry will undergo sweeping changes. Carriers are swapping spectrum in hotly sought markets, smaller competitors will disappear, and most experts expect four or five major carriers to emerge. Analysts have already begun to speculate on possible combinations: Sprint and Verizon Wireless make obvious bedfellows, given that they are the only two carriers that use CDMA technology in their networks. And Nextel Communications will almost certainly be swallowed up by a major carrier looking to cash in on the company's lucrative business-customer base, though there are some technology evolution issues with iDEN. Consolidation can take many different forms, including full-blown mergers and acquisitions, marketing alliances and shared infrastructure agreements. Most recently, as wireless companies face enormous network upgrade costs to offer the next-generation services, infrastructure sharing is gaining in popularity. Wireless carriers also develop partnerships and affiliate relationships to realize economies of scale. For example, Sprint started with ample spectrum but no network. Sprint created a handful of smaller affiliates to speed the build out of a national system that could be marketed under one brand name. Ideally, the major wireless players want to own nationwide footprints, otherwise they tend to pay significantly higher roaming fees to other carriers. This, in turn, reduces profit margins and makes it more difficult to compete with service offerings. In many cases, a wireless company may be spending more to provide roaming coverage than it can charge customers, thereby subsidizing roaming traffic at the expense of on-network profitability.

Many wireless companies are looking beyond their own borders for new customers. Among European wireless carriers, Vodafone has been the most aggressive, with major investments in Verizon in the USA, Iusacell in Mexico, Japan Telecom in Asia and many carriers in Europe. Germany's Deutsche Telecom T-Mobile unit recently acquired US-based VoiceStream. Likewise, France Telecom is branching out with its purchase of UK-based Orange and participation in Germany with MobileCom. Scandinavian operators Sonera (Finland) and Telenor (Norway) are also managing international investments.

Exceptions to the primarily Eurocentric focus are Spain's Telefonica and Italy's Telecom Italia, which have major stakes in Latin America. In Asia, NTT DoCoMo is the only major operator with significant international investments with a stake in US-based AT&T Wireless and Dutch-based KPN.

Following Vodafone's acquisition of AirTouch, major US wireless operators have been largely absent from the international scene; one significant exception is BellSouth with investments in several major South American markets.

US-based operators' limited international efforts have focused primarily on other North American markets. Verizon invested in Mexico's Lusacell and Canada's Telus, Sprint invested in Mexico's Pegaso, AT&T Wireless increased its stake in Rogers of Canada, and VoiceStream owns a percentage of Canadian partner MicroCell. As wireless operators prepare to offer 3G services, the issue of spectrum allocations will loom even more prominently. Companies will need a significant amount of additional spectrum to cost efficiently offer 3G services. US carriers are hampered by the FCC's allocation of spectrum bands to the military, government agencies, TV broadcasters, educational institutions and other public groups. Although earlier discussions raised the possibility of reallocating some of that spectrum for commercial operators, the current military situation makes it unlikely that this will happen in the near future.

In any case, 2003–2005 will see several shifts in the carrier space. In other parts of the value chain, consolidation has occurred in different segments; for instance, Ericsson and Sony have joined forces for handset manufacturing as it is becoming very expensive to carry out the burden of research and production on one's own. Similar agreements are being fostered between other handset manufacturers such as Motorola, Siemens, Mitsubishi, and others.

9.8.1   Is consolidation always the right answer?

A consolidation makes sense if the industry in general can gain some economies of scale, through synergies between two or more entities that are rolled-up in one. In the USA, the main issues that have dragged the sector in the past few years are a low ARPU (under $50 for most, except Nextel), and a very high level of debt (except for Alltel) necessary to finance the heavy capital expenditures to upgrade the network and maintain it, as well as front-end expenses to acquire FCC licenses in auctions rounds.

Increasing ARPU is one possibility: the risk, however, is that the regulatory bodies may see it as taking advantage of a monopoly or duopoly position. The only way to increase the ARPU is to offer more value-added services. To offer more value-added services, a better sharing agreement with solution developers needs to encourage the burst of applications. Consolidation is not necessarily the answer.

A consolidation will not bring down the level of debt of a merged group of entities, unless cumulative cash flow grows faster than the financing requirements. Many seem to believe that investments in infrastructure would be spared by a consolidation. It is a fair statement as long as technologies are similar and a switch over to one network is seamless and not costly. Analysts seem to forget the cost of acquiring customers: in a merger, how many customers will walk out and sign on with the competitor? As the present authors believe that the FCC will not allow less than two competitors at least in each CMSA market, the consumer will always have the choice to leave. What would be the cost of merging two network? Is there a potential saving? Maybe, but it has to be thoroughly studied case by case.

9.9   Market saturation and search for new market strategies

As we discussed earlier in the book, major markets are nearing saturation. In the early days (early 1990s), it was more about land-grab – trying to get the bigger piece of the pie – but now the challenge has turned to finding new segments and customers to sell to. In Japan, the consumer market is pretty much saturated not only for voice users, but also for wireless Internet users such as i-mode who are almost approaching saturation, so DoCoMo, KDDI, and J-phone are trying to tap into the enterprise market – traditionally an untapped segment for the Japanese carriers. Similarly, European carriers, especially Scandinavians, are making the most of the potential to get more customers. In the USA, where selling to enterprises is easier than attracting new segments, carriers are trying to seek new niche market segments. US carriers have mostly failed at tapping into the teenage market unlike their Japanese or European counterparts. Larger carriers are looking overseas to increase their consumer base. As countries enter the saturation market, typically starting at the inflection point of 45% penetration, key dynamics of the industry begin to change, often accelerating with each percentage point gain. These dynamics include slower subscriber growth, accelerating subscriber churn, competitive pricing pressure and a decline in subscriber quality. Within saturation markets, the nature of competition shifts and, with it, the critical factors required for success.

In North American markets, key industry metrics such as churn and subscriber growth have taken a turn for the worse. The US market is indeed changing, forcing key players to alter the nature and scope of their activities. There is some good news. Despite the apparent challenges, many foreign wireless carriers have thrived within a saturation market, providing powerful lessons for their US counterparts. In Europe, operators such as Orange, Vodafone and Telecom Italia Mobile have successfully implemented "battle plans" to cope with market saturation.

In phase one of these plans, successful carriers introduced initiatives to improve efficiency and increase earnings. Key wireless metrics are improved through innovative churn management policies, sticky portals and additional value-added services to stimulate voice revenues. But the most important phase-one initiative is a fundamental reversal of the handset subsidies which have plagued the industry since its inception. Efficient saturation markets tend to eliminate subsidies, as has happened in the UK and Australian markets within the large pre-paid segment.

In addition, selling, branding and the intense focus on benefits versus technology is the biggest single change required. As Orange's UK success illustrates, where it has moved from fourth-placed carrier to the clear market leader, branding that pervades the entire organization is a core requirement for success within a saturation market. Lastly, the delivery of services will also change dramatically. Pre-saturation carriers tend to "do it all" – that is both run an airline (the brand) and operate the airport (the network). Saturation players increasingly have strategies and organizational structures that differ for the network and the brand service operations. As 3G economics change and network capacity utilization becomes a key success factor, wireless networks will both consolidate and be opened to third-party brands as mobile virtual network operators (MVNOs). While all markets will experience some degree of saturation effect, the speed and impact will vary. One key determinant is the lag between the slowdown of the voice market and the emergence of wireless data and value-added services.

9.10   Privacy

Amongst the 3Ps of pervasive computing – Privacy of user data, Personalized user experience, and Protection of information assets – the most important issue is privacy. Lately, a lot of companies, organizations and politicians have been jousting for their stance on privacy. A battle is brewing between Passport (Microsoft) and Liberty Alliance (SUN Microsystems). There is talk of privacy legislation on various fronts – E911 and online – and organizations like EPIC (Electronic Privacy Information Center) have launched scathing attacks on Microsoft. So, besides appearing ethically and morally astute, what's at stake here? The answer is transactions and the user. The conventional wisdom is the more you know about as many customers as possible, the more dollars per user will flow towards you. The grab for consumer data is going to be an interesting battle in the coming years. AOL has five times more subscribers than MSN while it was estimated that 90 million PCs would have Windows XP by 2002 (source: Business 2.0). All of this lays the groundwork for interesting times ahead.

Privacy is all about "trust". If someone masquerades as someone I should trust with my "information", then turns around and sells that information to the highest bidder, without users' consent, it is WRONG – plain and simple. They can muddle around with verbiage all they want, but the fact remains that it is wrong. Unfortunately, it is a common practice. The Internet is here, and collecting user information is absolutely essential to provide a usable user experience. Otherwise one is bound to get junk more often than not, and people do not have time to sort through it. It might be hard to believe, but an amazing amount of information can be collected, stored, and mined to build a pretty good user profile. If one gets value for what information one provides to the vendor – for example, purchasing airline tickets and books online with minimal clicks or viewing customized news and sports scores and so forth – it is worth it. Who has the time to enter addresses and credit card info, again and again? Amazon and Expedia have made our lives easier.

However, problems arise if these companies go beyond using the information to improve the user experience, and then do underhand dealings with spammers. That is a betrayal of one's trust and needs to stop. We will now look at some of the technology solutions being introduced to address privacy.

9.10.1   P3P

Good progress has been made on the Platform for Privacy Preferences Project (P3P)[3] project at W3C, which allows automatic, computerized reading of a Web site's privacy policy by browsers. There are two key components:

---

[3] http://www.w3.org/P3P/

- the client side that allows P3P clients to automatically fetch and read P3P privacy policies on Web sites, and
- the server side (Web site) component that allows Web sites to translate their human-readable privacy practices into a standard, machine-readable format that can be retrieved and read by browsers.

However, the problems with P3P include the following.

- It is not a complete solution: we need more customized capability. Instead of generically stating, "I don't want mining of my data," you can say, "I don't want mining of my data from XYZ and there needs to be a way to track my private information from changing hands on demand."
- Users need a way for their information to go into hibernation and have the ability to delete it from at any website, if that's what they choose to do. (Vendors could also consider destroying post-transaction personal data, depending on non-repudiation requirements.) P3P is largely unknown to consumers and businesses alike. There is no automatic way to equip legacy browsers with P3P capability.
- It's a chicken-and-egg dilemma: companies will not make the translations until customers have the tools and demand P3P capability on the server side, but consumers will not bother to download and configure tools just to interpret a privacy policy that they do not read anyway.
- Also, P3P categorizes the types of information handed over by the user in the following ways:
  - the purpose for which it is collected,
  - the recipients of the information, and
  - the duration of the information's retention.

These categories can be misused. So, if the purpose category is <current/> or <stated-purpose/>, the user has to dig for more details of the vendor's meaning.

- There are very complex user scenarios with the class of devices that most need privacy – wireless phones. This is especially important in location-based data sharing scenarios.
- The most serious problem is of course the inability to enforce privacy. There is a need to devise mechanisms so the policy agreement between consumer and vendor is legally binding.
- They are not deployed for any major wireless application.

How many users will actually change the browser default? When users visit a site that uses P3P, they can click on the privacy icon in their browser to "privacy check" the site. This brings up a window that explains any areas where a site's policy conflicts with a user's preferences. Users can also use this window to jump directly to a site's privacy policy, as well as to see whether the site has a privacy seal.

However, P3P is a good first step. A Web-agent-based approach is best for a privacy handshake, but there is a need to keep working with organizations like EPIC to make progress. Because of the pressures from such groups, Microsoft has now limited the information required to use Passport. Additionally, strict enforcement of privacy laws is needed, so there is no doubt in anyone's mind about the repercussions. Unfortunately, legislative bodies move too slowly for the information age we live in. We need to design legislation keeping the next decade in mind. Then, we might have a faint chance of getting it right.

9.10.2   XNS

In 2001, XNSORG (eXtensible Name Service Public Trust Organization) took the P3P concept to the next level by introducing a new open protocol and open-source platform, XNS. Based on XML and using Web agents, XNS is designed as a global solution for automatically exchanging XML data between two devices with privacy, security and synchronization controls. XNS uses the concept of an XNS business agent first to negotiate a legal contract between user and the business before doing the P3P step. In addition there are anonymity and pseudo-anonymity tools, encryption tools, filters, identity management tools, and other devices available. However, they can not be used as generic solutions to address privacy concerns, because the average consumer will not go through the trouble of learning the nuances. For any solution to gain widespread adoption, it needs to be part of the browser.

9.10.3   Regulations

There are current US regulations that protect consumers' financial (The Gramm–Leach–Bliley Act) and medical (Health Insurance Portability and Accountability Act) information from being sold to third parties for purposes of telemarketing or other direct marketing. The Children's Online Privacy Protection Act, enacted in law in 1998, requires that Web sites visited by children under age 13 post a privacy policy detailing any personally identifiable information collected from those children.

In Europe, most of the EU member states have implemented the EU 1995 Data Protection Directive that seeks to protect consumer data. There are also some laws in place for data that cross borders. A "safe harbor" arrangement exists between the USA and the EU. It declares that personal data about EU citizens may be transferred to the USA only if adequate protection is provided, such as obtaining consent for any sensitive information used for purposes other than originally stated in the privacy policy. Because of the importance of the issue, we will see more regulations in the future.

9.10.4   Wireless advertising

As we earlier discussed, wireless advertising has a huge potential, especially if it is coupled with location technology. However, unless we develop UI standards for devices that allow complete control over what comes to them (when and how users want it) wireless advertising is just not going to be successful. For a carrier to monetize location services, they must develop end-user subscriber applications, giving users the control over what application gets which personal information and which personal information is off limits. The Wireless Advertising Association (WAA) has made progress in defining standards and measurement definitions, but more needs to be done to cover a range of devices and technologies. Privacy filters are critical to the success of any consumer wireless application or service. It is an absolutely critical element that needs to happen before any location-based services see the light of day.

9.10.5   Dos and don'ts of privacy

It is also in the user's interest to get audited by third parties like PwC, E&Y, Truste, and others. Everybody in the value chain should work proactively on the privacy issue.

Protect consumers from your partners as well. It is not permissible to ship user information to your partners without legally binding contracts that adhere to your privacy standards so that your partners will not misuse the data. User data should be guarded or anonymized for partner usage, just as you would guard any other sensitive information such as user ID and password lists. Also, you should stick to your privacy policy and not change it frivolously to suit your business needs.

The privacy issue is not just a desktop issue. It is an AORTA (always on real-time access)[4] issue – anything connected to a network can potentially transmit personal information about usage habits. Recent issues with TiVo and other broadband platforms such as GPSs in rental cars, biometrics technology in stores/malls/companies to identify shoplifters, raise this point. Aggregators such as Axiom, Experion, and Engage collect a broad range of customer data across various channels to build a lifestyle score – which could be used either to provide useful services or discriminate.

It is not in the best interest of product and service companies stealthily to record information without consent. It leads to PR nightmares, bad press, and a black mark on their privacy records. Ask Real Networks, Doubleclick, Microsoft, Intel, and others. Somebody somewhere is bound to figure things out, so why waste time and effort in questionable activities?

Another risk that should be avoided is mixing collective (aggregate) data with personal information. As Mary Modal of Forrester once commented, "Companies should think of personal information and collective data as the church and state of Internet business. Keep the two separate."

We also need to keep in mind that when it comes to privacy, we can not generalize. We have to pay attention to both sides of privacy concerns. People who do not feel

---

[4] AORTA was coined by Mark Anderson.

Table 9.3. Securing the information value

Steps in place	Information value			
	Low	Medium	High	Highest
Authentication	✓	✓	✓	✓
Authorization	✓	✓	✓	✓
Policies and procedures	✓	✓	✓	✓
Encryption		✓	✓	✓
Monitoring		✓	✓	✓
Auditing			✓	✓
Fraud prevention				✓

comfortable trusting the Internet need to be accommodated with consumer-friendly technologies and policies.

Privacy can be used as a competitive advantage mechanism. Companies will build up great brand value propositions around having absolutely reliable "privacy", and consumers will flock to these brands, as consumers will not have to question the reselling of their data. A new "brand value" will soon be total and unassailable end-user privacy. Computer makers can also help by bundling personal firewall products with their PCs.

9.11   Security

Most analysts will make you believe that security is a technology problem. However, historical data prove that this is far from truth. It is as much a business problem, if not more, as it is a technology problem. First, the amount of security needed is directly proportional to the value of the information that needs to be protected (Table 9.3). For example, under normal circumstances the value of a user ID and password list is high and should be protected at all costs while items such as news releases or executive bios are not as critical as damage risks in the event of a security breach are small. Paying $100 to the janitor (to wipe some confidential papers and passwords stuck on computers) is still cheaper than breaking complex algorithms and codes, yet people worry more about how many bits are being used for encryption rather than examining their policies and procedures or employing effective monitoring. Your security is only as strong as your weakest link. If you believe that authentication and encryption are enough, your e-security will e-vaporate at some point.

The most fundamental shift when moving from 2G to 3G (via the intermediate 2.5G GPRS stage) is the way in which the platforms are managed – and here new security problems are clearly identifiable.

The GSM environment relied partly on proprietary standards and components like SS7 and X25 to keep out external influences, and such protocols were shared between roaming partners. GPRS and 3G, by contrast, follow the global trend of mapping nearly all network use to TCP/IP. This is great for standardization, but it does introduce new risks that require attention to ensure that extra problems, such as denial-of-service attacks and loss of confidentiality, are not created. On the plus side, integrating data streams with roaming partners or corporate clients is thus made considerably easier as it can be carried via an industry standard VPN connection. The conversion is not completely across the board of course – circuit switching (the standard GSM voice call) will follow the original route. Here, however, we are mainly concerned with the packet-switching side.

At the handset level things are likely to change too. The handset itself is expected to become more powerful, with built-in PDA or other computing capabilities. This presents challenges for both operators and end-users. From the operators' point of view, the more closely a user's computing component and the air interface processing are coupled, the more likely it is that one will be able to influence the other. It will be up to the handset manufacturers to maintain the separation and stop unintentional or malicious modifications to the way the handset communicates with the radio network.

But for the end-user the exposure goes a lot further. The new-generation handsets will, in effect, have power approaching that of common laptops and desktops. They are likely to run the same software – and thus find themselves with the same problems. Wideband 3G will give access to high levels of bandwidth per terminal, and thus to many more data services, making an end-user an attractive target for virus infections and a potentially unwitting participant in distributed denial-of-service attacks.

The billing model will also become a security challenge as we move to the wideband 3G future. There could also be a threat to revenues. As operators will be charging for both inbound and outbound traffic, end-users could face unexpected bills as a result of practices that are currently acceptable on fixed-fee connections. We will continue the discussion on security in the next chapter.

9.12   Position location rollout and privacy

The FCC required wireless carriers to report their plans for implementing E911 phase II, including the technology they plan to use to provide caller location, by November 2000. The top US wireless carriers – Verizon, Cingular, AT&T Wireless, Sprint PCS, and Nextel Communications – account for roughly 76 million subscribers. In reviewing their plans and progress made for E911 deployment, apart from Sprint PCS, these carriers are way behind in their implementation schedule. For some, the service won't start until later in 2003. The position location industry in the USA is very complex. Since there is such diversity in wireless standards (AMPS, TDMA, CDMA, GSM, GPRS, iDEN), wireless carriers have been struggling for the past four years to come up with a strategy that will suit both short-term needs and long-term goals. It is clear that in the next five years or so, GPS-based solutions will become the norm for the industry, but it is the short-term compliance that is causing the carriers to split hairs. The majority of the carriers are being forced to consider a dual-implementation strategy: network-based solutions for their antique AMPS and second-generation networks, and GPS-based solutions for GSM and next generation networks. For GPS-based solutions, since they are handset-based solutions, the existing handsets in the market need to be recycled and, for the existing handsets, the only way you can locate them is by using network-based solutions, which means infrastructure costs in millions, if not billions.

Another conundrum is recovery costs. Carriers want to see their investments recovered as quickly as possible; they would ideally like the state to fund the roll-out. There are still several states that are unclear on cost-recovery schemes. NENA (National Emergency Number Association), APCO, and other public-interest groups have been arguing on behalf of consumers for a speedy implementation of position location solutions so that they can start providing E911 services – the domino that was supposed to enthuse the industry with all sorts of applications and services. The issue is very evident at AT&T Wireless, which has filed for a waiver as it concludes that the network-based technology (E-OTD – Enhanced Observed Time Difference) that it has tested is not meeting FCC ALI accuracy requirements for its TDMA network; AT&T is proposing Mobile-Assisted Network Location System (MNLS) technology for its TDMA networks. The most significant problem facing this technology is its accuracy, especially in rural markets. AT&T gave an expected MNLS accuracy estimate of 250 m for 67% of the calls and 750 m for 95% of the calls, which is short of FCC requirements. According to the FCC definition, MNLS is a network solution because it does not require modifications to legacy handsets and, thus, it is subjected to the less stringent accuracy requirements of 100 m for 67% of the calls and 300 m for 95% of the calls. So, even though the solution is less expensive and has some nice features, it probably will not make the grade.

For handset-based solutions, handset manufacturers are wary of production capacity risks. There are no rules for the evolution of technology in the USA and there are different technical innovations occurring simultaneously. Manufacturers are asking customers (carriers) to prioritize but nothing has been forthcoming according to Nokia. Nokia believes that it would take two years from an order date to get to 50% activation of ALI-capable handsets for a single carrier and for that to happen, GPS needs to be available into the lowest end handsets to meet the numbers. Also, since accuracy requirements are statistical; there is no general agreement amongst carriers regarding accuracy of the various technologies, because solutions accuracy varies depending on the topography – accuracy of a solution is quite different in the plain fields of Topeka, Kansas, to the urban canyons of Manhattan, NY.

There are also issues regarding interoperability. In this situation there is a need to establish a global forum to address the complexity and multiplicity of current solutions and the market situation. With this purpose in mind, Motorola, Nokia and Ericsson established the Location Interoperability Forum (LIF – www.locationforum.org) in September 2000. LIF's purpose is to define and promote, through the global standard bodies and specification organizations, a common and ubiquitous location services solution. The forum's goal is to define and promote an interoperable location services solution that is open, simple, and secure. This solution should allow user appliances and Internet-based applications to obtain location information from the wireless networks independent of their air interfaces and positioning methods. Let us move past the location technology discussion to delivery issues. It is safe to assume that it will be another two years before position-location-enabled applications and services are available widely in the USA (owing to implementation and handset delays). Although APCO and NENA have done an admirable job in pushing the issue, things just have not moved fast enough. Once position location of any mobile device is a given, E911 application infrastructure (PSAPs, ALIs, ANIs, Routers) will be ready to make it work end-to-end. However, that is not true for non-E911 applications and services. There are three key missing components that need to work in the position-location-enabled application architecture:

- location middleware and business intelligence components to integrate location information into applications and have the capability to do analysis to leverage trends and usage;
- user-awareness components to update a user's presence information provide privacy filters (controlled by the user) and configuration options, and usability filters and transcoders to customize and render on need/demand; and
- operations and management components – somebody has to track and bill for all this information.

Amongst all these, providing privacy filters (opt-in NOT opt-out policy) is absolutely essential for the industry, otherwise there is going to be a backlash against position-location applications and services. We will probably see legislation of some sort to protect consumers' privacy and unauthorized use of data. These three areas will also be fertile ground for new, different business models and strategies – who owns and stores this vital user information?

Opt-in or opt-out are only one simplistic area of what should be termed "shared personalization". It is very doubtful that opt-out will be accepted by consumers in an LBS (location based system). Opt-in is the more likely and most basic point of enabling personal data to be shared. Taking this one major step forward – "shared personalization" is what information a consumer can specify that can be shared from vendor to vendor within a carrier infrastructure. Storage location is almost irrelevant in this context. In effect a consumer can say "this info can be shared by all vendors" and "this info can only be shared with my permissions first". That way the consumer can share the data only when specifically interacting with a vendor. Put another

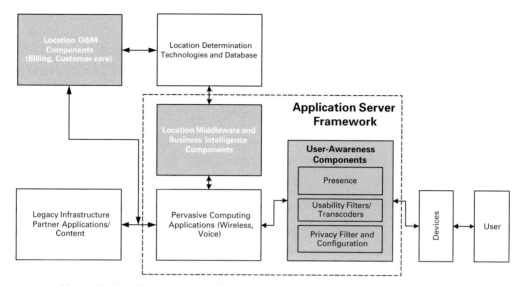

Figure 9.4. Position-location application framework.

way – these data are off limits for data mining and target marketing, etc., from a carrier perspective or a vendor perspective.

The second area is "synced personalization data". Consumer-defined data sets can be automatically shared with a new vendor application and it is made immediately available. But the data sync term here just means "shared in near real-time" and not necessarily a Fusion 1 heavy overhead approach. This is especially critical in a WAP UI where any keystroke activity to "pull" data from one system to another is a usability showstopper. Each time a consumer adds or modifies a data set, it is updated automatically across all the systems it feeds.

The third area is authentication of an identity to change the consumers data set. This could be a sign-on password, fingerprints on devices in the future (available on guns today) or other methods. This has to be part of the sunshine platform. In this scenario even if the personal data were lost the consumer would still have access.

These are the three main keys to providing secure, consumer controlled personalization and enable vendors to provide "1-to-1" services and applications customizations and leverage location based data. Consumers have the control here and this is necessary before LBS applications take off.

Also, some new application and services companies are springing up to get ready for the position-location application boom (Figure 9.4). First, the wave is going to be delayed, and secondly, start spending some time with the carriers. It is naïve to think that carriers will just give you location information or just because a user is using your application (and driving airtime), you can assume a share of airtime revenues from carriers. It just does not work that way. Carriers are investing millions of dollars in network and handset rollouts and are going to do their best to sell location information most appropriately. They should be working with content providers to enable applications and services to be location sensitive. An increase in such applications will lead to an increase in airtime – the real source of their revenues.

All this discussion begs the question: what can I do today or in the short-term? It will certainly be worthwhile to start experimenting with cell-sector-ID based position-location prototype applications and services, especially for applications that do not need to pin-point location with high accuracy and start planning the road-map for the future – as position location based revenues are bound to sky rocket.

9.13   Mobile fraud

The mCommerce world combines good and bad components of the wireless, Internet and credit card industries. For the following reasons, there is more potential for fraud:

- mCommerce: more reward for risk
- more information available
- stolen credit cards
- over-the-air activation of mobile accounts and services
- stolen identity
- new technologies, new security holes
- authentication for each session
- latency in authentication
- intermediary gateway architecture.

Billions of dollars of revenue are lost every year from mobile fraud. The GSM Association, a mobile industry body, says that 250 million euros are lost each year from mobile roaming fraud alone – and this is just one of the most popular types of mobile fraud.

There are many different types of mobile fraud (see Figure 9.5), with new variations being developed all the time. The most common by value according to Cerebrus, which produces fraud-detection software, are as follows.

- Internal fraud, which accounts for 40.3% of losses from mobile fraud. This may involve an employee altering the telecoms switch to give a free mobile number or numbers to friends – who use them for commercial gain.
- Premium-rate services fraud, which accounts for 13.1%. Here bogus companies set up premium-rate services and collect revenues from the mobile operator knowing their services have been called up by fraudsters who will never pay.
- Subscription fraud, which accounts for 11.6%. This involves registering for a service under a false name or signing up for a service with no intention of paying for it.
- Roaming fraud (running up bills outside the home network with no intention of paying), which accounts for 11.4% of losses.

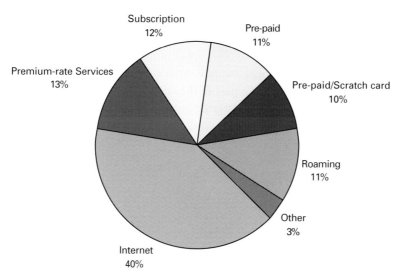

Figure 9.5. Mobile fraud distribution by type (source: Cerebrus).

- Pre-paid fraud, which accounts for 10.8%. This includes cloning pre-paid cards from newsagents, so one card's details can be used by 100 or so people.
- Pre-paid/scratch card fraud accounts for 9.5% of reported cases; handset fraud 5.4% (stealing handset shipments and changing their SIM cards or identities); and internal fraud (which is probably under-reported) 8.2%.

Many of the same types of fraud crop up again when looking at the most frequently reported cases. Subscription fraud and roaming fraud are the two most frequently reported types, accounting for 40.8% and 16.3% respectively, according to Cerebrus data based on telecoms industry association figures. In practice, subscription fraud and roaming fraud are often related. Criminals will often subscribe to a mobile service under a false identity, then go abroad to use the phone, running up bills they never intend to pay. Roaming charges for using a network outside the subscriber's home network may not appear on the bill for several weeks, so considerable sums have usually been spent before the operator realizes there is a problem.

Another type of roaming fraud occurs when unethical resellers in emerging markets inflate the number of subscriptions. They do this by submitting the same application forms twice, leaving an interval of a few months. If the mobile operator does not pick up on this, the reseller will receive a second bundle of handsets and SIM cards (the tiny smart cards that make a mobile work). Typically these are split, with the SIM cards sold to organized crime, while the handsets are sold to another country such as South Africa.

A bootlegged SIM will be worth $200–300 as it can be used to offer long-distance calls on street corners in areas where there are high numbers of migrant workers. A mobile used in this way can easily run up a bill of thousands of dollars over a weekend.

Criminals also have software that enables them to piggyback a call from a second mobile on to a call record made using another mobile. The first mobile will ring a number that is not answered, creating a call record. But as the call is not answered most of the record will be deleted, making it unlikely that the call will be traced. Some operators, however, leave part of the call record open, enabling criminals to piggyback off this call record undetected.

Another type of roaming fraud occurs when people run up bills abroad that exceed their credit limit and then do not pay. This is believed to be a problem for some Netherlands operators, which experience high call traffic levels from people on holiday in Greece in the summer months calling home. Or for other European operators, which experience surging holiday traffic from budget sunspots.

The operators do not always get details of these calls immediately, hence it is hard for them to prevent them being made if they exceed users' credit limits. The US-based Communications Fraud Control Association says that fraud costs its (fixed and mobile) telecoms members 3–5% of gross revenue annually. Revenue leakage from roaming fraud is believed to account for 0.3% of airtime revenues and this is an area that Cerebrus is specifically targeting.

It is likely that with the rise in mCommerce solutions, fraudulent activities will also rise. For hackers, pranksters, and organized criminals, mCommerce provides more reward for the risk. Not only can they steal airtime, but they can also perform financial transactions on your behalf.

To put up a unified front against mCommerce fraud, all the players (carriers, application developers, infrastructure providers, handset manufacturers) need to work together to fend off this emerging yet unacknowledged threat. The following data sources already exist and can be used intelligently to create profiles of users that can help in detecting and preventing fraud as it happens:

- subscriber data
- server logs
- wireless billing records
- wireless usage records
- online history
- behavior patterns and personalization data
- device transactions
- user profile, device capabilities
- location.

Fuzzy logic or artificial intelligence (AI) engines can be used dynamically to analyze and approve transactions. Also, gateway architecture introduces another point of failure. You should either have fairly tight SLAs or, for confidential data, host gateways within your corporation.

The impact of mCommerce, supported by 3G networks, can perhaps best be illustrated by a fledgling service already being introduced by several existing 2G

GSM networks. The trial allows a caller to send an SMS message to a telephone number advertised on a soft drink vending machine. On receipt of the SMS (which by default includes the caller's identity) a management system tells the machine to vend a can of Coca Cola or some alternative, the cost of which is automatically billed to the customer's telephone account. This last is the crucial point. In the above scenario, the caller has effectively received goods up front (cans of Coke) by using a credit mechanism – their post-paid GSM account. The SIM card within their mobile handset, which identifies them to the network for billing purposes, has effectively become a credit card. If the caller commits fraud and, with pre-meditation, fails to pay their telephone bill, then the value of that fraud will not merely be the value of the SMS messages sent (a paltry sum) but the value of all of the cans of Coke, as well as any other items taken in a similar fashion.

The additional commercial applications already being proposed for GPRS and UMTS networks certainly imply that this exposure to new frauds, where the value of the transaction exceeds the value of the message many times over, will increase exponentially during the next two years. It is worth noting that during 1999 the UK witnessed a 53% increase in the value of traditional credit card fraud, suggesting that a large and talented community of criminals stands waiting in line for the first 3G offerings.

The extent of this possible increase in fraud, and its true impact on 3G networks and service providers, will ultimately depend on a single factor – will networks follow the current trend and allow credit to be obtained against the telephone account for non-telecom goods and services? If they do, they will almost inevitably bear the resulting financial burden of fraud. If they do not, they must fight a much tougher battle than they have perhaps envisioned, increasing 3G call volumes to a level that recoups their already massive investment.

9.14   Transfer of technology

The term "transfer of technology" usually refers to the transfer from research center to development and engineering department. In this context, the successful transfer of technology is the technology born in the research labs that is then moved to the development department and finally leads to a successful product launch using the technology. Many technology-based companies have a research division and a development division too. It is like "ping-pong" between the research division and development division. There are several key factors to make the "ping-pong" succeed and there are many good books to explain it.

What we want to discuss here has a different viewpoint. It is the issue of technology transfer in a global market. This is the case when one (wireless) technology is born in one market or area and an attempt is made to move to another market or area across the border. This is a very important question for those who succeeded tremendously in their markets and plan to extend the successes into new geographies. So, what are some of the challenges? What are some of the key factors for such technology transfers? What issues do we need to overcome? This is the theme of this section.

Technology transfer is a new area for the wireless industry since the transfers were not that successful in the era of non-global standard systems such as 2G cellular. NTT tried to export its PDC system in Asian countries in the early 1990s, but was not successful. Asian operators were reluctant to employ non-global standard technology such as PDC. GSM was successful because it was designed considering the regional roaming in Europe from the start and the technology is easy to deploy into areas other than Europe. However, even GSM is not a global system.

Next, we will pick up a technology transfer example that uses a wireless Internet access service. Most of the wireless carriers have steadily been recognizing that the next revenue stream will come from wireless data access services by cell-phones, such as i-mode.

In the USA, major wireless carriers were not active in developing these kind of services until 2001. However, in 2002, some major wireless US carriers started to move forward. US carriers have been putting in a lot of resources trying to understand the market situation in some Asian markets (Japan, South Korea, etc.) where Internet access services have become major revenue sources for carriers. US carriers are just trying to recover from a voice-only business model era.

As we previously mentioned, the most important aspect of the i-mode business model is the BOBO (Billing On Behalf Of) mechanism, which allows a wireless carrier and not a content provider to collect the information fee (see Chapter 8). BOBO is the key mechanism to achieving Internet-like ecosystem in the cellular world.

It is well known that AT&T Wireless made a strategic tie-up with NTT DoCoMo in early 2001 and has been working hard to learn the technology and business model of i-mode. AT&T Wireless started its PocketNet service several years ago using HDML (Handheld Device Markup Language) but it was far from being successful. AT&T Wireless established a new organization called MMS in the Fall of 2001 and released the m-mode service in June 2002; m-mode is now providing various wireless value-added services using GPRS and BOBO functions. The service is available from $2.99 per month. This price setting is very similar to i-mode (300 yen per month).

Also, Verizon Wireless announced a program download service using BOBO in June 2002. This is a typical example of the transfer of technology into the global wireless market. It should be noted that technology here means not only the technology itself but it also includes the related-business model and pricing strategy.

Just as the USA (and to some extent Europe) is trying to learn from Japan, Japan and Europe are trying to learn from North America when it comes to the enterprise market, as the USA is a much more mature market when it comes to enterprise solutions. Mobile enterprise solutions are the first to take off in the USA. Other successes like the popularity of Blackberry as an email device are being tested in foreign markets.

The successful transfer of technology across the borders in the wireless world can only be achieved by the transfer of the total ecosystem in which the technology can be fully utilized.

9.15   Conclusion

In the 1980s, the slogan of most fixed telecom carriers was "one telephone per family". This slogan was the driving force behind the telecom industry's development and investment in the 1980s. Major fixed telecom carriers like AT&T and NTT spent enormous amounts of money to build out the telephone network nationwide. In the past decade, the mobile communication service has been intensively developed and expanded under the slogan of "one telephone per person". In other words, the business chance of mobile communication has been aiming at personalization of telecom lifestyles. The challenge was for the mobile industry to create services and devices that would satisfy this goal. This strategy fits well with the public's instinctive desire and was tremendously successful all over the world irrespective of culture, race and language.

What will be the slogan in the coming decade? One of the keywords might be "multiple devices per person". A user can use multiple telecom devices depending on the purpose and situation. Mobile communication in the next decade will accelerate the diversification of telecom lifestyles. In this context, the challenges in the next decade will be how mobile industries can respond to the trends of diversification. Diversification is important not only for device design, but also for network and services. The recent sudden upsurge in the wireless LAN market around the world indicates how strongly the public and markets need new horizons that promote and enable different wireless lifestyles.

In this chapter, we have reviewed some of the business issues and challenges that the wireless industry needs to contend with, both short- and long-term. In the next chapter, we will take a look at some of the technology issues and challenges for the industry.

10 Technology issues and challenges

In the previous chapter, we discussed the major business issues and challenges facing the wireless industry. In this chapter, we focus our attention on technology issues such as limitations of Moore's Law, spectrum issues in the USA, the WWAN vs WLAN debate, security, transcoding, roaming and billing for services.

10.1 Technology development and the laws

Both the computing and communications industries are impacted by the various laws, theorems, and principles that have become part and parcel of the scientific community. Two laws that are most commonly used to predict progress with respect to the wireless world are Moore's Law and Shannon's Law. Let us discuss these two briefly to see how they have shaped technology development.

10.1.1 Moore's Law

In an article[1] titled, "Cramming more components onto integrated circuits" for *Electronics* magazine, young Gordon Moore of Shockley Semiconductor presented an idea that the development of smaller transistors at lower unit production costs would translate into exponential progress in silicon chip complexity at minimum cost. In other words, performance would double every 18 months. "Integrated circuits will lead to such wonders as home computers," wrote Gordon Moore, "...automatic controls for automobiles, and personal portable communications equipment." The year was 1965, three years before he would co-found the largest silicon chip manufacturer now in existence – Intel. In 1965, chips could hold 60 transistors. Moore predicted that by 1975 as many as 65 000 transistors could be crammed onto a single chip. To put this into perspective, Intel's Pentium 4 chips now hold over 42 million transistors. The ability to fit more transistors onto a silicon chip simply means that information can be processed faster.

---

[1] http://www.intel.com/research/silicon/moorespaper.pdf

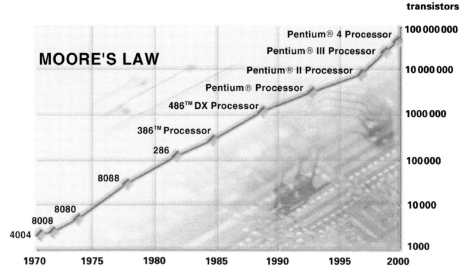

Figure 10.1. Moore's Law (source: Intel).

The press dubbed the observation "Moore's Law" (Figure 10.1), and it has become a widely accepted and universally acknowledged precept that has come to account not only for chip complexity but also for faster performance, translating into computer performance that doubles every 18–24 months. In the late 1990s, the digital world started becoming increasingly anxious about the supposedly inevitable demise of Moore's Law. Chip manufacturers had two major problems: physical barriers blocking the continued diminution of transistor size, and financial barriers in fabrication costs.

While computing power was doubling every two years, fabrication costs were doubling every three years. This situation would inevitably lead to the slowing or stopping of chip manufacturers' progress. If Moore's Law holds true, manufacturers will need to place atoms individually to manufacture silicon chips throughout the 2010s. At the 0.10-micron stage (each transistor would be composed of less than 100 atoms), small silicon chips containing millions or billions of transistors would no longer be able to control the flow of electrons.

With each generation of increased performance, the size of computer-based products becomes smaller. If manufacturers simply made larger chips to fit more transistors on them, the products would also have to become larger to accommodate them, or our smallest electronic products and applications would not be able to be improved upon.

Moore's Law has been a corner stone of the progress we have made in both the computing and communications fields. Chips continue to become smaller and faster. Additionally, more-complex devices can now be manufactured. For the designers of wireless systems, Moore's Law has enabled designers and system engineers to ignore complexity implications of their designs. It should be noted that Moore's Law

Table 10.1. Semiconductor manufacturing roadmap: key targets

Year of first product shipment	Feature Size	DRAM capacity	Cost performance (microcents/transistor)	Functions per chip (Gbits)	Microprocessor cross chip clock frequency
2001	130	512 Mb	107	2.15	1.6 GHz
2002	115	512 Mb	124	2.15	2.4 GHz
2004	90	1 Gb	62	4.29	4.0 GHz
2007	65	4 Gb	22	17.18	6.7 GHz
2010	45	8 Gb	4.71	34.36	11.5 GHz
2016	22	64 Gb	0.59	68.72	28.8 GHz

Source: ITRS, 2001 update.

applies only to microprocessors and associated DSPs, and not to the efficiency of radio communications where the number of voice calls per MHz of spectrum has changed at a slower pace. If Moore's Law holds true for the next couple of decades, then the processing devices will be nearly ten thousand times more powerful than they are today (in 2003). Increase in processing power does not imply improvement in radio efficiency. Enhanced processing power does help in (say) better codecs, which will reduce the bit rate and hence increase the system efficiency, but on a far smaller scale.

The International Technology Roadmap for Semiconductors (ITRS) puts out a forecast annually for semiconductor technology, specifying targets for every aspect of the semiconductor manufacturing industry. In each area addressed, the ITRS identifies the key challenges that the industry must overcome to continue the technical progress predicted by Moore's Law. Table 10.1 identifies a few targets set in the latest version of the ITRS[2].

The ITRS identifies many problems to solve to keep Moore's Law on track through to the end of its forecast period in 2016. These problems include the design, manufacture, and test of advanced semiconductor devices, with both material and equipment requirements in mind. In each area, the ITRS notes the status of solutions: they exist, are being pursued, or are not known. In almost all key areas, known solutions are available to solve near-term problems. Many areas of the ITRS state that no known solutions exist for requirements beyond 2005, and even more categories bear this designation for post-2008 requirements. Given the rapid pace of developments in the industry, however, this lack of clarity beyond a five-year horizon is not surprising. We might hit the limit of Moore's Law for several reasons; for example, manufacturing might not be possible, or it might not be possible to design or test a chip with more than a billion transistors.

---

[2] For updates please visit http://public.itrs.net/Files/2001ITRS/Home.htm

10.1.2   Shannon's Law

Claude Shannon's creation in the 1940s of the subject of information theory is arguably one of the great intellectual achievements of the twentieth century. Information theory has had an important and significant influence on mathematics, particularly on probability theory and ergodic theory, and Shannon's mathematics is in its own right a considerable and profound contribution to pure mathematics. But Shannon did his work primarily in the context of communication engineering, and it is in this area that it stands as a unique monument. In his classical paper of 1948 and its sequels, he formulated a model of a communication system that is distinctive for its generality as well as for its amenability to mathematical analysis. He formulated the central problems of theoretical interest, and gave a brilliant and elegant solution to these problems.

Shannon's Law defines the theoretical maximum rate at which error-free digits can be transmitted over a bandwidth-limited channel in the presence of noise, usually expressed in the form $C = W \log_2 (1 + S/N)$, where C is the channel capacity in bits per second, W is the bandwidth in hertz, and S/N is the signal-to-noise ratio. Note that error-correction codes can improve the communications performance relative to uncoded transmission, but no practical error-correction coding system exists that can closely approach the theoretical performance limit given by Shannon's Law.

The digital baseband processing power required for 3G has reached the level of many thousands of MOPS (million operations per second) per channel, and this is growing by a factor of 10 every four years. This is because of the use of increasingly sophisticated, digital transceiver algorithms needed to achieve increased channel capacity and network coverage. This rate of increase is driven by the pursuit of Shannon's Law. But this growth rate is much faster than the increase in processing power that can be achieved through a single-threaded, Von Neumann-based architecture: transistor density, and therefore processing power, only increases by a factor of 10 every six years. The traditional DSP + ASIC architectures used for 3G baseband processing are limited by this bottleneck: the fact that there is a central DSP that is shared across the entire architecture for multiple tasks.

Unlike Moore's Law, where things keep getting better every few months, with Shannon's Law, we can only approach a maximum limit (Figure 10.2). In a perfect case when Shannon's limit is reached, if we have 1 MHz of spectrum and each user's call requires 10 kbps of data, then we can accommodate approximately a maximum of 142 voice calls per cell in a clustered environment where spectrum is reused from cell to cell. At present, we are still far from the limit. Proposed 3G systems are expected to realize around 60–100 calls per carrier per sector. Each carrier is 5 MHz wide, so the number of calls per sector per MHz is around 12–20. By adding more sectors, we can reach 30–50 calls per cell. To get more capacity, the trend for operators is to install more smaller (or micro) cells. But these smaller cells lead to other issues such as the management complexity of frequency reuse and handover.

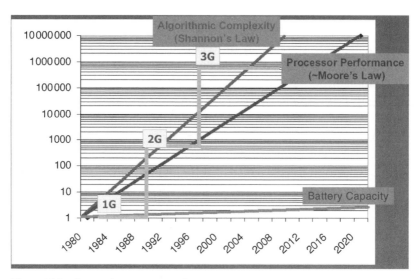

Figure 10.2. Shannon's Law vs Moore's Law.

10.1.3 Gene's Law

Gene Frantz's theory holds that "power consumption of integrated circuits decreases exponentially" over time, and because of that the whole system built around chips will get smaller and batteries will last longer. And there is ample proof that Gene's Law (named after Texas Instruments engineering guru Gene Frantz) holds water. Since Gene's Law was introduced in 1994, the power required to run an integrated circuit, or IC, has declined 10-fold every two years. In tandem with Moore's Law, Gene's Law just may be the basis for the advances we have seen, and will continue to see, in miniaturization. By 2005, according to Gene's Law, we should be able to carry around in our front pockets a device with the power of a Sun Ultra SPARC III workstation. By 2010, we should be able to wear on our fingers the computing power of IBM's ASCI White, the most powerful computer in the world today.

We know from our previous discussions that improvements in battery life are key to future wireless devices and applications that can run on these devices without being a major drain source. Figure 10.3 depicts Gene's Law.

10.2 Spectrum issues

Spectrum, in wireless communications, refers to the ability to transmit signals at a specific frequency, and the bandwidth of that frequency. The ability to transmit signals at a particular frequency is generally conferred by a national government, in the form of a license. In the USA, regulatory responsibility for radio spectrum is

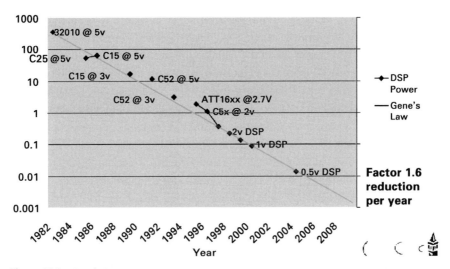

Figure 10.3. Gene's Law.

divided between the Federal Communications Commission (FCC) and the National Telecommunications and Information Administration (NTIA). The FCC, which is an independent regulatory agency, administers spectrum for non-Federal government use and the NTIA, which is an operating unit of the Department of Commerce, administers spectrum for federal government use. One of the primary missions of the FCC is to provide advice on technical and policy issues pertaining to spectrum allocation and use in order to manage the country's radio frequency (RF) spectrum most effectively. Demand for spectrum – from both the private sector and from government agencies – has rapidly outstripped its very limited supply. This is especially true for spectrum that is suitable for mobile wireless communications. Faced with a spectrum shortage, the difficult job of efficiently rationing the available RF spectrum has become one of the FCC's most important tasks.

Spectrum is the fundamental requirement for entering the mobile wireless communications industry, and is the key shortage in the industry. The available mobile wireless communications spectrum in the USA is rapidly becoming filled up, as the number of subscribers and their minutes of use continue to grow rapidly. Additionally, more spectrum will be required in order to accommodate data transfer and offer third-generation (3G) services. Capping nearly two years of intense negotiation among the US military, government agencies, and the wireless industry, the Bush Administration unveiled a compromise plan that calls for the allocation of 90 MHz of spectrum for third-generation (3G) wireless services in the USA. The agreement – which provides the wireless industry with half of the spectrum it had sought originally for 3G use – covers 45 MHz of spectrum in the 1710–1755 MHz band and an additional 45 MHz in the 2110–2170 MHz band. Prior to this agreement, the Defense Department (DOD) vigorously opposed any plan that would have required the military to vacate spectrum for commercial 3G use or to share that spectrum with 3G licensees. As a result of the 9/11 terrorist attacks, the 1770–1850 MHz band – which encompasses much of the DOD's operations and had also been earmarked for 3G – was removed from consideration for 3G services. Under the compromise, the DOD agreed to relocate military microwave systems in the 1710–1755 MHz band (as well as airborne telemetry, munitions operations and other systems in 14 of 16 "protected" sites) upon receiving reimbursement from the wireless industry, but no later than December 2008. The plan, which is expected to clear the path toward 3G license auctions in 2004–2005, calls upon the FCC to allow DOD ground systems to remain in that band on a secondary basis and on a primary basis in two protected sites in North Carolina and Arizona. Despite receiving only half of the 3G spectrum desired, Cellular Telecommunications and Internet Association President Tom Wheeler praised the accord as "a clear win for the economy, a win for consumers, and a win for national security".

The ability of the government to make more spectrum available is one of the key features of the industry. The government can and will make more spectrum available. They have both a public policy and a financial motive to do so. This is a potent combination in a political process, and should not be underestimated. Making new spectrum available is generally good for consumers, since prices usually come down and new services are made available. However, it generally makes the existing spectrum relatively less valuable as the industry becomes more competitive. This has broad implications for carriers and for investors. The existing mobile wireless communications spectrum in the USA is rapidly filling up. Greater capacity, in the form of spectrum, will be required to enable the future growth of the industry and the evolution to 3G wireless data services.

10.3   Challenge for high-performance devices

Device technology is one of key factors for the future wireless systems. As the access technology is becoming more complex such as WCDMA and OFDM, the technical requirements for 3G/4G devices are also increasing. The biggest challenges for next-generation devices are the battery and linearity of the amplifier.

The lithium-ion battery was initially developed in Japan in the 1990s and is now widely used for mobile devices, since this type of battery has a higher energy density than other battery technologies. The next targets for battery development will be (1) higher energy density (from 150 Wh/kg to 300 Wh/kg), (2) low power voltage (from 3.6 V to 2 V or less), (3) compact and flexible (less than 1 mm thickness), and (4) re-cycle material.

The lithium-polymer battery has advantages on energy density, flexibility and low power voltage, and the solar battery has advantages on cleanness and re-cycleability.

Table 10.2. Comparison of WLAN/PAN standards

	IEEE 802.11b	IEEE 802.11a	Bluetooth	HomeRF	HomeRF/2	HiperLAN/2	IrDA
Use	LAN	LAN	PAN	LAN	LAN	LAN	PTP
Maximum theorectical physical rate	11 Mbps	54 Mbps	1 Mbps	1.6 Mbps	10 Mbps	54 Mbps	1 Mbps
Maximum physical rate (reality)	11 Mbps	–	400–700 kbps	1.6 Mbps	–	–	1 Mbps
Range	100 m	100 m	10 m	50 m	50 m	100 m	1 m
Spectrum	2.4 GHz	5 GHz	2.4 GHz	2.4 GHz	2.4 GHz	5 GHz	Infrared region

Such breakthroughs in battery technology may dramatically change wireless lifestyles. The development may achieve "paper-like" batteries and "super safe" devices.

Another challenge is the power amplifier design issue, especially for OFDM-based wireless systems such as IEEE 802.11a wireless-LAN. In OFDM it is well known that the system requires very severe linear characteristics for a transmitting power amplifier since OFDM has higher PAPR (Peak-to-Average Power Ratio) than other modulation systems like PSK (Phase Sift Keying). So, a transmitting power amplifier of OFDM needs to have wider dynamic range for linear operation than other systems.

10.4   IEEE 802.11 vs Bluetooth vs 3G

Here, we will consider the typical evolving wireless technologies, wireless LAN (WLAN), Bluetooth (WPAN) and 3G systems (WWAN). Originally, these technologies were designed and proposed for different purpose and have different histories (see Table 10.2).

The definition of WLAN in IEEE 802.11 is "a wireless version of a local area networking protocol such as ETHERNET" and "its behavior must be compatible with other portions of the LAN infrastructure such as TCP/IP". This definition clearly says WLAN must behave like ETHERNET using wireless access.

The definition of WPAN in IEEE 802.15 says "effort focuses on the development of consensus standards for PAN or short distance wireless networks". Also, it indicates the distance is typically less than 10 m as a short range cable replacement.

The definition of WWAN is "a network that spans a relatively large geographical area" and "the largest WWAN is the mobile telephone system".

Table 10.3. Issues: LAN versus WAN

	LAN (802.11)	WAN (3G-WCDMA or cdma2000)
Locations	Hot spots	Metropolitan and suburban regions
Business Apps.	Computer-centric (full PC) applications	Integrated LAN/WAN access with mobile devices
Consumer Apps.	High-speed video, audio, and Internet access	MMS, gaming, location-based services
Range	Short-LAN 70–100 m Rapid linear decline over range	Wide-WAN 1–3 km Slow gradual decline over range
Throughput	Max. 54 Mbps	Max. 2 Mbps
Bandwidth	20 MHz	5 MHz
Mobility	Low mobility, nomadic use	High mobility, full roaming
QOS	Private network QOS	Public network QOS
Power Consumption	Uncontrolled power	More power control
Network Integration	Viral network $(n + 1)$ growth	Systematic and regulated approach
Major Backers	PC Vendors	Communications OEMS

Source: Interdigital and US Banoorp Piper Jaffray

The question is "should these technologies compete or converge?".

The answer is very clear: they will converge, not compete. Or at least they will play a complementary role for wireless data solution (see also Table 10.3).

(1) WLAN/WWAN (2G/3G) convergence

There is the strongest need in the market for this combination because each systems has different merits as shown above:

- high-speed data rate for hot spot or corporate applications
- increase new revenue stream
- enlarge the coverage area by WWAN
- reduce the huge investment of WWAN, especially for indoor coverage
- global roaming capability
- new services, especially for PC users to provide ubiquitous access.

Some wireless carriers are interested in the concept of convergence of WLAN and WWAN. For example, KDDI and CISCO announced in May 2002 that they jointly started the WLAN/3G (CDMA2000) roaming trial.

Also, NTT DoCoMo announced the concept of the FOMA/WLAN combined service using a dual-mode wireless PC card. Dr. Tachikawa, CEO of DoCoMo, explains that "a PC user can automatically enjoy high speed wireless data services unconsciously with a dual-mode PC card." WLAN/WWAN convergence is a very promising stream from both business and technical points of view.

(2) WPAN /WLAN convergence

This scenario looks less attractive than the WLAN/WWAN (2G/3G) convergence above since there are some overlaps in the target market. Another concern is the interference issue since IEEE 802.11b standard (WiFi) and Bluetooth technologies are both in operation in the 2.4GHz ISM band. (Also, a microwave oven uses the same band.)

IEEE 802.15.2: Coexistence SG is working hard to assess the issue.

We can conclude that the trends towards the convergence of multi-standard wireless systems such as WLAN/WWAN (2G/3G) convergence will match the requirement of the market, which is diversity of the service and usage scene.

To achieve this concept, we need to find a solution in many technical areas like billing architecture, security enhancement, roaming capability between multiple networks, radio resource management and protocols. Sunil Jain, a leading wireless industry analyst, aptly summarized the differences in an interview: "I think wireless LAN is mostly about 'portability,' wireless WAN about 'mobility,' and wireless PAN is primarily about 'convenience.' I consider them more complementary than competitive because each of these technologies represents a unique tradeoff in terms of range, data rates, power consumption and cost. Wireless LAN and WAN might compete against each other in some situations, however, fundamentally WLAN is about 'IP going wireless' while WWAN is about 'wireless going IP.'"

Inside offices and buildings, the utility of wireless LAN is clear and analogous to having cordless phones at home. They extend the reach of office LANs without requiring extensive wiring. Outside in the public space, wireless LAN is analogous to using pay phones. Pay phones were very successful but their popularity disappeared as wireless phones became more pervasive, reliable, and affordable. Initially wireless LAN would be the preferred choice for wireless data service that required high bandwidth and QoS at affordable prices. As wide-area data networks start to close this gap, the traffic would migrate to these networks. However, the two networks (WLAN and WWAN) will continue to coexist for a long period of time.

10.5   Transition from 2.5G to 3G

Some wireless operators have already started 3G commercial services and some operators are providing 2.5G services and are planning/preparing to offer 3G services in near future. Typical 2.5G systems are GPRS and PDC-Packet with wireless packet communications, and 3G systems are WCDMA and CDMA2000-1X. At this point, there is no other country but Japan and South Korea where large-scale commercial 3G services are being launched, so we will look at how the transition has been made by Japanese and Korean operators.

Initial market hype disguised the technical, financial, and logistical challenges of implementing 2.5G and 3G network infrastructure. However, wireless operators are embarking on the migration of their 2G network infrastructure, as these challenges are becoming better understood; and companies are learning to weigh such factors as their incumbent technologies, implementation costs, market conditions, regulatory constraints, and the generally preferred migration strategies. The five major 2G technologies adopted globally will converge to align with the GSM or cdmaONE migration paths. The deployment requirements for each migration path depend on the inherent capabilities of 2.5G and 3G technologies, and their backward-compatibility with incumbent 2G networks. In some cases, the network migration merely involves software and hardware upgrades to existing infrastructure, such as the migration from GSM to GPRS and EDGE, or cdmaONE to CDMA2000. However, other upgrades, such as from TDMA to GSM or GPRS/EDGE to WCDMA, require separate base station cabinets, which occupy additional cell-site ground space and, in many cases, separate antennas.

10.5.1 Type 1: Overlay approach; PDC to WCDMA (NTT DoCoMo and J-Phone)

As there is no technical compatibility between the two systems, DoCoMo's network strategy is to have overlay nationwide networks of PDC and WCDMA. DoCoMo had 478 000 FOMA subscribers by May 2003 and this number is relatively modest. The reasons for these figures are the results of combination of

- expensive terminal price ($300–500),
- relatively shorter battery time than that of 2G terminals (PDC/PHS),
- network coverage.

These are typical issues when a brand new wireless system is introduced. "It is easy to solve these factors as the time goes by", a DoCoMo spokesman said. In order to provide number portability between 2G and 3G networks, DoCoMo started a "dual network service" in June 2002. The user of the service can access from his/her PDC or FOMA phone to the PDC/PDC-P/FOMA network using one telephone number. Once a FOMA user has subscribed to this service, the SCP (Service Control Point) of the network has set service profiles of both PDC and FOMA. When the dual network service user selects the "activated" network by his/her terminal, the service profile of the "activated" service becomes available. Not only voice service, but email and voice-mail are available with a single access number. This is a network-based solution for the transition period when the 3G service coverage is not wide enough compared with existing 2G coverage.

Many of the 2.5G and 3G configurations require separate radio spectrum allocations for each of the overlaid technologies. Normally, when technologies use the same channel structure and modulation scheme, such as cdmaONE and CDMA2000 1xRTT, or GSM and GPRS, radio spectrum segmentation is not necessarily required.

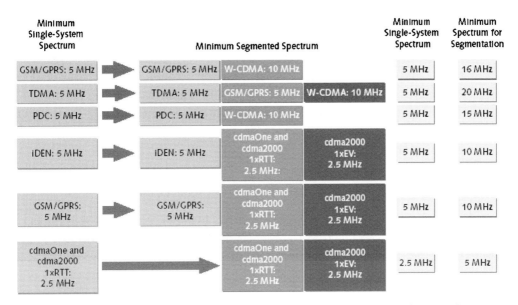

Figure 10.4. Spectrum requirements for technology migration (source: Yankee Group).

However, because TDMA and GSM, or iDEN and CDMA2000, do not use equivalent modulation schemes or channel structures, overlaid configurations of these technology combinations require radio spectrum segmentation. Segmenting radio spectrum effectively lowers the overall efficiency of a system that has an overlaid configuration, particularly when the overlaid technologies have low subscriber penetration. The minimum radio spectrum requirements for the planned 2.5G and 3G network deployments are approximated in Figure 10.4. The actual radio spectrum requirements for these deployments vary greatly, depending on factors such as subscriber distribution, geographical characteristics, and network layout. Some overlaid configurations are more spectrally efficient than others, especially when the systems have low subscriber penetration. In particular, the migration from cdmaONE to CDMA2000 offers the greatest spectral efficiency, requiring a minimum of approximately 5 MHz for the implementation of CDMA2000 1xRTT and 1xEV. Conversely, the TDMA migration to GSM/GPRS and, ultimately, to WCDMA requires a minimum of approximately 20 MHz.

The radio spectrum allocations in North America do not align with those in Europe; and in other regions, such as Asia–Pacific and Latin America, a combination of the western European and North American allocations have been adopted. This effectively limits the benefits that infrastructure and device vendors can lever from economies of scale associated with technology standardization. For example, modifications to base-station infrastructure and handsets are required to implement GSM overlays to TDMA systems in the Americas. Another approach is a terminal-based solution; that is, to have a dual mode terminal for 2G and 3G.

As DoCoMo is reluctant to introduce a dual mode phone for 2G/3G, most GSM operators in the world are planning to introduce a dual mode terminal for GSM/WCDMA. The aim is similar to DoCoMo's dual network service.

Dual mode phones are already a very popular solution for the interworking of different 2G networks. For example, AT & T Wireless has a GAIT phone (TDMA/GSM dual mode) and NTT DoCoMo has a Docchimo phone (PDC/PHS dual mode). Thanks to recent developments in device technology, these dual mode phones are not bulky or heavy.

10.5.2   Type 2: Upgrade approach; cdmaONE to CDMA2000-1X (KDDI)

KDDI has been using the different radio access technologies of cdmaONE and PDC for 2/2.5G. As CDMA2000-1X is an upgraded version of cdmaONE, network coverage of 1X is nearly the same as for cdmaONE and the terminal price is nearly $150. Because of this, the 1X service in Japan made a good start, with 1.15 million subscribers as of June 2002. However, it should be noted that some existing 2G users of KDDI (both cdmaONE and PDC) have just switched to 1X, so the number of KDDI's 2G users has been reduced by cannibalization of the 1X service.

KDDI announced that the company will start the datacentric EV-DO in late 2003 to compete with WCDMA's high-speed data service. GSM operators will be categorized in this way since the GSM and WCDMA infrastructures have many compatibilities with KDDI. Some operators (e.g. AT & T Wireless and Cingular) have announced their intention to migrate firstly from GSM to EDGE and secondly to WCDMA. In this case, terminal complexity must be considered since operators may need to prepare triple-mode and multi-band terminals. This approach may increase the terminal and operational costs.

10.6   Battery life

Estimated at $9.4 billion in 2002, the battery industry in the USA, according to Battery Council International, will have grown by 22% to $11.5 billion by the end of 2003. However, such a forecast may remain unfulfilled as there are simply no revolutionary technologies to improve significantly the performance of batteries in the near future.

Unlike the devices they power, batteries have lagged behind the times. While processing power – measured in MIPS – has increased 3000% over the past ten years, battery technology has increased by only 80%.

The basic concept of battery technology has not altered since the 1830s, and that is a pity because, along with the commercial necessity to keep RF power outputs adequate to sustain always-on high-speed data, the combination of wireless and non-wireless feature sets within next-generation terminals places absurd demands on today's under-equipped battery technologies. Many of the 3G terminal designs promise power-sapping feature sets, such as high-resolution color screens, radios, huge memory reservoirs to run Java applications, MP3 players, WLAN capabilities, dual-SIMs, GPS services, barcode scanners, Bluetooth interoperability, full-motion cameras, storage modules, and voice recognition.

The use of many of these features in a smartphone affects the longevity of the battery power, and means dispensing with the traditional notion of talk and standby times of current devices, as these relate primarily to the time the devices will spend being connected to a network. The use of legacy battery technologies, like the ubiquitous lithium-polymer in smartphones, creates a very simple conundrum: if consumers sap power from their smartphones by using popular non-wireless feature sets like MP3 playback, then they are unlikely to be constantly connected to a network. No connection means no revenue for operators. That is not exactly a sure-fire route to recoup billions spent on 2.5 and 3G systems. No wonder many telecoms analysts are keeping a sharp eye on developments in battery technology.

Efforts to improve power capabilities by increasing the size of the device battery are doomed to instant market ridicule by consumers now long accustomed to tiny form factors. But compromise units that force an unacceptable trade-off between size, cost, speed, feature sets and usage times could create a consumer backlash.

However, several strong methodologies have emerged to bridge the divide between power, size and the price of terminals. The first strategy is to improve the energy efficiency of displays and chipsets by integrating disparate chip components into single low-consumption chips that sip rather than suck power from batteries.

The second strategy is a move away from batteries that store a fixed amount of energy, to lightweight new-generation fuel cells that can produce power as long as they are supplied with a replenishable fuel, much like butane lighter refills. Current prototypes revolve around so-called micro-fuel cell technology, an amalgam of new forms of carbon, methanol, new acids, or nano-technologies that promise to produce batteries with standby times of up to a month for current-generation mobile phones without recharging, or work a laptop computer for a full day. Some studies even concentrate on self-powering chips that do not need external power sources.

Whatever may be the technology or strategy involved, this research area will be closely watched for many years to come. We will continue this discussion in Chapter 13 where we discuss the future of battery technology.

10.7   Security

As we discussed in the previous chapter, the amount of security needed is directly proportional to the value of information that needs to be protected. For example, under normal circumstances, the value of a user-id and password list is high and should be protected at all costs, while items such as news releases or executive bios are not that critical as damage risks in the event of security breach are small. As such, it is critical to conduct end-to-end vulnerability assessments, application and service assessments, benchmarking and gap analysis, and finally to have frequent policy reviews.

10.7.1   Vulnerability assessments

Vulnerability assessments are intensive examinations of targeted areas of your technology infrastructure to determine the current state. Having security policies is an important first step, but actually implementing them and keeping system and network configurations current with vendor fixes and state of the art advances is where policies become meaningful. Strengths and weaknesses in the targeted elements are identified, and prioritized remediation recommendations are provided.

10.7.2   Enterprise application assessments

Enterprise assessments are broad-based assessments of data security or privacy programs. They examine policy, security or privacy program management, technology infrastructure, and operational effectiveness of enterprise security or privacy programs. Strengths and weaknesses are identified and recommendations for program improvements are described. Assessments can be tailored to address specific needs and concerns. Over time, assessments can track changes in the effectiveness of existing security programs. Enterprise assessments help spending to be prioritized on the most needed security and privacy efforts, greatly improving the program for a very small cost.

10.7.3   Benchmarking and gap analysis

Benchmark and gap analysis helps in reviewing a security or privacy program or elements of it. Results are compared against standards relevant to the business.

10.7.4   Policy reviews

Security and privacy policy reviews examine the current policy infrastructure and compare it against your business needs and policy practices in the world at large. Because some policies were written before personal computers, networks, and e-commerce, they often do not address the business risks of today's information technology infrastructures. Some policies are overly strict or technically unenforceable and can result in wasted human or technological resources as you aim for unrealistic objectives.

    With the above as a backdrop, we review some of the security issues for various wireless WAN, LAN, and PAN technologies in Tables 10.4–10.7.

Table 10.4. WAN – wireless technologies – CDMA, TDMA, GSM, CDPD, GPRS

Risk/incident	Control
Interception of data transactions	128-bit encryption combined with digital certificates
WAP gap – data in clear at the WAP gateway for few ms	Upgrade to WAP 1.2 Control data flows for wireless access by segmentation Introduce an intermediary step for information exchange so that user is not directly getting to corporate databases WPKI
Unauthorized access to mail and information servers	Limit access by IP address, UserID/password
Virus	Virus detection at firewalls, servers Anti-virus policy Application filtering Personal use restrictions (downloading)
Identity theft	Policies in place Fraud detection in place
Lost/stolen mobile device – in majority of the cases devices are not locked and the data on the device (PDAs) are not encrypted	Policies in place Depending on the sensitive nature of application, encryption of sensitive information is recommended Data back-up Device access controls and secure configuration
Man-in-the-middle attack Gateways are not checking that the domain name part of the SSL certificate presented by an HTTPS server to the WAP gateway matches up with the domain name part of the URL requested. Because of this the possibility of an SSL man-in-the-middle attack is raised when DNS spoofing is possible	
Session hijacking	First of all, you must take care not to sequentially assign any ID numbers, which form the basis of numeric session IDs (a better idea is to use MD5 checksums or similar) since this is both predictable and easily manipulated. An

(*cont.*)

Table 10.4. (*cont.*)

Risk/Incident	Control
	example of this would be if an attacking party were issued with a session ID in a URL and examining the ID, modified it to be a lower number. This simple step could potentially result in the attacking party successfully impersonating the user to whom the lower session ID number was issued
	If you are going to use numeric session ID numbers then rather than using an autoincrementing integer, consider using a good random number generator to create the ID numbers. Alternatively, use alphanumeric session IDs from something like MD5
Weak passwords/PINs Limited input facilities mean that there is a great temptation to use short numeric passwords where access-control is desired. While this may be appropriate for some systems you should be aware that at one request per second with no parallelism the entirety of a four-digit PIN key space can be searched in less than three hours. A more realistic attack would run at two requests per second with a parallelism of ten giving a 50% key space search in a few minutes	Rigorous password management policies in place Monitor logs and exception reports daily
WAP application development. WMLScript variables are global, so it is not just your application that can read the value of any variables that have been set, $(password) for example. Exposure is limited by the small cache in current WAP devices, but this will most likely change in the future as devices are equipped with more RAM	Use unpredictable dynamic variable names or explicitly clearing variable contents immediately after use
Denial of service attacks	Monitoring

Table 10.5. LAN – IEEE 802.11

Risk/incident	Control
Attack from within the network's user community	Use 128 bit WEP products such as ones from BreezeCom, Ornico (Lucent)
Unauthorized users gaining access	For a corporate intranet or internal network to be properly configured to handle wireless traffic, access to and from wireless access points, as well as to and from the Internet, should be controlled by firewalls. Intrusion detection and response sensors should also be in place to monitor traffic on each wireless segment
	VPN could be used to access the internal network when connected over 802.11
Eavesdropping (capture LAN traffic)	The best solution is probably to require the use of IPSec for all hosts on the wireless network. While this will incur a performance penalty, it will solve problems of impersonating users, monitoring user data, and so on. Various IPSec implementations support the use of certificates and other forms of strong authentication. Windows 2000 sports a combination of (integrated) Kerberos, IPSec and Microsoft authentication methods along with policy support. With almost universal support for IPSec, and the generally low speeds of 802.11 (maximum 11 megabits, probably shared with others), this plan should not be too difficult to implement or sell to management
	Isolate wireless and data networks
	Turn on WEP (Wired Equivalent Privacy)
	Every host should have its own encryption keys, and keys should be changed with high frequency
Other exploits that might endanger wireless LAN environments include jamming, which overwhelms the frequencies with illegitimate traffic; client-to-client attacks that bypass the access point; and encryption attacks that exploit well known weaknesses in the Wired Equivalent Privacy (WEP) encryption system	Frequency hopping
Symmetric encryption	Dynamic WEP keys
If keys of any user are compromised, all devices will need to be re-keyed which can become a unmanageable task	Extensible Authentication Protocol (EAP)
	128 bit key length

Table 10.6. PAN – Bluetooth

Risk/incident	Control
Bluetooth has three security modes: • non-secure – the device does not initiate any kind of security procedure • service level security – the service level security mode allows more flexibility in application access policies; this mode is especially useful when running several applications in parallel with differing security requirements • link level security – the device sets up security procedures before the link setup is completed	Use service or link level security depending on your situation
Default PINs (E22 algorithm) Often times the default pins (0000) are not changed	Have a policy to not allow default PINs

Table 10.7. Network elements (routers, bridges, firewalls)

Risk	Control
Network switching equipment, such as FDDI hubs and switches, often are not password protected or have default vendor passwords	All switches, hubs, bridges, and gateways should be password protected to prevent unauthorized changes to the configuration files and security. Never leave the default passwords in place
Vendors often have remote access to routers so that they can reprogram them and upgrade the software. Penetration of the vendor's security or discovery of vendor access capability by unauthorized persons could result in penetration on organization's network	Install authentication procedures on the remote access capability and dump the router tables to identify all addresses and investigate any unknown or unusual addresses Allow remote capability on an as-needed basis only
Router tables may contain unauthorized connections	Review tables periodically

10.8   Transcoding and usability – phones/apps/services

There are several aspects of application delivery to a device that differ, sometimes significantly, based on the manufacturer, model, revision, etc. The diversity found by the network configuration also adds to the complexity. Consider the various aspects where things could be different for an application or a service.

(1) Wireless Network Protocol (WAP, HDML, i-mode, etc.).
(2) Infrastructure servers (gateways, cache, compression, etc.).
(3) User interface (soft keys, touch screen, roller, voice, stylus, etc.).
(4) Content (WML, HDML, cHTML, XHTML, etc.).
(5) Multimedia interfaces.
(6) Browsers (HDML, WAP, XHTML, cHTML, SALT, VoiceXML+XHTML).
(7) Application environments (JAVA, BREW, SIM Toolkit).
(8) Air interface (CDMA, CDPD, TDMA, GSM, WCDMA, etc.).
(9) Operating systems (EPOC, CE, Palm, Linux).
(10) Physical device types (phone, pager, smartphone, PDA, watch, autoPC, etc.).

Add network performance conditions and some other factors such as device models and manufacturers, and the picture gets extremely complicated. One of the biggest challenges that developers have is to separate business logic and presentation layers of the application in such a way that the presentation can be customized independent of the end-device service and capability. As we discussed in Chapter 8, this desire to make an application or service device independent led to the term "transcoding", which applies to transforming the content to appropriate format. Although we have made great strides with transcoding engines in toolkits such as WebSphere Transcoding from IBM and Net Compact Framework from Microsoft, there still a lot of room for improvement. These toolkits focus on converting a piece of content into different markup languages and formats for different form factors, different device capabilities, and different user preferences. Although we have improved in doing this content transformation, what is missing is "adaptiveness" of content transformation. Instead of hard-coded rules on the server, the transformation of content needs to be based on real-time circumstances and events. The W3C and WAP forum, through its work on RDF, CC/PP and UAProf, has pushed for a similar concept, but it is still a work in progress. With the variables of configurations changing so rapidly and with the content becoming more complex by the day with the integration of audio and video, an "adaptive" approach is a must or else mass "one-on-one" customization and personalization of content would be a very difficult task and, as we established earlier, personalization of content is the corner stone of any application or service.

Application developers should both "proactively" design "adaptive" applications and "reactively" look for holes in the application and service by continuously testing and monitoring usage logs with tools such as the one offered by Argogroup and others (Figure 10.5).

10.9   Global roaming

Global roaming is a very attractive service, not only for business executives but also for tourists. The acceleration of globalization increases the demand for such a service.

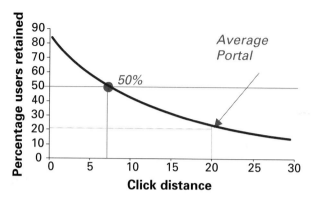

Figure 10.5. User retention vs click distance (source: Argo group).

For example, about 16 million Japanese go abroad each year and this number is nearly 15% of total population in Japan.

GSM operators are fully enjoying the benefits of the GSM standard. The GSM Association (GSMA) is specifying the network operation guidelines for international roaming and 377 operators have joined GSMA. For example, BT Cellnet in the UK is offering roaming service with 104 countries and Hutchison in Hong Kong with 103 counties. People expect the global roaming service for voice, email and also content access using GSM phones as well as for 3G phones. World travelers want to talk with their family and friends about their impressions of the visited countries and business executives need to check emails anywhere and at any time. Wireless Internet users will try to check the websites of the visited countries for local information using wireless and Internet-enabled devices. It is highly desirable to send a live video-mail from the World Cup final game in Yokohama Stadium to a Brazilian friend in Rio de Janeiro or to a German friend in Munich.

The global roaming service for mCommerce applications and services using a cellular phone makes shopping easier and faster anywhere in the world. The global location service informs parents in New York where their daughter is walking around in the park of London. A travel agency can provide a total package of an international tour through a global roaming terminal to a customer. A customer can fully arrange the tour by accessing the portal site of the agent anywhere on the globe. So, a strong demand to start global roaming using 3G systems indeed exists.

There are three major functions to be considered for 3G global roaming. They are: network function, terminal function, and UIM function.

3GPP has established the detail specifications for the interfaces between the elements, but legal, operational and billing rules among 3G operators should be discussed and decided for the global roaming service. Packet roaming has a lot of technical challenges. The challenges are optimum routing sequences of multiple connection types, contents design, matching between roaming terminals and servers, billing mechanism, human interface, quality management and tariff setting.

However, when all is negotiated, configured, set and ready to go, we can fully enjoy the 3G multimedia services anywhere on the globe and wireless truly becomes the best communication tool for the global village in the global era.

10.10 Billing and payment issues

As we have mentioned in earlier chapters, payment and billing issues are some of the most important ones that most of the wireless players need to contend with, yet they are among the most underestimated and misunderstood areas of the wireless value chain. As the success of future-generation applications and services, and in turn the carriers and the industry, becomes more dependent on content and successful implementation of business models around content-rich applications, billing and payment issues are important considerations for any application developer.

One of the biggest issues that wireless carriers will face once their networks are ready to support high-value content revolves around pricing and settlement. Consider that the target audience is probably already locked into a contract with a set price point. The challenge comes in migrating the customer to a more comprehensive service package, most likely at a higher price. For virtually every customer, the wireless carrier will encounter the following customer-facing issues:

- up selling existing data customers to higher-value, usage-based billed content services;
- offering attractive service bundles that demonstrate value without risking sticker shock; and
- providing easily understood billing, where the value of the content service makes sense to the end customer.

Wireless carriers face even greater issue complexity when they look at network and content delivery issues. Considering that the majority of wireless carriers continue to own the customer by offering a reliable end point for content delivery, maintaining that reliable end point requires:

- upgrading of existing billing and customer care infrastructure to support the transaction-based usage of data services and quality of service;
- a robust partnership strategy with both content aggregators and content providers (ASPs) to ensure smooth delivery of service, and settlement of revenues based on customer usage; and
- investment by content providers and aggregators in a billing system that supports, at minimum, transaction processing that can accommodate flat-rate subscription billing and credit card processing. Expectations should be to ensure support for more complex billing as services become increasingly complex and as companies move closer to the customer along the value chain.

As carriers ponder their next steps related to wireless data networks, all agree that the services, partnerships, advertisements, and – ultimately – content will be critical elements in determining the success or failure of this multibillion-dollar investment. To a large extent, the capital outlay is something of a leap of faith driven not only by market hype, but also by market hope as it relates to the uptake of services delivered over this very expensive infrastructure.

For billing and customer care vendors addressing the needs of these carriers, the investment in 3G (and beyond) functional requirements will be closely tied to the evolution of the mobile content itself. Regardless of the technology or bandwidth used to deliver mobile data content, the complexity of settlement and partner relationship management begins today. Few carriers have the resources to create and cultivate compelling content and services, and they will increasingly look to their partners for compelling services to be offered over the mobile device.

Also, integration remains one of the toughest and most daunting challenges, and vendors offer a million and one new answers for solving ongoing integration problems. In the past few years, carriers have almost always attacked integration from a technology perspective, with the goal of making systems talk to each other. The industry has invested millions in enterprise application integration (EAI) technology – only to have vendors turn around today and point out all of its flaws, after spending years selling its long-term benefits. At the heart of this problem is a lack of commonality in data models. In simple terms, different systems think about different elements of a carrier's business in different ways. Naming conventions and relationships are handled differently, which is what makes integrating two or more disparate systems a nightmare.

The extreme cost of integration is crippling service providers. More importantly, this cost can cripple any vendor's return on investment, because application-focused numbers often go out the window once the actual environment and integration costs come into play. To help close the gap in data modeling, TMF has launched its Shared Information/Data (SID) model initiative. This initiative focuses on documenting the relationships and roles that business entities and processes play, driving toward a model and methodology dictated not by technology, but by business. The core idea behind the model is to come up with nomenclature and models for relationships that can deliver sufficient commonality but remain generic enough to allow differentiation. Every process or interaction involves specific information exchanged among organizations that play specific roles. These roles and common interactions need common definitions.

From a very practical point of view, a shared-data model could be useful in implementation and integration. OSS applications can do their jobs only when loaded with the right data, rules, policies and processes that dictate how they interact with or manage the carrier's business. Carriers spend millions for systems integrators to bring in armies of inexperienced consultants who do little more than map data from one system to another. They bill thousands of hours creating the data element and process documentation necessary to the people who actually implement and integrate systems. Often, this arduous process is poorly managed and can not keep up with the speed of the business. By the time the documentation is done, requirements change. Turning to common data and process models could help to shorten this process immensely, because the process and data ingredients for the most part would be modeled, leaving implementers to focus just on the pieces that are unique to the carrier in question.

Additional complexities arise as more bandwidth-intensive content becomes available, creating more network-oriented issues such as quality of service and network utilization while delivering those bandwidth-hungry services. Ultimately, the evolution to data will demand a level of cohesiveness between the network and the business (one that was never required for voice services), and billing and customer care systems will be instrumental in orchestrating this cooperation.

The mobile device represents the first content delivery experiment in which telecommunications companies are instrumental in bringing services to the end-user. Lessons learned by billing and customer care vendors helping these carriers with their business infrastructure needs can and will be ported to other verticals – such as cable and DBS – where we feel the next content delivery chain is waiting to be unleashed.

For settlement solutions to meet the demands of this content value chain, both operational and fiscal elements must be considered in the solution design. Just as on the retail side, carriers may have "rate plans" for consumers and business customers, so they will eventually manage "contracts tables" on the wholesale or settlement side of the business. A prime example would be the delivery of QoS-sensitive information where contracts may dictate certain timeliness or latency/degradation provisions. Although many dispute the applicability of these QoS-based agreements to the end-user (more so on the consumer side), the applicability to settlement is absolute and is a potential stream of cost savings if content delivery is properly managed to the terms of the contract.

From the business management aspect, settlement solutions and functionality have the opportunity to evolve into a wholesale business operations platform. Settlement will play a role not only in managing the financial aspects of partner settlement, but also in providing financial insight into the true cost of delivering services and content. And like least-cost routing within interconnect platforms, settlement platforms will also help to identify more efficient ways of delivering content.

There are many challenges that mobile payment systems face (Figure 10.6). They need to address the following issues.
- Cost
  - to the user for using the service
  - to the content provider for integration

Table 10.8. Mobile payment solutions around the world

Mobile payment solution	Geography	Description/type
Banko.max	Austria	Virtual POS
Bibit	Belgium, Holland, Germany, Scandinavia	Virtual POS
Cingular DirectBill	USA	Virtual POS, Limited to micropayments under $10
EMPS	Finland	Virtual POS with plans to use Bluetooth to offer real POS
GiSMO	Sweden, UK, Germany	Virtual POS
iCash	Sweden	Real and virtual POS, limited to micropayments
i-mode	Japan	Virtual and real POS, by far the most successful model with over 35 million users
Metax	Denmark	Real POS at gas stations
Mint	Sweden	Real and virtual POS
MobilPay	Germany, Austria	Virtual POS
Orange Mobile Payment	Denmark	Virtual POS
PayBox	Germany, Austria, Sweden, UK	Real and virtual POS, P2P
Paypal	USA (fixed internet solution in many more countries)	Virtual POS, P2P
Sonera Mobile Pay	Finland, Sweden	Real POS

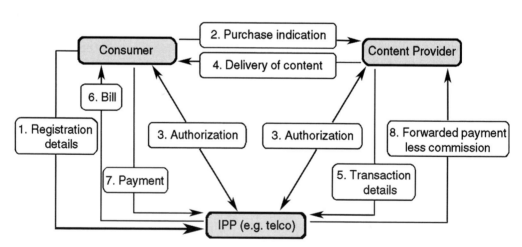

Figure 10.6. mCommerce payment model (source : CGEY).

- Security
    - confidentiality
    - authentication
    - integrity
    - non-repudiation
    - authorization
- Accessibility
    - convenience
    - speed
    - ease of use
- Lack of standardization across geographies (see Table 10.8).

Cost and accessibility are important but security is the key issue that service and application providers need to grapple with as consumers who do not have confidence in the payment method just will not use it.

Various initiatives and consortiums are working to address the above-mentioned challenges and issues. MeT (Mobile Electronic Transactions), a consortium led by Nokia, Motorola, and Ericsson, is seeking to establish a framework for secure mobile commerce transactions.

In the past two chapters we have looked at various business and technology issues that the wireless industry is facing. It is critical that players in the value chain come together to solve these problems. As new innovations are introduced, as technology breaks new grounds, we will continue to see the morphing of the value chain with new players coming in and older players adjusting to the realities of the changing world. Those who are fast in adapting to change will succeed in monetizing technology. In the next chapter we take a look at a few case studies of companies who have embraced wireless solutions and used them to enhance productivity or improve revenues or gain mindshare of their customers.

11 Case studies

In the past two chapters, we focused our attention on issues and challenges facing our industry. In this chapter, we will discuss four unique case studies that show how wireless technology can be used to enhance productivity, improve revenues, or gain mindshare.

11.1 Southern Wines and Spirits

11.1.1 Problem description

Like a lot of industries, the wines and spirits industry is in the midst of consolidation. Distributors are searching for ways to cut costs while maintaining customer relationships crucial to increasing market share and revenue. Southern Wines and Spirits of California Inc. (SWS), the single largest distributor of wine, spirits, beer and other non-alcoholic beverages in the USA, saw an opportunity to save by automating its sales processes. Five thousand reps in the field were generating half a million sheets of paperwork per week. They placed orders in a "Brick" that sent data back to the home office through a 300 bps (remember, that is *bits* per second) acoustic coupler. The error rate was high, and delays in orders were an all too common occurrence. This information was fed into the enterprise system, a group of about five IBM AS/400® servers and an IBM S/390® enterprise server.

11.1.2 How wireless was used to solve the problem

"We knew there had to be a better way," said Steve Borrows, VP of Spirits at SWS, which represents more than 300 suppliers and nearly 6000 individual brands. "Once information came into our systems we had applications that worked. But we needed to link these assets more closely with our sales force and customers in order to keep costs down, share the benefits with our customers and gain market share. We knew that if we didn't do this, somebody else would."

Traveling from customer to customer to take orders, company reps do not normally have access to a PC with a modem link, and building that kind of network did not make sense anyway. It did not bring information into the field where it was needed. Wireless communications seemed to be the obvious solution, but that brought its own challenges. One was the intermittent nature of wireless communications. If a rep was in a basement or in a remote area, there was no connection, and without a connection, there was no server to supply the web pages used for order entry or order status checks. A wireless system would not provide any advantages over the old system if reps were dependent on a precarious wireless link to complete orders while in the field.

A second problem was related to the device and OS. SWS did not like the idea of making a commitment to one device or operating system when new technologies were becoming available almost daily, especially in the wireless world. If SWS bought into an elaborate dedicated solution that locked them into a particular device and operating system, their solution would be obsolete in months.

SWS found its answer in Pocket Strategi™, a suite of programs that enable browser-based access and native IBM® iSeries® 5250 emulation on wireless devices. Developed and sold by IBM Business Partner Advanced BusinessLink Corp., Pocket Strategi installs a replica "server" on remote devices. Using the device's browser, SWS reps are able to access this server when they are not linked to the iSeries server through a wireless connection. A small Java applet of about 500 K provides a GUI or "green screen" view of the 5250 emulation and a replicated version of data specific to the SWS rep. Pocket Strategi is complemented in the SWS solution by Strategi™, a desktop-enabled version of the solution that allows customers and sales reps to access needed information and processes.

11.1.3 Solution architecture

The sales information and enablement system is powered by five new iSeries servers running OS/400, version 5.1. Written specifically for the iSeries platform, the Strategi and Pocket Strategi suite is made up of four separate components. The first is a pure Java client applet that provides browser-based remote access. Second, Strategi's unique "Push Technology," actively delivers files directly to the appropriate persons (and, more importantly, not to inappropriate persons). Third, Strategi and Pocket Strategi are built on the world's only AS/400-specific HTTP server. The Strategi server is composed of unique template-based architecture that provides direct access to iSeries data through the HSM architecture. Fourth, File Cabinet provides secure, manageable access by end-users to any file, whether it is an iSeries file, spreadsheet, word processing document, or presentation located in the entire defined system, including information residing in the mainframe's data warehouse.

"We designed and architected Strategi and Pocket Strategi to address the business issues faced in the development, deployment, and management of an e-business initiative," said David Tassie, director of technical consulting, Advanced BusinessLink, Austrialia. "By combining all the Strategi components, we are able to provide a single environment that enables you to build a bridge to the past and deep into your systems – to your legacy applications through the Java applet – and a bridge to the future of HSM e-business applications."

Resting on the bedrock of the interoperability and application flexibility of the iSeries server, Pocket Strategi from Advanced BusinessLink is another example of how IBM and its partners are making e-business look easy.

11.1.4   ROI/lessons learned

SWS is in the process of rolling out the Pocket Strategi on iSeries solution to its entire sales force. So far about 2000 of 6000 reps are using the new tool. Although it is too early to say just what effect the Pocket Strategi is having on sales, SWS knows that sales people using the solution are able to make on average one more sales call per day, meaning that SWS is making 2000 more sales calls per day with the same number of people.

"We are very happy with the Pocket Strategi sales automation tool," said SWS's Steve Borrows. "We decided on this solution because Advanced BusinessLink and IBM applied their technology to our specific situation and methods. They didn't come in telling us that we would have to change the way we do business. Their system adapts to what we do, basically by employing our AS/400 applications without major modification. Wireless sales automation is not only giving us new capabilities; it's also adding value to the investments we've made over the years in enterprise systems. That's how an e-business deployment should work."

11.2   British Petroleum

11.2.1   Problem description

British Petroleum, a global energy provider, has thousands of natural gas wells, many in remote locations throughout North America. These wells are monitored and maintained by field technicians who communicate regularly with field offices. These technicians needed a way to collect data at well sites and capture HSE (health, safety and environmental) and operational data. Although wireless phones were available, the client needed better communication between on-site technicians and field offices, including the ability to receive and submit work orders, forms and maintenance procedures.

11.2.2   How wireless was used to solve the problem

The solution was to give technicians handheld computers containing well-production information, maintenance procedures and forms for collecting data. To manage in-

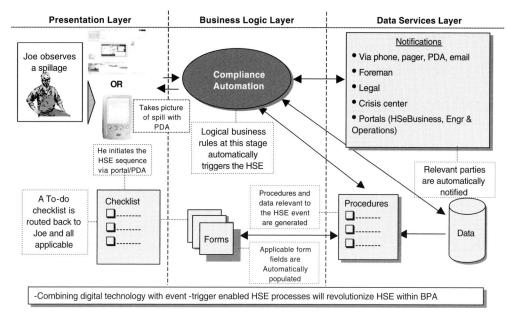

Figure 11.1. Process flow of HSE application.

formation exchange between field offices and job sites, a wirelessly accessible portal was created and deployed. Office personnel can review production data reports, issue HSE bulletins, and maintain team schedules and task lists (see Figure 11.1). Field members are able to receive this information on their mobile devices by synchronization or via a wireless connection.

11.2.3 Solution architecture

The solution was developed using Plumtree Software, a corporate portal software, AvantGo, a provider of infrastructure software and services, and Microsoft's WindowsCE technology (Figure 11.2). Technicians can now exchange and process accurate well information with field offices. This reduces the time spent traveling between well sites and offices, cuts down on paperwork through one-time data capture, and ultimately enables field personnel to make better decisions on the job. Additionally, drop-down menu features improve the reliability of the information gathered. Email and calendar tools promote better communicate with field offices.

The benefits of the architecture are as follows.

- Works disconnected.
- Thin client.
    No software distribution (other than AvantGo client).
    Easy development (ASP vs. C++/VB).
- OS independent (Palm or WinCE).
- Hardware independent (HP, Symbol, Itronix, Casio, Handspring, Palm, etc.).

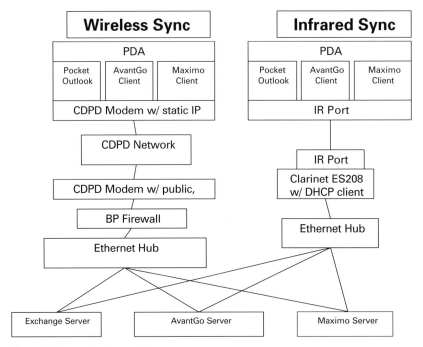

Figure 11.2. Application architecture.

The limitations of the architecture are as follows.

- Forms must be simple to be usable.
- No lookups or wizards.
- Wireless network coverage is sparse in remote areas.
- Messages are not guaranteed.
- Devices have limited CPU, battery life and screen real estate.

11.2.4 ROI/lessons learned

This solution increased field technician productivity by 7.7% (60 technicians, each saves 3.08 hours in a 40 hour working week owing to a decrease in the need to drive in to the office in the morning to pick up work orders and reports, and intra-day trips to get the latest reports of under-performing wells to visit next). This solution also increased the accuracy and timeliness of reporting.

Since technicians were able to service more wells per day, downtime decreased and well output increased, reducing the difference between actual and forecasted volumes by a corresponding 7.7%. This translated into a $1582 daily saving, based on a $2.12 unit (kcf) cost (Table 11.1).

In addition to daily savings, shifting from paper to electronic forms led to a one-time decrease in administrative expenses by $60 000.

Table 11.1. Solutions for the BP problem

Daily flow volume (kcf)	Forecast volume (kcf)	Difference (kcf)	7.7% change in difference (kcf)	January 2002 Avg gas price	Daily savings
320 373	330 065	9692	746	$2.12/kcf	$1582

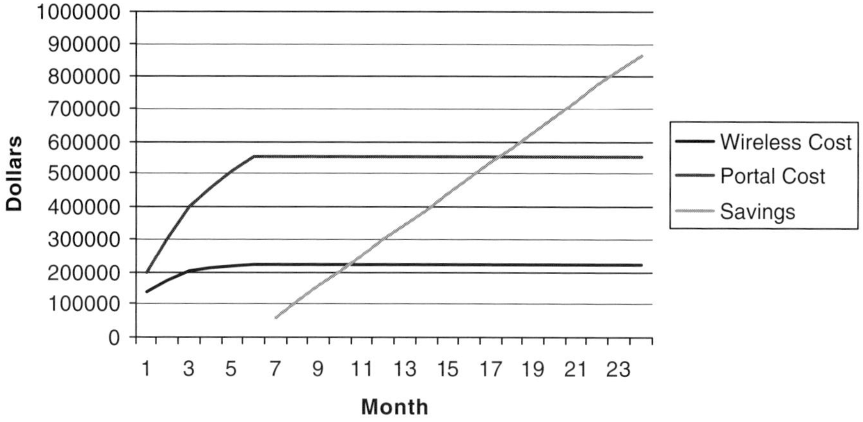

Figure 11.3. ROI for the BP project.

Over time, these daily returns not only cover the costs of the wireless investment, but also lead to profit (Figure 11.3). The client broke even on its wireless investment four months after deploying the solution. The 12 month ROI on the $224 000 investment was 160% (assuming investments to be made at the launch of the solution).

11.3 SEGA

11.3.1 Problem description

SEGA is a world leader and top brand of entertainment business. The SEGA Corporation, founded in 1951 and incorporated in 1960, is engaged in the comprehensive business of consumer products, amusement machine sales and amusement center operations called JOYPOLICE. SEGA has a vision to become *the world top content provider* and the contents business has three different areas: amusement machine, network games and consumer games.

The competition in the entertainment business has been one of the most furious areas and the recent participation of computer giant Microsoft only accelerated it. So, SEGA needed a brand new strategy to create a new business opportunity and of course to win the war.

Figure 11.4. SEGA mobile games.

11.3.2 How wireless was used to address the issue

The strategy of SEGA is to provide seamless entertainment services on different platforms in the home, on the street and on mobile phones (Figure 11.4). Different platforms will give customers different spaces in which they can enjoy entertainment games.

SEGA has made alliances with DoCoMo and Sony Computer Entertainment (SCE) to realize its vision. In mobile spaces, SEGA will provide a variety of game contents through an official i-mode site called *SEGA mode*. SONIC CAFÉ is a game content service on the i-appli service using Java. SONIC CAFÉ provides dynamic and interactive game contents such as Samba de Amigo.

In the home, users can enjoy SEGA contents with Playstation 1 and 2 and XBOX and also on PCs. In the street spaces, which include amusement spots, convenience stores and also vending machines, SEGA and DoCoMo jointly provide wireless access methods to the contents using i-mode.

SEGA mobile Friends is the name of the membership service that allows users to enjoy email coupons by i-mode. Once a member accesses the service and pre-registers on the *SEGA mobile Friends* site by i-mode phone, he or she can download the barcode membership card on the phone. Additionally, when users go to SEGA amusement spots, they can get points by scanning the barcode on the display of the phone. SEGA has already deployed the service in 350 shops in Japan.

Another service called VF.NET synchronizes games machines and i-mode phones. When a person plays the game *Virtual Fighter-4* at an amusement center, the game results are transmitted to the VF.NET site in the *SEGA mode* sites. Users can then check the game result and national ranking, and enjoy online games with the i-mode phone afterwards.

11.3.3 Solution architecture

SEGA's solution architecture is based on the principle that entertainment solutions from SEGA should be available irrespective of location, network or device

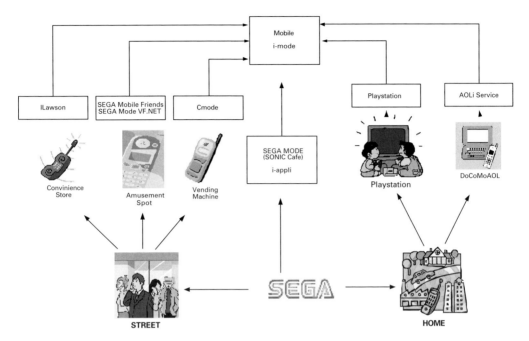

Figure 11.5. SEGA wireless universe.

(Figure 11.5). As such, they have Playstations at home that offer SEGA entertainment solutions. The same solutions can also be accessed via the AOLi Service
(DoCoMoAOL Internet Service). At various cafes (SONIC CAFÉS), instant access is
available as well. In addition, games and entertainment content can be accessed from
mobile devices, vending machines, and other devices. So, the concept is simple and
a very powerful one. SEGA wanted to be pervasive, and by using a combination of
technologies such as i-mode, Internet, etc., they have been able to create a compelling
brand that continuously attracts its followers to its content or, in other words, its
content morphs into something a customer wants rather than forcing a medium on
to them. Wireless by its very nature provides freedom of location.

11.3.4   ROI/lessons learned

SEGA is very aggressive in making collaborations with different market leaders like
DoCoMo and Sony. SEGA strongly believes that collaboration can make a great
impact on the entertainment business since the competition in the business is fierce.
Also, some national content providers try to create new markets beyond their borders.
So, the development of a global standard wireless system is a great chance for them
to make their business global. Such a strategy could not have been imaginable five
years ago when the wireless system was used for only voice services.

11.4   RIM in enterprise

Email is increasingly becoming the preferred means of communication. An independent Ipsos-Reid study of 1000 active Internet users in June 2001 found that 61% agreed that they preferred communicating via email than by other methods. This same study found that half of the active Internet users agreed that they could not live without their email. While online shopping and other activities receive the most attention, email remains the Internet's most important application. The key weakness with email for the busy professional is that keeping up with it can be an arduous task. The study quoted above found that 39% agreed that they can hardly keep up with all the email they received. This is not difficult to fathom. Professionals who find themselves consistently on the road or busy in meetings can become overwhelmed with the number of emails that have accumulated. Additionally, it is unavoidable that many of these emails are time sensitive and require immediate attention. Not being able to address these types of message in a timely manner can have severe consequences.

11.4.1   BlackBerry in the workplace – potential users

Let us look at the potential users of BlackBerry (see also Table 11.2).
- Executives/management. Many executives use email as their prime method of communication. Some may need to respond to a large number of emails in a timely manner. If an executive is in a long meeting or away from the office, decisions can still be made during moments of downtime, making themselves and those who rely on these decisions more efficient.
- Support functions. This group of people ensures that specific systems are operational. This applies both for the production system and desktop support. If there

Table 11.2. BlackBerry ROI

Source of value	Scenario # 1	Scenario # 2	Scenario # 3
Immediacy	$6630 (value per time sensitive email = $2)	$16 575 (value per time sensitive email = $5)	$33 150 (value per time sensitive email = $10)
Productivity	$5521 (salary = $50 000)	$8281 (salary = $75 000)	$15 459 (salary = $140 000)
Direct Cost Savings	$722.29	$722.29	$722.29
TOTAL	$12 873.29	$25 578.29	$49 331.29
BlackBerry TCO	$811.11	$811.11	$811.11
BlackBerry ROI	**1587%**	**3153%**	**6082%**

are any problems, they need to be notified as soon as possible, in as much detail as possible. Any significant downtime with these systems could lead to a major loss of revenue or productivity.

- Mobile sales. This group is out in the field, supporting customers most of the day. They are in their offices only rarely. While they primarily rely on their phones, some of them may use email as a main contact method, or would like to use email more often in certain situations when dealing with clients.

- Investment analysts/brokers. This group has a need for keeping track of portfolio information while away from the office. Like mobile sales people, they primarily rely on their phones. However, they may be able to communicate with each other more effectively during a busy time, and there are services in place where stock quotes or company information are available to the BlackBerry for a price.

11.5   Conclusions

Each of the case studies discussed in this chapter offers us a unique perspective on how wireless technology can be used to enhance productivity, improve revenues, or gain mindshare. Southern Wines and Spirits used wireless data technology to do things in a better way, empowering their field agents with readily available customer information. BP used wireless technology to improve productivity of their field force. SEGA wanted to be connected to their loyal customers at all times so they have been using wireless data technology to make their technology available to customers wherever they are. As discussed in the RIM case study, by introducing a RIM-based email solution in their enterprise, corporate customers can gain immense benefits. By being always connected, users can respond to situations that demand attention – there and then – rather than waiting to get back to the office or hotel. The time saved and productivity gained have great value to the enterprise. In the next chapter we will present our discussion with some industry thought leaders to get their perspective on the wireless industry and where they think the industry is heading.

12 Perspectives

12.1 Introduction

So far we have talked about various aspects of wireless industry – its peculiarities, the challenges and issues with business and technologies, case studies and a glimpse at where the industry is heading. For this chapter, we sat down with several industry leaders representing different segments of the value chain, operating in various parts of the world. Our interviewees have been around the industry for a long time and have witnessed the trials and tribulations of the wireless sector over the years. This chapter presents our one-on-one discussions with them on various aspects of the industry, their own organizations, and where they see the wireless industry in the future. We again thank our interviewees and their organizations for sharing their view and for their time for our project. The executives are:

- Mark Anderson, President Technology Alliance and Strategic News Service, LLC,
- Frank Yester, Vice President – Communications Lab, Motorola,
- Jon Prial, Vice President, Pervasive Computing, IBM,
- Steve Wood, Chairman and CEO, Wireless Services Corporation, and
- Mark Tapling, President and CEO, Everypath.

12.2 One-on-one with Mark Anderson

Mark Anderson is president of Technology Alliance Partners (TAP), and of the Strategic News Service(tm) LLC. TAP was founded in 1989, and provides trends and marketing alliance assistance to firms leading the convergence of telecom and computing. Mark is a Seybold Fellow. He is the founder of two software companies and of the Washington Software Alliance Investors' Forum, Washington's premier software investment conference, and has participated in the launch of many software startups. A past director of the WSA, Mark chairs the WSA Presidents' Group. He regularly appears on the Wall Street Review/KSDO and National Public Radio/KPLU programs. Mark is a member of the Merrill Lynch Technology Advisory Board, and is an advisor and/or investor in Authora, Ontain Corporation, Ignition Partners, Mohr Davidow Ventures, and others. He is also a principal in the investment advisory firm Resonance Capital Management LLC, which manages the accounts of institutions and high-net-worth investors, focused on technology markets.

You have been credited to be amongst the select few who have forecasted trends accurately in the computing and communications industry. What's your assessment of the wireless industry today across Europe, Japan, and the USA?

As you are aware, a great harm was done to the legacy players who were involved in 3G auctions, this doesn't directly affect Japan as it does Europe and the USA though the USA hasn't had 3G auctions yet. With those caveats in place, it is increasingly clear that taking off the $180 billion in cash from the legacy operators who bid has caused an extreme amount of damage to the balance sheets and their ability to bounce back. I don't think anybody really appreciates as to how much damage. It looks to me that the companies which went through the auction process fully aware of the damage and risks have done best to come out of it. Two of the obvious ones are Vodafone and AT&T Wireless. It is a pretty grim situation for most of the players. This sets up the stage for Japan and we should add South Korea to that. If you look at Japan, Government has continuing interest in building broadband infrastructure in general, very supportive environment. The result is going to be that significant advances will be made in handset and infrastructure over there and this will lead to what they really want, which is strategic global advantage. My guess is that though applications is the important question that people are facing, i-mode is not the end game for the Japanese. They want to get involved in the infrastructure itself, especially with the contractual agreements or incentives with minority investments and partners for using WCDMA that NTT finds acceptable. South Korea is a story midway between Europe and Japan. High penetration, funding is good and positive. Look at Samsung. It was the largest supplier of handsets in the USA. Last year nobody would have guessed. They are succeeding, though their market share remains small. There are very large opportunities for handset manufacturers in Japan and Korea in terms of unseating the current leaders. Over the next five years a significant rise in their fortunes could be predicted. In infrastructure, Ericsson is hurting because of lack of infrastructure (80% of revenues come from infrastructure). It is not clear to Ericsson and it is not clear to me when it is going to rebound. But, it is going to happen with competition coming into the metropolitan areas and the carriers who haven't thought about spending will hurt.

Where would the competition come from?

Exhibit A is Hutchinson Wampoa. Companies who don't have cash problems and have current infrastructure holding them back, new companies will move fast. That is the plan HW is following. When that happens companies like BT will change their timetable. In fact, there was a complete revamp of management at BT because of this kind of pressure and their new marketing push is broadband. It will happen in other European countries as well and the wild card in all of this is the US market. It is unclear to me that the current administration has any interest for rational spectrum policy though new bills are being passed in congress for new spectrum policy. It is not clear if the Department of Defense is willing to relinquish spectrum.

You coined AORTA long before most of us thought it was possible. Could you please discuss your vision of AORTA and how do see it impacting our daily lives?

I guess it was 4 years ago when I started this idea and what occurred to me was that there is going to be a discontinuous step function like change in human behavior when they are given access to the Internet is an instantaneous way. Unlike dialup, where you have to wait for dialup and response, you get access to the net with the speed of thought. You just get it, the result would be that you would be in touch all the time. You turn to those devices and information reflexively for transactions and communications. That is such a huge change from previous thinking that it would create a revolution over the years. We have seen acceleration in communications, commerce, and results, etc., all that is happening now.

What are some of the key components for a business model in the wireless industry to succeed? If you could illustrate by using examples of successes and failures.

Operators who come to the market with wall-gardened approach even in low-bandwidth space didn't succeed. Partly because of bandwidth, but even more important as i-mode shows that it shouldn't be a walled garden. The primary success of i-mode is the intelligence of Japanese operators to allow even renegades to use their network . . . you had certified ISVs who were allowed to use NTT's billing system but half of the ISVs were uncertified, this is just great! Bill Gates model. You want to have as many developers as possible and let them figure out what is going to be popular and what's going to catch on. That business model is brilliant. That will be the basic biz model for all operators in the future. Operators should provide a platform, provide value-add through billing services, QoS services, and infrastructure benefits and then let ISVs come and take the risk.

Where do you see China in all of this?

China is a difficult market to assess as it is so driven by politics and that will probably stay true. It makes for a difficult investment environment; it is hostile even though a lot of people are vying for that business. It will get better with time. Whether you are Qualcomm and have been hurt a few times in public and saw a drop in stock price because contracts suddenly disappeared, or anybody else, you are nervous. Market is growing faster but there are impediments. Actually, most of Chinese see this and want to make changes and make the problems go away. They want to make their own and export infrastructure equipment. It is not hard to imagine that in 5 years China will be the hot-bed for wireless.

Which carriers/countries do you think get AORTA and are progressively working towards that goal?

A lot of this is happening with teenagers first. It started with Nordic teens, then to Japan with i-mode rollout, then again in Japan with teenagers and 3G rollout there. I am sure that people in Europe who are rolling out the infrastructure hope the same kind of demand will pick up there. The wireless world will be led by teenagers and they will use short messages and pictures as their primary means of communication. I think gaming is misunderstood still and is going to be huge. If you have to look at applications that will drive this, I always believed it will be sex and gambling. You can take it the wrong way but the taking advantage of romantic impulses is an opportunity and business driver.

In your experience in the wireless industry, what has consistently surprised you the most?

The slowness in the US market. The only light I see at the end of the tunnel, ok, two lights that I see at the end of the tunnel. One is foreign companies, not US companies, purchasing US assets and making things move faster and that's happening. The other light is from the ground up, widespread use, decentralized use of things like Wi-Fi and increase in investment where government has no control and monopolists have no sway and therefore path is clear for the users for new applications and that's starting to happen.

How soon do you see telematics becoming part of our everyday life?

I increasingly believe that car value-add is increasingly electronic. The car should be like a living bus. The human relation to car in terms of security, entertainment, emergency services, and 3Cs (common constrained comfortable) things you do with the car is changing. The wrong approach was initially taken by the Japanese by putting things that required lot of attention from the driver, then they put in special switches when you would use them only when you are parked. But now with GPS systems in the car, the real uses will be beyond that and are less attention requiring.

If you can figure out these 3Cs ahead of time, there is opportunity of real revenue. If you think about it, you are trapped in the car and you can accomplish a lot of low-attention requiring tasks.

What are your thoughts on wireless LANs (Wi-Fi)? Are they a threat to 3G?

There is no greater threat to 3G than Wi-Fi or any successor of Wi-Fi because of unregulated use of spectrum as it becomes attractive to investors. It is easy to put together a scenario to stitch together a network with hotspots across the nation and brand them and sell them against 3G. If this happens, some estimates are that operators are going to see as much as 30% drop in their gross revenues.

Do you see carriers getting involved in this more seriously now?

Yes. Carriers have been playing with wireless LANs in secret for over 12–18 months now because they see this as a threat and are concerned about the potential impact on the revenues. They all have an answer or their own wireless LAN solution today as to how to stitch together hotspots in airports, etc., through a 3G system where you can do a handoff and be complimentary for the user.

What I have not seen yet and have been arguing for vociferously is that from a user perspective consider building out 3G network in a different way and that is not to consider handsets attached to its 3G systems but consider laptops at the end of the network. Then the argument why you need this bandwidth completely goes away. That's so clear to me that I am completely convinced that either wireless LANs are going to beat out 3G for revenues or operators are going to have to rollout their network laptop focused to support small screens too. And if they did this, they will create IPv6 VPN network, you are always connected no matter where you are in the world. Unlike the talk about people buying 3G services for photo messaging, I see business people buying data access for say $100/month for a large amount of data.

Are we quickly heading down the road to a GSM roll up of AT&T Wireless, Cingular, and T-Mobile? On the CMDA side, Sprint and Verizon. With the cost of build outs, the elimination of cell duplication has to be a significant saving. Couple the savings with the FCC refunds, and less need for additional spectrum, wireless could really improve its overall financial position?

I don't have quite the same view that you see in the media. When you see things in the market, they are: a bunch of monopolists. One thing about monopolists is that they run the entire market, so they don't see the sense of urgency or immediacy as the media or analysts see. The exception is of course foreign companies where DT, BT are looking to enter the market quickly. So from that perspective there will be more of it. The most prized targets are of course GSM properties, so AT&T is a target. Nextel is a good target but has some technical issues associated with it.

The new E911 system for wireless phones will bring into play the fact that wireless phones can be constantly tracked by the carriers and means that anyone with a wireless phone can have their movements tracked at all times. Do you believe that there is a need to legislate to force all wireless to be on a GPS system where the navigation system is at all times in the wireless phone itself and that only at times when the owner of the wireless phone needs help via E911 will the location of the user be known by the user making the decision to disclose their location by activating the phone only for that or a similar reason?

I expect people will just use it. The benefits far outweigh the privacy issues and risks. Hopefully, there will be legislation to turn off locating and tracking features when you don't want to be tracked. The application issues in my mind are greater. I am not convinced that people would want to get ads splashed on their screens in the malls. We will probably see dating services such that you go to Starbucks and want to talk to the cute girl in the corner, and you use the service to send a message. It might sound stupid but there might be a large market for such offerings. If you look at the classified business, the largest revenue comes from the dating services. Wireless is just a different platform, just a new way of doing it.

With the advent of Virgin as an MVNO, some interesting models might emerge? Do you see this trend continuing and what if any are the risks to the carriers that support these MVNOs?

I think Virgin is a special case. I don't think anybody has a better understanding of sense of brand as they do, so things they can do are not easily repeatable. So, Richard is just brilliant in understanding customers and I won't speak against him. In general, I don't expect that this kind of model will proliferate, rather, a model, where services are offered à la carte by third-party companies instead of the entire umbrella of services of operators and that's how they will make money.

What about wireless in home?

It is natural for the transition of technologies we use in the office to the home. The existing fight between the wireless standards for home will go away and Wi-Fi will become standard in both office and in homes, because people will use one laptop with one card. It is unclear to me the vision that is held by Sony and some other manufacturers. That these devices will be talking to each other, I am relatively unconvinced by that. The vision put forth by the Moaxi team that Paul Allen owns is more compelling to me. The idea that with drop in prices of flat screens, and TV are suddenly flat screens (dumb flat screen) room by room and there is a transceiver some place, enough bandwidth is there, that makes sense to me. I don't think that my refrigerator wants to talk to my washing machine.

How will 4G/5G wireless differ from 3G?

At some point, the issue isn't speed of transmission, it is the addressing space. As applications become more sophisticated, it is less important if you are getting 1 Mb/s or 2 Mb/s, flexibility in addressability and VPNs are more important. The infrastructure changes that I am suggesting operators make right now will probably happen with 4G.

Will speech recognition be an important part of the device–user interface? How long before speech processing is done on the device vs on the network? Are there advantages of one model vs another?

I believe it will happen in ways in which we almost don't notice. So, it is happening now. New techs and companies are coming out that make it happen faster. Increasingly you will see call centers as a target. There was a time when I thought it will happen as a revolution but it is happening more as an evolution. We will have all forms of queries being done just by speech rather than "dial 1" interfaces. It is going to be so much better.

I think a more interesting change will come when operators will understand that they should offer this as server farms. They are doing this for directory services which have become pretty advanced. But they could be providing similar services and access to their customers' databases and they don't understand that today. Revenues will be huge.

Are there certain products or concepts that are not mainstream on the industry's radar screen but have the potential of having an impact in the future?

Couple of things that come to mind. Historically, the governments tell you what spectrum you are going to use because of interference. If it is possible to create two new technologies. One, if we can create such devices that the problem of interference goes away, it will create a revolution for all concerned as there will be no government control and there will be a power shift from the government regulation to device interaction. I don't think it will work but if it did, you will see it in the 2.4 GHz spectrum in the USA. If that power shift could occur, there will be pressure to release additional parts of the spectrum. We will move from a different society from government regulated in the spectrum to device regulated. Another potentially large change will come from UWB. As you know, it is highly regulated. But there are a lot of people who want to make use of it – not simply to rescue people but many more commercial apps. If you look to dolphins or killer whales in the way they use their sonars is that they do a broad sweep first across all spectra and then they send out a focus. That makes sense. It is not too difficult to figure out a day when you will combine Bluetooth, 3G, Wi-Fi into one device, you figure out 2–3 people whom you are interested in and get your own VPN setup and refocus the power in your device to those kind of APIs that you want to use. That will be a smart thing to do.

12.3 One-on-one with Frank Yester

Frank received his BSEE degree in 1966 from Pennsylvania State University and his MSEE in 1968 from the University of Pennsylvania. Frank joined Motorola in 1968 and has been a member of various Motorola research organizations since that time. He has contributed in a number of technical areas including custom integrated circuit design, RF power amplifier modeling, analog and digital microwave system design and A/D converters for digitally implemented radios. In 1988, he became a manager of research where he had responsibility for the development of the RF system design, modulation, RF hardware architecture and custom integrated circuits used in iDEN and other digital cellular systems. Frank became VP & Director of the Wireless Access and Applications Lab at the inception of Motorola Labs in 1999. Frank is currently VP & Director of the Communications Research Labs in Motorola Labs. He has 18 issued patents and lives in Illinois with his wife Lucy and two children.

What kind of role do you play at Motorola labs?

I am in the research arm of Motorola supporting the handset and infrastructure side of businesses along with some other Motorola businesses as well.

Motorola brought wireless phones to the market in the early 1980s. You have seen the evolution of the wireless industry. From your vantage point, how has the transition taken place, has it been faster than you expected?

I don't think too many people expected the rapid growth and increase in percentage uptake that we have had over the years and some of the later things that came to the market. Now, the market is becoming saturated in different parts of the world. Many people, and certainly I, didn't expect this to happen. There were a few other surprises along the way though certain things did happen the way we expected.

Over the years, has the research process changed?

It has varied quite a bit over the years. Part of it was due to the maturity of the market causing changes in the organization. We have been flipping between centralized and decentralized research organizations depending on the market. Earlier, when I started with Motorola in the late 1960s, it was a centralized research organization because we weren't that large. As we grew, each of the individual sectors, portable, infrastructure, each had its own research organization, so we became decentralized. Then in the late 1990s, we went back to a centralized organization to coordinate efforts between groups under Motorola Labs because we were getting bigger. Over those time periods we have worked differently with the customers. Initially we were focused on particular problems in the sectors – short to mid-term. In Motorola Labs, we have tried to cover the long-term issues while being connected to the short-term issues in the various sectors. So, it has changed quite a bit over the years.

What are some of the most challenging issues that you guys are working on?

Communication technologies are what I am going to focus on. One of the main challenges is in dealing with convergence. Mobile, software, and content industries convergence is really causing a lot of questions as far as how we develop technologies and what technology to develop and how we go about most effectively commercializing these. One of the big trends that we see for the future is the value the end-users are going to perceive in our devices. We are going to be associated with what applications and services the users can get on these devices. It is no longer going to be voice only as you know. As that happens, we are faced with the issue of content coming from content providers, information coming from other areas, people are building wireless LANs or computer side of things. All this means, there is a lot more technology to be knowledgeable on and be judicious about picking the right ones in what you want to do to maintain competitive advantage.

SMS is one technology that most agree will generate revenues for carriers. How do you see this phenomenon morphing with the advent of next generation messaging?

SMS has been very successful, we are moving to a point that standards for MMS are becoming solid. Several companies have products in that space. We see MMS as a key technology in the next phase of evolution including GPRS. We see it as providing more compelling content to the user. Text is one of the things that is supported but also photos, video clips, streaming video, are all important in next phase of messaging. More content will be sent over MMS than SMS.

Does the transition from thin-client to thick-client give handset manufacturers more advantage to capture the segments of the value chain, and are you getting involved with the application side as well?

You are probably aware that Motorola is a leader in Java, J2ME for handset; we want the applications on the handset to be richer in content, effective, easier to use. In terms of putting more processing power on the handset to do all this stuff, it is a challenge for handset manufacturers to get this additional processing to minimize power drains, so it is an area of differentiation for handset providers to provide that kind of processing power.

What's your view of the industry debate on wireless LAN vs 3G?

I think each has its place (my view). Wireless LAN technology is very effective in giving high bandwidth at low cost in concentrated areas. Now, if you want to talk about broad coverage – with wireless LAN it is difficult therefore there is a need for wireless WANS for high speed. I very much take the view that wireless LAN is a key aspect in enterprises and hotspot areas (significant population) for mobile users but over the long run where the population density is not that high, you can't afford to put large numbers of wireless LAN, that's where wireless WAN will have a significant play. How does it help or hurt? Wireless LANs are going to have a positive effect on wireless WAN. That's where you will see the evolution of high-power mobile applications first because it allows you to directly extend what you do in your office or on the Internet in a mobile environment. On the wireless WAN side of things, we have a limitation on user interface or bit rates. But it will help us in getting the view as to what people are going to be using high-bit rates for and help in evolution of wireless WAN applications and services.

Because of this do you see a trend where the handset will choose what is available?

Absolutely, this will be a key element for future generation networks. The system should provide choice as to what technology is most appropriate for the situation you might find yourself in and the next-generation system will provide a way to accomplish that. Identify what's available, what's most appropriate and choose the right one for the type of things you want to accomplish.

Billing systems have often lagged behind and they sometimes restrict carriers to implement new billing strategies. Do billing systems hinder progress in the marketplace or don't they matter in the overall scheme of things?

I am not an expert in billing systems but there are certain elements in billing systems that are very relevant and important. As we talk about moving applications like micropayments where the billing has to be a low amount, billing system become very important. You need to have a cost-effective way to deliver that capability. A lot of billing systems out there don't seem to have the capability of low-cost payment schemes.

We are already seeing 3G implementations. How would 4G be different than 3G in terms of what it will provide?

We are in the early stages in the industry sorting this out. The view in a lot of different parts of the world is that 3G is going to evolve into beyond 3G kinds of systems that will integrate wireless LAN, PAN and WAN and broadcast systems and integrate in a seamless fashion. What will be after 3G is integration of these different concepts and technologies. Beyond that I think there are some views on 4G as well – 4G as wireless WAN beyond the integration phase – technology development – what's happening is continuing demand for higher bit rate that is needed in the future. We will continue to see bandwidth in the home and this will provide new apps, etc., and these things will naturally spill over into the wireless world. Both wireless LANs and WANs will continue to have higher speed.

Can you give some examples of the kind of human-to-machine or machine-to-machine applications that you expect to become important? How would these applications work in the home?

If you want to look at 5–10 years timeframe – some of the key aspects that are being discussed in various research forums have to do with personal agent types of capabilities. These agents will know your preferences not only by you telling the agent but actually them learning as you continue to interact with them. It understands the types of things you are trying to do. Remember birthdays, schedule meetings, update calendar. Things that you do manually, agents can do them automatically. They understand some of the limitations of your current environment that can help you do things that you want to do. It also understands what you want to do for fun – types of games. What games have become available, music, videos and alert you to what's available. The view is that agents will understand your preferences, environment and make appropriate suggestions and arrangements in a non-interfering way to help you and help you enjoy your life more.

Have we made progress with battery technology?

I think there have been significant changes in battery technologies over the years. The transition from NiCad to Li ion has been a significant change over the last 5 years. Looking forward, there is the opportunity for fuel-cells on the way we think about battery and mobile devices. It is not going to be in the next couple of years but a bit longer.

What's your view on how we are doing on display technologies?

With regards to display technology – we have seen a number of advances – larger screens, not only for cell phones but also PDAs with larger screens and colors. Virtual displays have been around for a while, but slow to take off. There are some interesting possibilities there. Display technologies are not evolving as fast as we would like them to. For some of the applications, it will be nice to have some of the foldable displays a little bit faster, displays that can project, technologies are coming along, a little bit faster would be nice.

Motorola has been involved with ITU's DSR work? Can you speak about the future of DSR in handset technology?

Speech recognition is a key future technology. There are a number of types of SR that are implemented into handsets today. The opportunity for DSR is that you can offload the processing to a centralized server from the user device. It makes things easier in terms of capability and processing power you need on a device. Our view is that this is a key technology for mobile devices and we continue to push that in our organization.

Can you talk about some of the technical issues for QoS-based services and protocols in heterogeneous networked multimedia systems?

QoS is an important issue going forward and transition of QoS as apps evolve is very important today and in the future. Things like multimedia, video conferencing, real-time and simultaneous video, and audio on IP kind of systems (and the industry is moving in that direction) are not resolved yet. Progress is being made but there is still quite a bit to do. If you take a look at how wireless connects with the Internet, we have approaches of some versions of QoS on the Internet but it is not resolved. We are still ironing out these issues and not quite there yet. We will be learning from how QoS evolves in the Internet environment to provide real-time.

And wireless poses some interesting problems like more limited bandwidth, it is not useful to just extend.

Wouldn't it be simply great if you could pick up a mobile phone and know that it could work with virtually any air protocol, no matter where you were in the world? What are your views on software-defined radios (SDR)?

We are going to see more and more in the way of SDR. There are a couple of ways you can think about SDR. If there is a piece of hardware that can handle any kind of air interface or radio access technology you happen to be in – wireless PAN, LAN or WAN – and clearly some elements of these are going to be very important and you are already seeing that in the market place. Motorola already offers a cellular phone with Bluetooth, WAN connectivity. We are starting to see that already. There is lot of discussion about adding wireless LAN to mobile devices. With FCC mandating location determination, many carriers have chosen GPS as the technology for location determination, so we are already seeing mobile phones being capable of doing several tasks (apart from plain voice) for multiple systems. Capabilities are evolving and we can see more to come. Handset manufacturers are constantly thinking about how to integrate a set of capabilities against the market requirement at a low cost.

The other aspect of SDR is to download configurations and apps into the handset and certainly J2ME is a great way to download, and this capability will improve over time.

What are your views on UWB research work being done?

UWB has been talked about for a number of years, we will start to see it used very soon. There are a couple of applications that we are sure to find relevant – for some of the location technologies, UWB has the capability to do very accurate location identification, very short-range, high-bit rate kind of applications, for instance a video camera with a little card that saves this information, and we might get to a point that video gets transferred automatically from video camera to home server instead of us doing it manually. Same with digital cameras to reduce the inconvenience and time it takes to transfer information. So we will find a lot of applications. There is potential for new applications – personal devices that we want to use to download content like video from some access point in a rapid manner, so I think we are going to see UWB – high-bit rate/local area kind of environment. There is a lot of potential for applications in public safety and commercial applications as well.

Where does telematics fit into all of this especially in the USA where we are in our cars most of the time?

Telematics is an interesting area in which Motorola has a good position in the telematics market space. Telematics has been evolving. We can certainly expect services in the entertainment area to be very useful. We are seeing a lot of that with video systems in our cars and vans. What we will see these systems do is that in local content centers like gas stations you will be able to download stuff into your car or games. Another area that gets a lot of attention is the potential for remote diagnostics. A lot of personalized services that we talked about for wireless LAN are also going to be available in your car. Emergency services will continue to be an important application for the car.

What about P2P wireless?

That will be a key element of the future. We are already seeing wireless LAN and PAN have P2P capability. Some of the applications will be in a business context – how to automatically share information with business partners. P2P gaming is a key application area. Get a few people together to see who is available to play games in that small geographic area.

Over the years you have must seen several technologies come and go. Were there any surprises for you as to what took off and what didn't?

That's a really good question. I think early on, we go through a cycle of extreme hype, then it goes away for a while, and often they come back a bit later in a typical fashion when reality sets in and you find out what this technology is really good for. There are a lot of technologies that have followed that kind of trend. Like wireless LAN – we are still in the hype stage, it will eventually find its very useful place in the wireless world. In some ways it seems like wireless LANs have been around for a while and suddenly more and more people are getting interested in them.

The big surprise is that it has been around and now it is taking off so rapidly and I wonder what phase are we in now with that. PAN – we have seen that sort of thing. Bluetooth was the poster child of the new and wonderful with wireless but it is having a slow uptake, but we are starting to see that take off again in a much more solid and continuing way. If you take a longer term view with new technologies – hype – exciting – everyone jumps on the bandwagon then we go to the reality phase where nothing seems to be happening and then we see strong uptake later on. We continue to see that kind of behavior all the time. WAP went through that. Now 3G is going through that. Advanced intelligent services on mobile devices went through that.

You stand back and look at them and say jeez these things are taking off. Actually technologies find their places in the industry in due course of time. It is not much of a surprise, we continue to go through these, and it is kind of surprising that it continues to happen.

We have talked about several challenges that the industry is facing. Are there any other challenges short-term or long-term that the industry needs to sort out?

There are going to be continuing issues and problems and challenges that we face in a number of areas. We talked about convergence. Not only from the point the of view of certain industries but also from other aspects like security and privacy – how do you provide the right tradeoff between services and invasion of privacy? That is a key issue – where customers view they are getting value and where they view their privacy is invaded. Security is a key issue. We need to be flexible enough to provide different ways of using applications. We talked about P2P, sharing information amongst small groups of users who want to share. How do we enable people to share content in a most appropriate way that maximizes the value for content owner as well as the end-user? There is a role for different players to help determine that content management ownership as well as provide flexibility to the end-user. We have started that in the Internet realm but it is going to continue to be a challenge.

12.4   One-on-one with Jon Prial

Jonathan L. Prial, Vice President of Marketing and Strategy, IBM Pervasive Computing Division: as VP of Marketing and Strategy for IBM's Pervasive Computing Division, Jon Prial leads strategic planning, marketing communications and worldwide marketing programs for pervasive computing technologies, which extend real-time information and interactive Web-based services to smart devices (e.g. mobile phones, personal digital assistants (PDAs), networked vehicles, home appliances). Before his current role, Prial was Director of Integrated Solutions and Linux Marketing for IBM's Software Solutions Division, with worldwide marketing responsibility for IBM's cross-platform software strategy and the delivery of integrated solutions. Prial also led IBM's marketing strategy for Linux. Prial joined IBM in 1978 as a programmer developing communications software products and in 1981 entered the sales organization holding various technical marketing, regional support and managerial positions. In 1990, Prial began marketing IBM's document management products, launching IBM Digital Library in 1995. After working as Executive Assistant to the General Manager of the Software Solutions Division, Prial was appointed Director of NT Solutions Marketing. Prial is currently on the Board of Directors of the WAP (Wireless Application Protocol) Forum. Prial has a BS in Mathematics and an MS in Computer Science from Rensselaer Polytechnic Institute.

IBM is one of the pioneers in the wireless industry. Could you please elaborate on the types of solution IBM is providing to the marketplace?

There is a wide range of wireless solutions that we provide. Some of the basics are extending our clients' Internet infrastructure to wireless LAN and WAN and we help our customers do that as we have a tremendous services capability all the way to extending e-business applications to PDAs and phones (enterprise). We help service providers (carriers in particular) to deploy wireless solutions. We are also working on the embedded end helping the device manufacturers with the software stacks or hardware technology, Si-Ge chips, software stacks running on say Linux, JavaOS, smart cards, SIM cards, so it is across the board. We are across all the major industries. Financial, hype shifted from B-C, B-E, access automation, inventor applications, CRM applications. We help customers solve their problem.

How has IBM been working with the carriers? Have there been any shifts?

IBM has billions of dollars in revenue from carriers. We are helping them in a number of ways. Voice is becoming more IP and IT oriented. Telephony applications are becoming more IT oriented. A few years back WAP applications were being done by the Nokias of the world, now they are being done by SUN, IBMs of the world.

As carriers tightly integrate with enterprises, they want to become more than bit pipes, application models and programming models extend quite nicely to carriers. Java model will extend out nicely to the enterprise.

IBM has the strongest services arm in the industry. What are some of the key system integration challenges you encounter that are unique to wireless? With time, will system integration get harder or simpler?

Clearly it is going to get easier. Things that are making it harder – skills need to integrate to the backend and network side of the house. Serving and managing the clients is a challenge. Software management, sync, and series of challenges that need to be addressed. The C/S model extends to pervasive computing. A lot of system integrators have point solutions but we are focused to the horizontal middleware layer. It will simplify integration.

Does telematics fit into your short-term strategy?

For us what we call the device-led solutions we see as the in-hand market, we talk about in-home, in-machine, in-factory market spaces. Telematics needs to mature. We have been working with car manufacturers. Next year we will be launching something with Johnson Controls – a Bluetooth-enabled phone kit for the car, it is going to be for Chrysler – embedded voice, email system; the car is becoming an interesting computing platform. It can act both as server to remote applications as well as client accessing connecting to remote channels.

What are some of the challenges and lessons learned in launching your solutions for consumers across different geographies and cultures?

Clearly there has been a difference. Part of the success for i-mode has been less Internet access at home, a lot of free time on the trains. DCM did a fantastic job building a business model around it. Tremendous amount of WAP take up in Europe. They didn't have vision of some carriers – throw away your PC, surf the Web on the phone.

What are some of the key components for a business model in the wireless industry to succeed?

It has something to do with more than just the wireless world. It takes a long time to get there. I am a user and advocate as well. People want added value. I signed up for a $14.99 baseball service. There are hundreds of games in a year and I want to follow them. I get great service from no matter where I am. I am willing to pay money to get rid of the spam that I keep getting. So it comes down to the value proposition to an individual. A business model is about getting down to value. Amazon gives you a shopping experience; some are starting to pay off. You asked about the value chain – in the pervasive space there are a lot of elements there, emerging pieces – home network market setup box, game machine, a lot of control points in the house – the value chain is huge, with several control points in the chain. It is not cost effective for the power company to put a gateway in-house to do meter reading. If all the players got together – phone, power, appliance manufacturer got together and built an ecosystem and value chain around it, then you own this space. The power company can say, I will give you a dollar off if you let me read your meter wirelessly. However, convergence needs to happen before it takes off.

In your experience in the wireless industry, what has consistently surprised you the most?

What has consistently surprised me is that people haven't figured out that over-hyping is not a good thing. We continue to over-hype everything. WAP was over-hyped. The same is going on with 3G as opposed to highlighting practical benefits of the technology. This has a cause and effect relationship. When people have spent billions of dollars on something, they don't have patience.

SMS is one technology that most agree will generate revenues for carriers. How do you see this phenomenon morphing with the advent of EMS and MMS? How soon before they become relevant?

There are some challenges that make it a bit different. SMS is pretty cheap, MMS is not. I was talking to some of my colleagues in Europe as to what they do for their kids' cell-phones. They pay the base fee and the rest comes from their kid's allowance. Well, w.r.t. they can get quite a number of transactions they are pretty happy. What I don't know is how MMS fits into the picture as you won't have an equivalent number of transactions with MMS.

It is ok to push SMS, it is not ok to push MMS, maybe I need to download it on my own. MMS is going to be like the email system with a to list and cc lists and what not. Let me reiterate. The model will change a bit. There will be some adoption delay. It is going to happen; it is just not going to happen overnight.

How important is the transition from circuit-switched data to packet-switched data in the mobile environment?

Without getting into packet switched we are not going to have a good data infrastructure. We are going to treat voice as IP, for WAP I don't want to dial in and then check stuff whereas in my house, as a normal human being, I use DSL/cable modem to access. I want the same experience of being always connected.

What are your thoughts on wireless LANs (Wi-Fi)? Are they a threat to 3G?

It could have been competition. Carriers are waking up that if we don't eat our children, someone else will. For any technology you have to do that. Recognizing that the transition of Wi-Fi to 3G has to be seamless and carriers are getting involved in this, it is going to be complementary. Because they don't want to compete. Heterogeneity wins. Networks matter. OMA initiative standards will help grow the market.

How will 4G/5G wireless differ from 3G?

Richer content. Apps are going to change. I have been a happy user – for running my mission critical applications like checking the METS score, I don't need bandwidth, 9.6 kbps works fine for me. I would pay big money to have a live stream of traffic on various bridges in NY – George Washington Bridge, etc., to help me decide what road to take. That means richer content.

How do you think the battle between Qualcomm's BREW platform and Java's J2ME platform is going right now, and what's your guess as to how it'll play out over the next year or two? Are there any other platforms (Symbian, OMAP, etc.) that have a chance to take a significant market share?

Microsoft is going to make a play. It is not a slam dunk for them. It is not quite clear yet. Java is big play. BREW did a nice job establishing themselves but it is not clear if they can do enterprise class applications, they have some nice apps, gaming apps. Good model. But at the end of the day, Java is more easily deployable and is going to spread in the enterprises.

For developers building apps that benefit from location and presence data, what are some of the successful business models going to be?

At some point location-based data is going to be available to the applications. This is the fundamental change in this new world where we will start having information based on context – on where I am, what I am doing right now. If I start my transaction at home on a desktop, continue the transaction in my car using a phone and when I leave my car, I still can continue. All along the system knows that it is me. So, it is all context. Intelligence of the backend system will help in creating new classes of apps that add value to the individual.

What in your opinion are some of the most challenging issues that our industry needs to address – in the short-term and long-term?

Security always comes up – people always worry about security. It is no different than the Internet world in 1997. People give away their credit card to strangers in stores but they are hesitant to give it over the Net. It is not a showstopper though. Identification is going to get stronger with biometrics solutions. Industry needs to do a good job of educating the masses.

Also, privacy issues are important. More and more information is coming in and it needs to be handled appropriately. We have a chief privacy officer. It is very important for us.

12.5 One-on-one with Steve Wood

Steve Wood received his Bachelor of Science degree in Computer Engineering from Case Western Reserve and his Master of Electrical Engineering degree from Stanford. For the past 25 years, he has focused on emerging technologies. In addition to being a member of the executive team at McCaw Cellular Communications, Steve Wood has either founded or has been part of the founding team for a number of companies including Microsoft, Asymetrix, Starwave Corporation, Interval Research, and Notable Technologies. Currently, Steve Wood serves on the Board of Directors for Telescan Inc., a publicly traded online financial services company based in Houston, Texas, and Sproqit Wireless, a developer of distributed systems for wireless carriers.

You are amongst the pioneers in the wireless data industry, starting with Notable Technologies. Could you please elaborate on types of solutions Wireless Services is providing to the marketplace?

Basically what we do is that we provide server software that handles messaging, advanced messaging, flexible messaging applications. So, we provide functionality that ranges from basic/SMS gateways to two-paging applications, IM, pop email, email, interfaces to outlook and notes and corporate email systems. And around that we have a variety of APIs and protocols that we support so we integrate with other applications that people have or are building. If you look at what's happening in a data network, everything boils down to messaging and messages that go back and forth – transactions, chats, email, etc., inside the system there are messages being handled by SMSC or WAP gateways or gateways. So, what we focus on is becoming the best and most flexible, high-capacity messaging infrastructure. It is part of the carrier infrastructure.

Application developers in the USA (and even in Europe) often get frustrated when dealing with carriers. What has been your key to success in securing carrier contracts?

There are some key things to recognize and some hard lessons you learn. The first thing is to recognize what the carriers are good at and what they perceive their core business is. The core business is acquiring customers and building a sustainable brand. The core biz doesn't include applications. They excel in running networks and infrastructure. The areas where I see companies fail most of the time is when they think they can sell applications through carriers to consumers or enterprises because carriers are not good at that. It is not part of their core business or competency. What they do understand and can sell them – you can go two different ways. They understand infrastructure – so you call sell them that – Nortel, Lucent, etc. The other kind of biz model that they understand is the revenue sharing model that DCM pioneered in Japan. If you go to them with apps or services and tell them they will pay carriers a percentage.

When you are selling infrastructure the important thing is the time frames. Sales cycles are very long – many years sometimes to work through a sales cycle. Once you are inside their network, you become a key partner and vendor with carriers – you can expand in other areas.

Two years from now, what kind of solutions do you see Wireless Services providing to the end users?

We are focused on a handful of areas that is clear to us that are going to be important. One of those is aimed at enterprises – functionality to introduce more advanced functionality to access corporate information from mobile devices – email, contact list, address in real time. Another important area is real-time text communications in a business context, so collaboration, dispatch, and those kind of applications. On the consumer side, obviously IM is a key area; there is a presence component to that. Another area is alternative billing interfaces – CPP paid billing schemes. The whole world has that but not the USA. Credit limits kind of billing schemes. More and more what we are seeing is that carriers are lowering the cost of customer acquisitions by offering pre-paid systems and other ways to bill.

Wireless Services seems to be focusing more on carriers as customers vs corporations. What are some of the thoughts behind this strategy? Also, is your pricing based on licensing or per transaction or something else?

It actually came from several years of selling to carriers and enterprises. And though we have several enterprise customers, it is not our focus and the reason it is not our focus is because a sale to an enterprise is a difficult sale, enterprises are relying on carriers for such applications, the other problem is that most enterprises are at the pilot state so what we found was that we could sign many corporations but only a few hundred customers at a time, it is difficult to make a business model. Working with carriers you can reach thousands and thousands of end-users at once.

We will do licensing but the bulk of our business is coming from capacity or transaction-based pricing. Our primary biz model is managed services – we install our servers and software inside the carrier and we sign a long-term operation agreement and we have the responsibility to run it. It is not an ASP model because it is not our hardware or data center and the reason we don't do that is that it doesn't fit with the carrier model. It is very difficult to operate and interact with carrier infrastructure without being tightly integrated with them. It is not practical. Most carriers want us to operate because they don't have application expertise and there are also capital constraints so they are trying to outsource as much as possible.

Does telematics fit into your short-term strategy?

We are doing a little bit of short-term but it is very important for us long-term. It is a large business. If you look at what GM is doing, they have lots of subscribers, not that much functionality but you can see the potential. I think the bigger market is in the verticals – shipping companies are putting transceivers on their equipment but you don't hear that much about it.

What are some of the challenges and lessons learned in launching your Wireless Application Delivery Platform for consumers abroad?

We have server installations in Mexico and Latin America, and the Philippines through our relationship with Nextel International. The USA is different from almost everywhere else. Outside the USA, carriers are more aggressive about data services, the usage is higher outside the USA, and revenue opps are higher. The US market is broken basically in a couple of ways. One way is that we have done such a good job of training everybody that everything is free, in this environment it is very difficult to generate revenue. Another aspect is that we have very good wired and Internet infrastructure. Revenue/sub is higher. In Latin America, people use phones for Internet access because there aren't other ways to access the Internet. They use text messaging much more.

Obviously we had to do localization. We designed our basic server platform with that in mind. We are now running at least 5 different language variations. We support the systems from here. It allows us to provide low cost, high reliability to the carriers.

You have been involved in wireless since the McCaw days. In your experience in the wireless industry, what has consistently surprised you the most?

The biggest surprise to me is how long it has taken for the mobile data services to arrive. Back in 1994, when we were at McCaw, we expected that there would be reasonable usage of digital network usage within a couple of years. In the USA only in the last couple of years, is it taking off.

Why do you think it took such a long time?

There are a variety of reasons. It takes people time to adapt to new technology. Carriers in the USA have been focused on voice and not data and that's because that's where most of the revenue comes from. It is difficult to generate incremental revenue with data services. In addition to marketing issues there are infrastructure issues. Until recently, most of the carriers here couldn't support data billing, couldn't support usage-based billing. When the data services came out, every one was charging a flat fee because they couldn't do anything else and it was easy to sell that. Now, where we are just now, carriers have billing systems in place that will allow them usage-based billing. Now they have to convert their users to the usage-based model. It has slowed things down. The fact that we have 4–5 standards in the market slowed things down.

SMS is one technology that most agree will generate revenues for carriers. How do you see this phenomenon morphing with the advent of EMS and MMS? How soon before they become relevant? Can GPRS operators deliver MMS?

We have a variety of data technology deployed in the USA like SMS, CDMA data technology, packet data networks, GPRS networks, etc. So we have a variety of ways people do messages. Most carriers don't use two-way SMS. Nextel has a very good two-way messaging. Not SMS but running on their packet network. More and more you have to focus on two-way communication and messaging. The network technology you use is not that important. More and more it won't be SMS because of its limitations.

I think MMS is cool but it is going to be a long time before anybody but DCM makes money off that. The Japanese market is way ahead on that. Japanese culture is very adaptable for such services. It will take a long time for the European and US markets to catch-up.

What are some of the key security implications of wireless data applications and services?

Some, but not a whole lot. The issue comes up on a regular basis but if you really drill down as to what people are concerned about, it turns out they haven't thought through it very much and often times they are focusing on the wrong aspects of security. The digital networks have some level of security built into them. It is not bullet proof, you could crack it using enough computing power but it's sufficient for most enterprises. If you look at what kind of information is being sent back and forth to mobile devices it is very limited due to bandwidth constraints, so it limits exposure. As we get advances in mobile devices, it is going to get much better. What we find over and over again is that when we answer the security question with – oh! It is going to cost you 20% to add another layer of security; it really isn't that big of an issue to most enterprises. There is a lot of talk about it but not willingness to spend money on it.

How important is the transition from circuit-switched data to packet-switched data in the mobile environment?

> It is critical. Nobody is going to use large amounts of data on circuit-switched networks – it is too expensive and too slow. Nextel has a very functional 2.5G packet network, so we can see the capabilities, it is very clear that it is the direction everyone is going to go.
>
> High wireless bandwidth would make sense for access using laptops and maybe PDAs but the billing models need to be different.

Is the industry geared to adopt adaptive billing models that mix technology (Wi-Fi and 2.5/3G), usage scenarios (critical transmissions, event-based data exchange vs casual web browsing), and customer history and profile?

> It is very complicated right now. Most carriers are struggling with that. All around the world there are different billing models that vary in pricing quite a bit. Clearly, you have to drive the cost down for higher usage. Threshold for early adopters in enterprise has been $40–50/month. Pricing has to come down in the future.

How can carriers expect to achieve any reasonable ROI considering the huge cash expenditures for 3G licenses? With a capital crunch in place and with a fickle consumer response to wireless data, what do you see compelling migration to 2.5 + or 3G?

> Some of the carriers who paid large amounts for 3G spectrum are not going to be successful. Governments are coming back in to loan money, defer costs. If you have paid so much money, you can't support the business.
>
> Given that, there is clearly going to be a delay in migration. Countries that will have 3G services are going to be the ones who allocated spectrum via non-auction frameworks. The other problem with 3G is that nobody knows what applications need all that bandwidth. The only obvious ones are access from laptop kind of devices to the network but you don't need 3G for localized high-bandwidth services.

What are your thoughts on wireless LANs (Wi-Fi)? Are they a threat to 3G? What implications do wireless LANs have on software platforms such as WADP?

> Carriers will step in. We have seen VoiceStream stepping in and other carriers are looking at either building or acquiring hotspot networks. Dual phones are also getting built within the next couple of years – iDEN and 802.11 or CDMA and 802.11. In the long term, you can do the same kind of things with 3G – building localized hotspots. Right now, 802.11 is well engineered technology available at low cost.

How do carriers utilize 3G bandwidth?

> I haven't found a good reason yet for having a 3G bandwidth on my handset. I don't know what I am going to do with it.

Are we quickly heading down the road to a GSM roll up of AT&T Wireless, Cinqular, and T-Mobile? On the CMDA side, Sprint and Verizon? With the cost of build outs, the elimination of cell duplication has to be a significant savings. Couple the savings with the FCC refunds, and less need for additional spectrum, could wireless really improve its overall financial position?

> There are two different ways it can go. I actually think that the likely way we would see is proliferation of more standards. The underlying standard is becoming less and less important for compatibility. Other network components and radio are becoming much better and much

smarter for doing transmission so it is very easy for us to do translation and interoperate between GSM, CDMA, and TDMA networks and to the customers it clearly doesn't matter.

As technology advances, software is going to enable carriers to install a wide variety of standards.

How do you think the battle between Qualcomm's BREW platform and Java's J2ME platform is going right now, and what's your guess as to how it'll play out over the next year or two? Are there any other platforms (Symbian, OMAP, etc.) that have a chance to take significant market share?

It is going to make the apps easier to use hence easier to sell. One of the disadvantages of current phones is that they rely on WAP, it is not very good, the limitations on the user interface are pretty severe. Email apps on J2ME are widely distributed. Client apps on phones allow you to do more on the client rather than across the network. Right now there isn't much of a contest, J2ME has the majority of the market. But, some carriers are rolling out BREW and Symbian. Microsoft, I think, is going to be a big player in all of this with their next-generation platform they are promoting. It will be really nice for the customer.

How would E911 and position-location technology impact the wireless industry?

A couple of things on location: legislation is already in place to mandate E911. So, carriers have to eventually conform to that. Locating is very different than tracking. For E911, we mainly need to locate the phone. For most of the other applications that people are talking about, it requires you to track the user/phone in the network – a much more difficult problem, a network-intensive problem. A problem that doesn't have much consumer demand. We are seeing some generalized location-based applications that make sense and they are all enterprise related – dispatch – I want to know where my service people are – clearly people are willing to pay for that. Most of the cool commerce-based location applications – like – the classic one is I walk by a Starbucks and a coupon comes flying in – that's not going to happen anytime soon, no demand, difficult to monetize, nobody will pay for it. Don't think those services are going anywhere. Other things like "where is the nearest ATM?", I don't see a demand for that. If it is available free on my phone, maybe I will use it but I am not going to pay for that. So, areas where location-based services are going to get traction is with the enterprise applications. And location integrated with communication apps. The only consumer app that has a chance is location tied in with IM and chat and scheduling functionality. If I can know if people on my buddy list are close to me, that is valuable to me.

How about presence?

Presence is quite different from location and it is an easier problem to solve in the network and it is easier to see how people will use it. It is an extension of IM on the desktop. In order to do that effectively on wireless devices, you need presence, but all you need for that is same level of presence information available on the desktop. It is not that hard.

What in your opinion are some of the most challenging issues that our industry needs to address – in the short term and long term?

One of the critical short-term issues is just the balance-sheet and capital problem that carriers have. Too much debt relative to what their stock prices are, which means they are constrained in their ability to spend money on anything new. Even carriers who haven't made commitments for 3G have that problem. It is a constraint worldwide. The other constraint especially in the USA is regulatory – not just the frequency issue, but FCC in the USA, which has traditionally been good at promoting installing new technology, clearly failed with the spectrum auction and telecom regulation in 1996. It was a disaster and it hurt a lot and is responsible for the state of the industry today. Because what happened was that all of the incumbent regulated RBOCs ignored the legislation, and used their monopoly to drive everyone else out of the market. All the DSL providers failed. Guys who were trying to build fixed wireless, none of that took off. The whole cable industry is growing more slowly than it should. And that is largely down to the USA.

What do you see as some of the challenges the carriers face as they consider implementing mobile advertising and other value-added services to their subscribers? What changes (legislative, technological, business) do you think we will need to have before wireless advertising can be accepted by the masses?

We have started to work with some companies who are working on targeted opt-in marketing on mobile devices. There is clear opportunity there. It is attractive to carriers. Some of the things that I am hearing that might take off are opt-in interactive kind of apps. If you answer these three questions, we will send you a coupon at McDonalds or if you play this game, we will send a coupon for a free Budweiser. So these advertising-targeted services will take off. And the large consumer companies like McDonalds are interested in that. Key requirement is, however, that you should be able to deploy these kind of marketing programs across multiple networks and devices. If you are just limited to ATT, it doesn't work. You need to be able to do that across all carriers.

12.6 One-on-one with Mark Tapling

Before joining Everypath, Mark Tapling served as President and CEO of ServiceWare Technologies, Inc. During his time at ServiceWare, Tapling led fundraising efforts and a successful initial public offering (IPO), raising $67 million. Under his leadership, the company tripled revenues in less than 24 months, including six consecutive quarters of 100% growth and five consecutive quarters of new customer orders worth more than a million dollars. Previously, Mark Tapling held management positions with IBM's Lotus Development division. He also held a variety of management and executive positions with Comshare, Soft-Switch and Softlab GmbH. Mark Tapling holds a BS in economics and management from Michigan State University, and he has completed post-graduate work at both the University of Virginia's Darden School of Business Administration and Cornell University's SC Johnson School of Management. A recognized industry leader, Mark Tapling was named one of Software Magazine's "100 People to Watch in Knowledge Management" in 2001 and PNC Bank's "Entrepreneur of the Year" in 2000.

Could you please give your assessment of the industry? How does Everypath fits in the value chain and what kind of solutions do you offer to the marketplace?

Everypath is an enterprise software company that delivers applications to be installed on-premises behind the company firewall to companies who are looking to mobilize their enterprise applications. We have a full suite of products that support both wireless and disconnected mode of computing. We have hosted clients where we run the infrastructure. This market has evolved significantly.

One of the evolutions we have seen over the years, is that companies have invested literally billions and billions of dollars trying to fund companies with wireless solutions. In the end, much to the chagrin of VCs, what has been discovered and we agree with it, is that wireless is a transport for application, not an application in itself. It is just like the Internet, it is not an application in itself but really a means to develop applications on. As a result, it has created segmentation in technologies, customers, solutions and in the market. It is healthy for the market and the transition is creating robust solutions. So, for the enterprises, it has evolved from a wireless problem to a mobile computing problem. So, companies are looking at business applications and figuring out my mobile computing requirements and the spectrum. On the left we have disconnected devices like iPAQ, Jornada, etc., and on the right side we have fully synchronous wireless devices like phones or PDAs, and in the middle sync is happening through a cradle. As enterprises determine where on the spectrum of mobile computing they need to build their mobile computing applications, they are building targets for these applications. We are seeing that b2e and b2b are most focused in North America and even in Japan on a disconnected model and consumer market they are focused on pure wireless interaction.

There have been large companies like Aether and Avantgo who built strategies being OEM providers to handheld and other companies. Though they have aligned themselves very well to device providers, unfortunately, they are losing money on every customer, so there has been a lot of pain for companies who have a solution purely focused on wireless.

You have been doing business in North America and Japan and some in Europe. Have you seen clear differences in how enterprises have adopted your solutions?

Yes, 100% difference. In Japan, there is a high penetration of wireless users, mature infrastructure. We see a high penetration of b2c applications which are interactive and reliable and our business maps that. We sell into companies who are trying to deliver those kinds of solutions; as an example, we have Tokyo Gas, delivering and handling questions and billing issues over cell-phones for 8 million customers. So that's a prototypical application for Japan. The other side of the spectrum is North America. Infrastructure is poor, coverage is spotty, we are not even doing packet here, speed is slow, etc., so, we find here, and customers mostly use their phones for talking and not much in the way of wireless data. There are some exceptions like E*Trade. Enterprises are more operating in a disconnected fashion rather than continuous sync fashion for handheld computers. There are a few reasons for that. Lack of cellular coverage and lack of compelling wireless plans, lack of high-speed infrastructure and finally, the mobility tax you have to pay for your handheld to become wireless, for instance you have to pay an extra $200 for adding a modem to your $300 PDA for it to become wireless. Enterprises see this as a heavy cost for supporting spotty coverage. Europe is very aggressive with SMS and WAP and they don't have widely deployed PDAs. Most of the wireless data apps are on cell-phones. They have been aggressive in rolling out applications and services to the consumers but not aggressive in rolling out applications to the enterprises.

What is your read on the migration from 2.5 to 3G? Do you see compelling reasons for carriers to make the transition?

In 2–5 years yes. Next year, no. The carriers have a problem. They built the infrastructure but they can't charge for it. Some consolidation needs to take place before the transition happens. When the enterprises start to make demands based on large-scale commitments to wireless data – like $100 million for the next 5 years – so for reliable high-bandwidth wireless transport then those carriers are going to use those commitments to raise debt and fund the infrastructure rollout of 3G. They can't count on inconsistent consumer demand to justify 3G rollout. But it is a 24–36 month problem.

What suite of applications do you see enterprises adopting that will increase the demand for wireless data adoption?

I think scenarios where accuracy matters, as an example field services for pharmaceutical companies who are doing electronic signature capture for the distribution of controlled substances. For example a pharmaceutical representative goes to a doctor's office and want to leave 100 samples of a regulated drug, it needs to have a documented transaction trail with signatures, and accuracy matters because penalties and fines are heavy. Or a field service situation where a field service technician needs access to customer information or equipment data, so scenarios such as where a technician walks into a room full of Sun servers to upgrade them all and later finds out only half were under a serviceable contract and the rest were done for free since he didn't have access to customer information. Or opportunities where there is extended selling opportunity where customers want an add-on product but the sales rep doesn't have the capability of taking and processing an order on the spot. Historically, value has been added at the platform level. For example, mainframe aggregated enterprise value, servers and miniframes aggregated departmental value, desktops aggregated personal value, handhelds focus on task automation. We find highly repeated redundant tasks critical to corporate business process highest on the list of candidates for mobile computing applications. Everything from accepting a credit application in the field to updating CRM data to managing inventory to handling field services interaction to capturing signature electronically are all very useful.

What are your thoughts on position location?

It has enormous benefits. If we had the capability to locate cell-phones and maybe remotely monitor critical health readings like bp, etc., we could have saved more lives out of the 9/11 disaster. There are some real mission critical applications that can be built from public safety to commerce to military. And some of these might be offered as additional features by the carriers and people would be willing to pay for it.

What have been the success factors for your applications across vertical and horizontal industries?

Two things – our message has been that of leverage. We are not going to enterprises to build out expensive infrastructure for wireless applications. We would mobilize the applications you have. From a technology perspective, we remove the technology complexities from protocol, to data extraction and abstraction to device support from the business and this is a huge advantage to our clients. Devices are changing so rapidly that businesses don't want to support a moving target. Our product removes the risks and complexities. We put that behind a drag-and-drop interface that generates software automatically, and it makes life simple for enterprises.

What's your view about future of wireless?

Handheld devices are creating a personal-centric economy. In other words, the economy wouldn't evolve around a place where you have to go to a computer to get access, but rather a computer will be with the person and the economy will evolve around that.

12.7   Conclusion

We also touched base with Satoshi Nakajima, CEO and Founder of UIEvolution, who pioneered the development of Windows OS and Internet Explorer during his 14 year stay at Microsoft. He has great insights into how users interact with technology, applications, and services. We asked him about his thoughts on the wireless industry. Satoshi believes that wireless is all about connecting people, it is about communication. He says, "I truly believe communication is a killer application – SMS, MMS, etc., other forms of application that help you to connect, customer database, email, any application that helps to share information with others is important." The reason he thinks the Japanese market has been so successful with wireless data is that data services were available from day one.

"The packet network, even though it was 2G, the business model with content providers, was solid. Good set of handsets. DoCoMo did such a phenomenal job. My son in Japan is completely into messaging. Teenagers communicate that way. That's excitement. People are hooked on various types of application but the phenomenon of wireless communication is applicable everywhere. I see two categories: communication (email access, etc.) and personalization (wall paper, ring tones, face, and painting) are key. In Japan, people use messaging but only 30% of the time, the rest of the time they use it to talk between classes, etc. My son uses it even from his desk. The USA is struggling because of the driving culture, plus the US teenager has access to Internet."

We hope that, via these interviews, we have been able to present a diverse range of views and opinions that are not only insightful but also intellectually stimulating. There are some industry issues discussed in detail as well views about what the future might look like. So, what inferences can we draw from the above? It is clear that there are some contentious issues like privacy and security that need to be addressed by the industry to assuage consumers' concerns about their personal information. There is a big problem with respect to the huge debts that some of the carriers have and, unless something is done soon to remedy the situation, the road leads to bankruptcy for most of them. Another interesting aspect is the shift in battles from regional to global. Carriers are aggressively pursuing foreign partnerships to continue growth. The USA, though mired in intense competition and spectrum woes, looks more and more attractive to investors and foreign carriers alike. Globalization's impact on the wireless industry has been great. There are pros and cons. The pros are that users enjoy cheaper and better products and services owing to the scale merit of the global market. As for the industry, it fosters immense competition. Initially, companies used to work only in their territories, but now to sustain and grow, they have to compete globally. The trend will continue in the coming years. Only two or three players can eventually survive and be leaders; others will be consolidated into the market.

During the past couple of years, the US wireless market has continued its substantial growth: from about 86 million subscribers at the end of 1999, the total number of subscribers had increased to 128 million by the end of 2001, a 22% compound annual growth rate (147 million as of June 2003). Carriers are all upgrading their networks to gain more capacity and efficiency for their voice networks as well as to introduce high-bandwidth data capabilities; 2001 was the year of experimentation with wireless applications and of developing various relationships amongst the players in the wireless value chain. It is becoming more and more evident that carriers are finally accepting the need to partner with application developers with unique value proposition and offerings.

The US market is ripe for future growth and the market potential is one of the most attractive in the world owing to the market size, its maturity and the opportunity to launch innovative applications and services. With network upgrades, carriers are reducing the number of standards in the market thus enabling new roaming agreements with worldwide operators as well as opening the market for application developers to transport successes from foreign markets into the USA.

Though the US market is attractive for many reasons, the "real" progress is actually coming out of Asia – from Korea and Japan. SK Telecom, Hutchinson Wampoa, NTT DoCoMo, KDDI and J-phone are all marching ahead with their next-generation rollouts and the rest of the world is following and learning from their experiences. China remains an interesting conundrum to all. The market is exploding, making it the no. 1 player in the world in terms of pure subscribers and the revenue potential, but penetrating the Chinese market remains difficult because of governmental regulations and control. However, with their recent entry into the WTO (World Trade Organization), things are bound to loosen up for the better.

Carriers outside Japan are quickly realizing the importance of alliances with content providers and developers. DoCoMo works rigorously with the developers to evaluate the technology and value proposition and, once they decide on working together, DoCoMo works closely with the developer to promote and enhance the application. The developer gets "DoCoMo Value" certified, which means the application is certified and tested. DoCoMo has already given 20 such certifications around the world including to people like IBM. Close partnerships with the right application developers is extremely important for carriers.

Another aspect that the industry doesn't seem to learn from is the "hyping" of technology and services. To their own detriment, the established players keep talking technology to the consumers who are least interested in hearing about it. Consumers only care about the quality and diversity of the services. Period. When DoCoMo started i-mode service in Japan, they never used the term "wireless Internet" to label, promote, or refer to the service. Otherwise people would have thought they would get same type of experience, tariff structure, applications, etc., as in the wired desktop world, so they never used the "wireless Internet" term. They didn't want to confuse consumers. They wanted a new concept, a new term, a new word to describe the service. Many operators around the world make the mistake of using technical terminology to describe or refer to their services. Technical terms should never be used to explain services. A lot of European and US carriers labeled their services as WAP or wireless Internet service. It's just stupid. It's a mistake. People expect too much if you use technology terms.

Gaming will continue to be an important application to attract youth. Characters from Disney or Pokeman are universal for kids, which means that gaming applications will be accepted not only in Asia but all over the world. Also, sophisticated technology such as J2ME will dramatically improve the capability for the application.

Another important aspect to consider is the local issues and circumstances. For consumers, each continent/country is different. For example in an Asian country, the young generation uses the devices to focus on entertainment; South America is sensitive to price as the average income is low. Different approaches are needed depending on the local market and demographics. With respect to the enterprise market, the US market is advanced, so other countries should learn from the USA as to how the enterprise market is supported and grown. In Europe, people are addicted to SMS messaging. So, a lot of the applications revolve around SMS. It will remain important for the future.

The 3G vs WLAN debate is taking its own course. WLAN has created an interesting disruption in the wireless industry especially in the USA, not so much in Japan. We should remember how the Internet pervaded our lives from no where and now it is everywhere. Such a phenomenon could also happen with WLAN. WLAN and WWAN are complementary because each system has a different purpose and the customer will decide what to use in a given situation. From a technical viewpoint, converging the two is not difficult; you can easily make dual mode data cards and chipsets. But challenges are security, authentication, billing, and roaming between the two systems. If such challenges can be overcome, the combination will provide a perfect solution for the end-users that they will really like.

Even though the telecom industry has gone through serious turmoil, the future remains bright. The cellular service and the automobile are the two biggest inventions of the twentieth century. People do not leave home without their car, and a phone is not only for business – that is common across cultures and languages. We need a cell-phone because it is a lifeline for communication. In the short term there are ups and downs but in the long run the future is very positive because so many people need a cell-phone. In the next chapter, we will take a look at some future technologies that are going to occupy our imagination in the next decade.

13 Future of wireless technologies, applications and services

The future isn't what it used to be.

Arthur C. Clarke

My interest is in the future because I am going to spend rest of my life there.

Charles Kettering

13.1   Introduction

This chapter aims to focus on the discussion of future of wireless technologies, applications, and services in the twentyfirst century. In Chapter 10, we discussed some of the technology issues and challenges for the industry. In this chapter, we will continue the discussion on the evolution and challenges of wireless technology in the context of the next 5–10 years. Several industry standards groups made up of manufacturers, carriers and academic institutions – including the IPv6 Forum, SDR Forum, 3GPP, Internet Engineering Task Force (IETF), and the Wireless World Research Forum – are helping to formulate a vision of a next-generation wireless world. Manufacturers and carriers are already looking to build on existing 3G specifications. AT&T has been developing a network prototype called 4G Access that combines Enhanced Data Rates for GSM Evolution (EDGE) with wideband orthogonal frequency division multiplexing (OFDM). Nortel has been working on software radio power amplifier technology needed to make higher wireless speeds a reality, and the streaming media research group of HP Labs has been working on systems for delivering multimedia content over next-generation networks.

Several companies are funding wireless research activities at the University of California. NTT DoCoMo's research labs are constructing a trial 4G network based on the ITU's proposals. The system combines variable spreading factor (VSF) and OFDM technologies. Japan's Ministry of Post and Telecommunications is shelling out subsidies of over ¥2 billion ($17 million) through Japan's Communications Research

Laboratory and Telecommunications Advancement Organization to develop core future technologies, such as software radios.

Research is not centered just on new network concepts and radio interfaces. There is a concerted attempt to identify how wireless technology can complement a more user-focused wireless world. One major shift already taking place in the wireless business model, and one the present authors expect to impact the 4G business and technology model for the future, is the move from a device-driven world to a service- and experience-centered world. Studies are now assessing new ways that users will interact with wireless systems, new services and applications that might become possible with the new technologies, and new business models that may prevail in the future, overcoming the traditional user–server–provider hierarchy.

At the most general level, next-generation architecture will include three basic areas of connectivity: personal area networking (PAN), local high-speed access points on the network, and cellular connectivity.

Despite reduced R&D spending, research will continue to be driven by concepts such as interoperability, security, system and device management and end-to-end connectivity. The Open Mobile Alliance is a good example of this trend. More specifically, R&D spending is now focused on system architecture, services and applications, radio design, signal processing, transceiver technology, network protocol and signaling, reconfigurable core design, interoperability and coexistence as well as spectrum sharing and bandwidth allocation.

The future wireless communication model will have to cope with different access technologies from user devices such as cellular, cordless, WLAN type systems, short-range connectivity, broadcast systems and wired systems. All these access technologies will have to be integrated into some kind of common, flexible and expandable platform to complement each other and to satisfy different service requirements in several radio environments. The interworking, mobility management and roaming would be handled via the medium access system and the IP-based core network. Interconnectivity will be important to the role played by small embedded computational devices in our everyday environment – enabling them to be operated seamlessly and transparently. These devices are meant to be active and aware of their surroundings so that they can react and emit information when needed.

Also, location and context awareness will play a major role for future wireless network services. There are several projects currently underway, mostly concerned with virtual tourist guides and involving tracking technologies such as GPS (Global Positioning Services). However, several companies are already looking at developing more sophisticated location solutions, such as providing location information in geographic contexts, as well as developing directory services for things such as personal profiles and preferences. Web services, and in particular UDDI, are expected to make a significant impact in this regard.

The Wireless World Research Forum (WWRF) in its *Book of Visions 2001, Version 1.0* (see Figure 13.1) describes the following building blocks for the wireless world:

(1) Augmented reality/cyberworld
  (a) wearables
  (b) deviceless communications
  (c) avatars
  (d) augmented reality
(2) Semantic aware services
  (a) context aware services
  (b) location aware services
  (c) extensive use of AI to assist in information retrieval
  (d) personalization
(3) Peer discovery
  (a) user addressing
  (b) service discovery
(4) End-to-end security and privacy
(5) Cooperative networks and terminals
  (a) reliable transport among heterogeneous networks and terminals
  (b) all IP
(6) Heterogeneous *ad hoc* networking
  (a) *ad hoc* networks among homogeneous and heterogeneous communication nodes
(7) 4G radio interfaces
  (a) spectral coexistence and frequency etiquette
  (b) positioning
  (c) multi-carrier
  (d) new air interfaces
(8) Smart antennas and base stations
  (a) High-altitude platforms
  (b) beamforming
  (c) MIMO, space time coding
  (d) radio heads and optical fibers
(9) Software defined radios
  (a) reconfigurable, downloadable protocol stacks.

We will be discussing some of the above in more detail in this chapter. Others we have already discussed briefly in previous chapters.

Generally speaking, it is very difficult to forecast the technology advancement in the next 10 years, especially in the area of IT and communications. Here is a good example. NTT published a report with the title *Telecommunication technology in 2005* in 1990 when the company established the "V (Visual) I (Information) & P (Personal)" concept as its service vision for the twentyfirst century. Reading the report

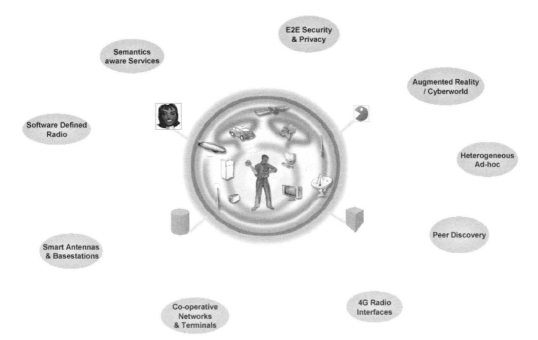

Figure 13.1. Nine building blocks of the wireless world (source: *Book of Visions*, Wireless World Research Forum).

now, it is hard to conclude that all the forecasts were achieved. Some technology has progressed beyond expectation and others did not do as well we expected.

Here are some typical examples.

Technology progressed beyond expectation (1990–2000)

Area; technology

Device/materials; optical disk, CMOS-chip, TFT-LCD

Human interface/information processing; voice recognition, CG, PC, work station

Access/network; WDM, wireless Internet, VoIP, car navigation

Field technology

Technology progressed contrary to expectation (1990–2000)

Area; technology

Device/materials; optical processing, GaAs logic-chip, color-CRT, micro-machine

Human interface/voice synthesis, fuzzy computer chip, 3G display, auto-navigation

Access/network; total operation system for multi-vendor, ATM-LAN

Field technology; space, marine, underground

Technology progressed as expected (1990–2000)

Area; technology

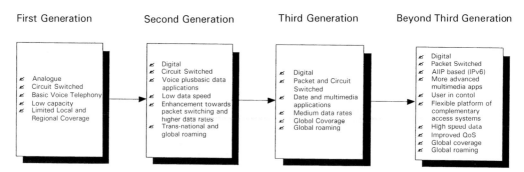

Figure 13.2. The paradigm shift from first generation towards beyond 3G (source: WWRF).

Device/materials; memory (256 Mbit), optical device, molecule (bio) device/switch

Human interface/neuro-computer chip, knowledge-base system

Access/network; cellular-phone (stand-by time), ATM-switch

Field technology

These are just some examples to show that it is hard to predict with certainty the trends of the next 10 years. However, we will consider some hot topics on the global trend of wireless technology evolution in next 10 years and also related challenges.

13.2 Systems beyond 3G (B3G)

One of the hottest topics for wireless engineers in the world now is the so-called fourth generation or "beyond 3G" mobile system. After this section, we will review the latest activities on this issue as a typical case study. Figure 13.2 shows the paradigm shift from first generation towards beyond 3G.

At present, there is no global consensus on the concept of beyond 3G mobile systems; however, we can find some important messages in Figure 13.3. One is all-IP based and the other is high-speed data and more advanced multimedia. Many international and regional standardization bodies have already started study programs both on entire system concepts and also on particular technologies.

Standardization the bodies on the entire system are:

• ITU-R WG8F, ITU-T SSG
• mITF (Sub-Committee of Japanese ARIB)
• IST (EU 5th RTD Framework Programme)
• 4G mobile forum (IEEE)
• the Wireless World Research Forum (WWRF).

Standardization bodies on particular technologies are:

• OFDM forum

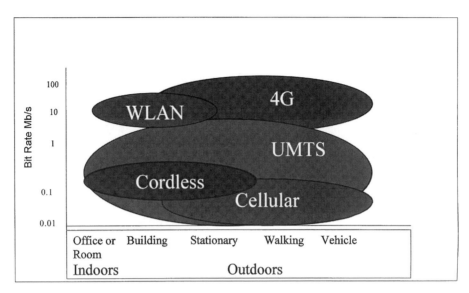

Figure 13.3. Mobility vs bit rate perspective of 3G–4G systems.

- IPV6 forum
- SDR (Software Defined Radio) forum
- ASMS (Advanced satellite mobile systems).
    Now let us look at the major global activities.

WWRF (www.wireless-world-research.org)

The Wireless World Research Forum (WWRF), launched in August 2001, is a non-standardization body and concentrates on the definition of international and national research items relevant to future wireless communications. As of January 2002, it contained 19 manufactures, 11 operators and 19 R&D centers and universities.

IST (www.cordis.lu/ist/home)

IST (Information Society Technologies), started in 1999, is concerned with European research on systems beyond 3G. It is a part of the EU 5th RTD Framework Programme and it has an indicative budget of 3600 million euros. IST will study the "System Beyond 3G Cluster" and its scope is the evolution of access systems, and IP in core and mobility management.

mITF

mITF (mobile IT Forum), established in June 2001, is a sub-committee of the Japanese standardization body ARIB and aims to realize an early implementation of "Systems beyond IMT-2000" and mobile commerce. Figure 13.4 shows mITF's view on Systems beyond IMT-2000.

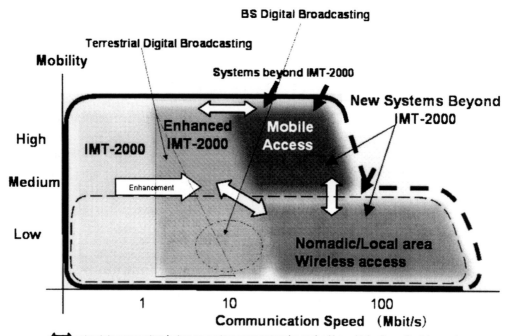

Figure 13.4. mITF's view on Systems beyond IMT-2000 (source: mITF, Agreed document in the 6th ITU-R WP8F, October 2001).

mITF proposes a development target in two steps.

Step 1 (by 2005):   enhancement of IMT-2000 (expansion stage)
                     about 30 Mbps (down-link)
Step 2 (by 2010):   systems beyond IMT-2000 (mature stage)
                     about 50–100 Mbps (down-link)

From the service viewpoint, mITF expects the Systems beyond IMT-2000 to be like a shopping mall, while IMT-2000 (3G) is like a department store. The concept comes from the idea that users of the future will prefer a flexible selection of optimum wireless services according to the usage environment. Also, on the frequency issue, mITF comments that in the year 2015 additional spectrum of 1.2–1.7 GHz bandwidth will be required (note that indoor traffic is excluded in this assesment) and the band below 5–6 GHz is suitable. The reasons for the consideration are two-fold: one is the increase in cell sites above 5–6 GHz, and the other is the loss caused by the human body above 5–6 GHz.

Liaison activities

Liaison activities on a global scale are indispensable for the success of wireless system development, because the wireless users are already living in a global economy and global society. For example, WWRF and mITF have started to exchange information and are seeking ways of collaboration. Readers can easily understand that this kind of activity is a natural conclusion of the theme of this book.

13.2.1   OFDM (Orthogonal Frequency Division Multiplexing)

The development of radio access technology for wireless communications has been enormous in the past two decades. The main challenge of the transition from 1G to 2G was to establish a stable digital transmission performance in a mobile environment, and the challenge of 2G to 3G was to increase the frequency utilization efficiency and strengthen the multi-path fading robustness of the system. In this long process of R&D, many great access technologies, such as TDMA, CDMA and WCDMA, were invented.

OFDM is one of the most notable access technologies going towards the post-3G system. Originally, OFDM was developed for the terrestrial digital broadcasting system in the UK and also was adopted as European DAB (Digital Audio Broadcasting) DVB (Digital Video Broadcasting) standard. OFDM has two main technical features: one is its super multi(sub)-carrier transmission with different center frequencies, and the other is orthogonal transmission between adjacent carriers. In cases where the number of multi(sub)-carriers increases, the equivalent transmission speed per carrier becomes low. The mechanisms can dramatically increase the robustness of the system against multi-path fading under high-speed digital transmission.

The adoption of OFDM technology for the WLAN IEEE 802.11a standard has made its status decisive. The typical parameters for the IEEE 802.11a standard are 52 sub-carriers (including four pilot carriers) with guard intervals of 800 ns.

The recent development of LSI, especially DSP technology on a chip, is a strong following wind for OFDM. The availability of the 0.18 μm rule CMOS-LSI and a 12-bit A-D/D-A converter enables the OFDM function on a chip. Some R&D organizations are now tackling the challenging issue of OFDM-based access technology for post-3G mobile systems. Multi-carrier CDMA (MC-CDMA, MC/DS-CDMA) is one attractive approach. This approach aims to combine the different merits of OFDM (strong multi-path robustness) and CDMA (high-frequency utilization efficiency due to code division).

NTT DoCoMo has proposed a new architecture called VSF-OFCDM (Variable Spreading Factor–Orthogonal Frequency and Code Division Multiplexing) based on the MC-CDMA concept. This system employs a variable spreading factor so that optimum system design can be possible depending on the different requirements for the operations such as WAN and LAN. Table 13.1 shows the major parameters of the VSF-OFCDM experimental system.

In the USA, intensive studies on OFDM are now underway, as described below. Flash-OFDM is a combination of frequency hopping of sub-carriers with OFDM technology proposed by Flarion Technologies (www.flarion.com). The company

Table 13.1. Major parameters of VSF-OFCDM experimental system

Transmission speed	max. 100 Mbps (down-link) and max. 20 Mbps (up-link)
Number of sub-carriers	768
Sub-carrier spacing	131.836 MHz
Bandwidth	101.5 MHz
Symbol length	9.259 μs
Frame length	54 symbol
Modulation	Q-PSK, 16–64 QAM
SF	1–256
Coding rate	1/3–5/6

aims to use this technology not only for a mobile service, but also for digital home electronics.

Many US makers propose MIMO-OFDM (Multiple Input Multiple Output) for BWA (Broadband Wireless Access) applications. These are "AirBurst" by Iospan Wireless. Inc., "Vector-OFDM" by Cisco Systems Inc., and "Adaptive Multibeam OFDM" by BeamReach Networks Inc. These technologies will use multiple transmit/receive antennas to improve frequency efficiency and transmission quality.

The most critical technical challenge to achieve OFDM as a commercial product is the requirement for linearity and power consumption of power amplifiers, because OFDM has a higher PAPR (Peak to Average Power Ratio) of dynamic range of transmitted signals compared with conventional modulation systems such as Q-PSK and FSK.

13.2.2   All-IP networking for 3G/beyond 3G mobile systems

The development of the Internet is tremendous and it has become a lifeline for daily and business activities in developed countries. Among the fixed telephone operators, the introduction of IP technology to fixed communications has started because of the following reasons:

(1) network cost reduction due to IP routing,

(2) operation cost reduction due to simple network architecture, and

(3) expectation for the new service deployment combined with Internet.

The increase of IP packet traffic is limited not only in fixed communications, but in wireless communications. The success of i-mode is one typical example.

Considering the great success of both Internet and cellular services in the past decade, the convergence of the mobile network and Internet is the mainstream of next-generation cellular network evolution. Introducing all-IP architecture into the mobile network is an effective approach to cope with the increasing wireless Internet traffic demand and compatibility with the Internet. Standardization bodies such as 3G-PP and UMTS Forum, Wireless Carriers and Vendors are promoting intensive study on all-IP networks for 3G and beyond 3G. The following items are major targets of future studies.

(1) Mobile-IP technology

The Mobile-IP concept was proposed by the IETF (Internet Engineering Task Force). This technology aims to achieve mobility control functions such as mobile subscriber authentication at IP layer. Using Mobile-IP, a roaming service between different networks (i.e. the cellular network and WLAN) becomes available.

(2) IP-Call Control technology and IP-QoS (Quality of Service)

These technologies are important to achieve IP-Multi-Media communication services. As IP-Call Control technology, SIP (Session Initiation Protocol) and H.323 protocol are proposed. These protocols are used to implement connection-type communication on connectionless-type IP networks. IP-QoS technology can satisfy different QoS requirements for multimedia communication on IP networks.

(3) Open API (Open Application Programming Interface)

This technology is effective for rapid service development of value-added services owing to open service control interfaces and is widely discussed in several standardization bodies like Parlay and JAIN (Java API for Integrated Network). Application of Open API technology to mobile network enables advanced services such as email and location-based application services.

Figure 13.5 shows the IP-based core network configuration.

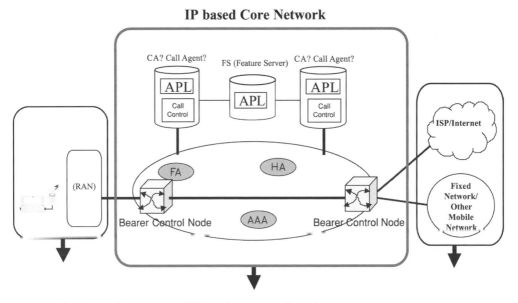

Figure 13.5. The concept of IP-based core network configuration.

NTT DoCoMo has proposed the concept of IP2: IP-based IMT Network Platform. The requirements for IP2 are:
(1) huge multimedia traffic handling
(2) diversified wireless access accommodations
(3) advanced mobility management
(4) seamless/ubiquitous services support.

In order to achieve IP2, separation of network controls and IP-based transport is important.

13.3   Future of devices and displays

Progress in devices is for the most part dependent on the evolution of networks, but is also affected by technical breakthroughs in areas of memory, display, battery, user interface, display, connectivity, interoperability, security, compression, OS, and APIs. As shown in Figure 13.6, future devices and displays will come in every imaginable shape, capability, and form factor. There will never be one specific device that dominates the market, rather, there will be increased segmentation in the market target groups, and new features and technologies will be embedded accordingly (plug-and-play to a certain extent) like an MP3 player, gameboy, GPS receiver, etc. Foldable displays will be available to communicate with transmitter and receiver units which might be physically separate. Displays that become your Wall Street Journal in the morning, writing pad in the afternoon, and game machine by evening will not be uncommon. Displays and devices developments are most exciting for consumers for they do not care about the underlying technology but rather the functionality and feature set of something tangible that they can use repeatedly and reliably. Table 13.2 shows the expected evolution of devices for the next 5–10 years.

13.4   Battery technology

In Chapter 10 we touched upon the challenges in developing batteries that enable longer usage of devices for both voice and data applications. It has been a challenge for the battery industry to scale down battery size, while at the same time obtaining high internal pressure resistance and meeting satisfactory insulation requirements. While there has been some progress, many prototypes based on micro-fuel cell technology are struggling to move from ideal lab conditions into commercial viability.

The idea is to exceed by orders of magnitude the power provided by the current cell-phone battery substrate champion, lithium, which at best provides six to eight hours of talk time from a fully charged battery. Initial test results from prototype fuel cells suggest the provision of up to 20 h of talk time for 3G phones.

Table 13.2. Evolution of devices

	Current	Future
Screens and input	Mostly B&W, small displays. Key pad and some voice input	Color, bigger, flexible and foldable screens. Input via speech, keypad, touch, gestures
Integration with PAN/LAN	Minimal	WAN/LAN/PAN capability will be integrated which will lead to enormous application capability
Security	SIM	SIM, biometrics
OS	Proprietary, Symbian, Microsoft, Palm, WAP, and other microbrowsers	XHTML, HTML, WAP will converge into the OS
Performance	Limited data storage and processing power	Significantly high storage and processing capability
Content type	Primarily text, limited video and audio	Primarily (adaptive) multimedia content

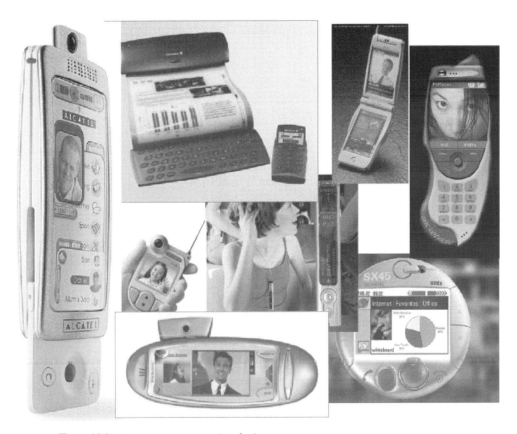

Figure 13.6. Concept next-generation devices.

Figure 13.6. (*cont.*)

Lithium-based battery technology is popular because of its high energy density, high voltage, excellent pulse capability, minimal self-discharge and none of the memory effect that plagued NiCd and NiMH batteries. Various incarnations, such as lithium ion and lithium polymer, will remain the focus for the next three to five years.

Lithium polymer is more expensive but is currently the darling of the battery industry because of its malleability, thin shape and modest weight. But lithium polymer batteries tend to swell about four times more than lithium-ion batteries when they are exposed to heat. Heat also reduces the number of times they can be recharged. So the development of cheaper and power-efficient lithium-ion batteries continues.

Samsung has developed an ultra-thin lithium-ion battery by employing a new sealing method. It is as light and thin as a polymer battery, and can store 355 Wh in a space of a 2.8 mm-thin pack. And Toshiba's Advanced Lithium Battery (ALB) derives its power from a polymer-like liquid that can be moulded to about 1 mm. Lithium Technology Inc. has developed a fibrous web substrate to create a composite battery structure that is said to reduce manufacturing costs. Developer Lightyear says that wrapping electrolyte components with its one-molecule thick TDM material will block internal corrosion, doubling battery life.

But it is the micro-fuel cells and the idea of self-powering chipsets that has grabbed most R&D attention. Fuel cells, the twentyfirst century versions of the bulky batteries used to power spacecraft in the 1960s, convert chemical energy directly to electrical energy by passing a fuel, such as hydrogen, across a special membrane to react with oxygen, creating an electric current (Figure 13.7). On the whole, the new fuel-cell technology is very efficient, converting almost 50% of the hydrogen fuel into electricity, compared with 25% for current battery technologies.

But because hydrogen is difficult to store and transport, modern fuel cells are instead designed to extract hydrogen from methanol, acids or other hydrocarbon fuels. Most of the current prototypes revolve around the splitting of methanol to produce hydrogen, a process called reformation.

In what is likely to involve a major change in socio-techno habits, users would snap in a new, inexpensive, diluted methanol cartridge rather than find the nearest power source to recharge the battery. The first fuel-based versions are not likely to be seen until 2004, however.

NEC, Toshiba, Sony, Motorola and a host of others are midway through prototyping clever variations of these new methanol-based fuel-cell technologies. The trick has been to miniaturize the methanol reformation process so that it can be used in cell-phone batteries. These reformed hydrogen fuel cells use catalysts like platinum and an internal membrane made from ceramics or newly discovered carbon polymers to split the diluted methanol. Several cells would be connected together in series electrically to increase the output voltage of the system, with the goal of five times the energy density of conventional lithium-ion rechargeable batteries.

Figure 13.7. Fuel cell from Polyfuel.

Several other innovations depend on improvements in battery technology before they see light of the day. It will be hard to convince consumers to run heavy multimedia applications unless better and affordable battery technology becomes available.

13.5   How to get on well with a limited frequency resource?

As everybody knows, frequency is the limited resource like a surface of the Earth. Once used, it is not an easy job to move it to another location. Because of this specific feature of frequency, national and international organizations have been working hard to establish a common way of frequency usage for national and international wireless services. Frequency allocation should be assigned with international consensus in order to avoid interference and increase the utilization efficiency. For example, the ITU has specified the basic principle of frequency allocation and usage and the RR (Radio Regulation) of ITU has specified in great detail the purpose of each frequency usage from 9 kHz to 400 GHz. In this range, the frequency band that is now mainly used for digital wireless communications is between 1 and 40 GHz. Expectations for the future usage of frequencies above 40 GHz are very high; however, the disadvantages of large propagation loss and expensive device cost are still to be solved.

Considering this, the frequency band for future wireless systems may be selected in the same range of the above 1–40 GHz band. As the band has been widely used and occupied by existing wireless systems, it will be very difficult to secure a continuous large-frequency band for future systems. With the above frequency environments, what technical solutions should we have?

One solution is intelligent and adaptive technology for high-speed wireless data. In the future, the increase of wireless multimedia traffic will make great impact on the frequency bandwidth requirements. So, the required frequency bandwidth should be secured by using multiple non-continuous bands as a unified one with adaptive technology. Adaptive technology includes, for example, optimum power control, adaptive modulation, adaptive antennas, software-defined radio, and equalization techniques. Some of these adaptive technologies are already developed and used for terrestrial digital fixed radio systems (adaptive antennas, equalization techniques) and for second-generation cellular systems (transmission power control). The challenge now is to establish the overall system design methods and criteria using the above technologies for future wireless systems.

13.6   Re-configurable terminal/software defined radio technology

The basic concept of software defined radio technology was proposed in the early 1980s for defense applications, and in 1996 the MMITS (currently SDR) Forum was founded (www.sdrforum.org). Now, SDR and related technologies are one of the hottest topics globally. The European Commission has initiated CAST and SCOUT projects with the IST Program (www.ist-scout.org).

The principal purpose of SDR is to achieve universal radio interfaces regardless of multi-band, multi-standard conditions and multi-access technologies (FDMA, CDMA, TDMA, etc.). There are some different motive forces of SDR depending on the market situation. In Europe and Asia, introduction of the same standard systems such as 2G (GSM) and 3G (WCDMA) has forced all wireless carriers to make service differentiation by the system itself. So, carriers seek ways to offer any services they desired quickly and flexibly. In the USA, on the other hand, there exist multiple standards and the transition between different access systems (from AMPS to digital, from 2G to 3G) increased the market chance for SDR.

The impact of SDR is not limited only to terminals, but is also for base-station infrastructures. SDR technology may achieve the flexible and multiple access function of a BTS under the beyond 3G environment where there is convergence towards all-IP core network and B3G radio access systems. Technical challenges for SDR are as follows.

(1) API (Application Interface)

API is a common application for multiple hardware by standardized API.

(2) RRM (Radio Resource Management) and spectrum allocation

RRM increases the efficient use of frequency by SDR.

(3) Software download mechanism

To increase the security, reliability and optimum protocol.

(4) Hardware and device technology

RF circuit, antenna, DSP, battery and other hardware technology for re-configurable terminals.

SDR may create a new value chain for operators, contents providers, terminal vendors and users where the open business model and platform are achieved.

13.7 Mesh networks

Mesh networks are networks that capitalize on the rise of Wi-Fi and other open wireless technologies. They shimmer into existence on their own, forming *ad hoc* out of whatever is in range – phones, PCs, laptops, tablet computers, and PDAs. Each device donates a little processing muscle and some memory. Packets jump from one user to the next – finding the best path for the conditions at any given moment – and finally skip to a high-bandwidth base station, which taps into the Internet. This boosts the range and speed of wireless signals. With the help of, say, 50 meshed PCs, PDAs, and phones, a typical Wi-Fi network with a 500-foot range can be transformed into one that extends 5 miles. In fact, the performance gains and cost savings are so great that these systems easily undercut today's wireless broadband service.

Traditional wireless systems are constrained by the old hub-and-spoke model. In a standard Wi-Fi network, for example, all the devices in range connect to a single transmitter. The wireless PC cards, which are flying off shelves at a rate of 1.5 million a month, can communicate with one another, peer to peer. The same goes for Bluetooth and the handful of proprietary technologies just bubbling up. An issue with mesh networks is that they are potentially really cheap, so there is no incentive to build one if you are a telco. But as mesh spreads, it will be hard to resist.

Each user device acts as a router/repeater for other devices. This means that users cooperate rather than compete for network resources. The device's ability to function as a router/repeater allows users to hop through other users to reach network access points. Hopping increases both the network throughput and the coverage area by leveraging users as part of the network.

This network architecture supports the ability to store MP3 files at home or on a network server, and then stream them wirelessly to cars, laptops, PDAs or wireless MP3 players. Users would also be able to share files directly with other users in an Internet-like wireless environment. The present authors believe that this technology will result in most MP3 players being wirelessly enabled in the future. It also means that expensive, single-function satellite radios could be replaced by much less expensive and much more personal wireless MP3 players.

Here are some important points about mesh networks (source: Mesh Networks).

- **Self-forming**. Subscriber devices discover, build and maintain their own routing tables in real time. Each device co-operates with its neighbors to create low-latency and high-quality paths throughout the network. New subscribers and infrastructure components are seamlessly integrated into the network routing tables in real time.

- **Self-healing**. Each element of the network, including subscriber devices, plays a part in making the network incredibly robust. Self-forming routes and Multi-Hopping™ capabilities mean the network is able to deal with point failures, congestion and other issues with little or no interruption in service.
- **Self-balancing**. Multiple network access points (IAPs) are available to subscribers since they seek out available resources in the network via the self-forming, Multi-Hopping™ routing process. This is possible because the network is made up of distributed, low-power, high-bandwidth radio access points and intelligent, self-routing, multi-hop capable subscriber devices. Essentially, a mesh enabled network turns the "Congested Cell" paradigm on its head; the more subscribers in an area, the more paths to alternative IAPs – thus increasing overall network utilization.
- **Supports high mobile data rates**. Subscribers get up to 6 Mbps burst rates and 2 Mbps sustained data rates. These are not per cell or per sector rates, but are real per subscriber rates – at speeds of up to 250 mph.
- **Spectrally efficient**. Subscribers do not have to "shout" at a centralized base station as in a cell-based network, but rather can whisper (via power control) to a near-by device that passes on the transmission to its destination (via Multi-Hopping™). Our technology enables subscriber terminals to co-operate, instead of competing for resources. Spectrum reuse increases dramatically, while overall battery consumption and RF output within a community of subscribers is drastically reduced.
- **Supports end-to-end IP**. By leveraging standards-based IP, any IP-based application and IP-capable device works seamlessly in the network. No modifications are required.
- **Built-in geo-location**. The network does not depend on GPS for subscriber geo-location. In fact, it is quicker and more accurate than consumer GPS, and supports location-aware services that will lead to next-generation mobile applications.
- **Quality of service management for voice, video and data**. Routing algorithms manage each media stream separately. This means it offers excellent cellular quality voice, high quality video, as well as high-speed data – all on a single network.
- **Supports infrastructure-less peer-to-peer networks**. Subscribers can form networks between themselves without the need for network infrastructure. This peer-to-peer capability has two major benefits. First, subscriber-to-subscriber communications can avoid using infrastructure resources – which significantly reduces network congestion. This increases the effective capacity of the network and increases subscriber service levels. Secondly, networks can be established where infrastructure does not exist or has been damaged. A high-speed broadband network will automatically form between subscribers, while Wireless Routers and IAPs will be re-incorporated when they come back on-line.

Mesh networks can address a wide variety of applications as a result of patented architecture and routing technology. This technology, found in mobile network solutions, uses a peer-to-peer network architecture instead of a traditional cellular architecture (i.e. one that uses tall towers that all communications must pass through). This eliminates the need for towers and their expensive radio equipment by making the subscriber device a major part of the network infrastructure. The result is a robust, high-bandwidth, mobile networking solution that can be deployed and operated for a fraction of the cost of next-generation cellular networks. However, these benefits are not limited to wide-area mobile data networks. Telematics, last-mile access and enterprise LANs are just a few of the additional applications that benefit from this technology.

Mesh radio replaces the conventional cellular tower with small "nodes" that have directional antennas beaming signals to neighboring nodes, creating a web-like mesh. PDAs, smartphones and laptops equipped with inexpensive 802.11b cards then become wireless routers capable of reaching any point in the network.

At the heart of mesh radio networks is a stubby "node" sitting atop a building, lamp post, or other stable structure. Inside these access points are four intelligent directional antennas seeking out the strongest signal and realigning should a transmission weaken.

The most immediate advantage of mesh radio is cost. Firms like Florida-based Mesh Networks estimate that the infrastructure investment required for mesh radio is a tenth of traditional cellular networks. Other advantages of mesh radio include the following.

- Use of unlicensed 2.4 GHz band. As the coffers of wireless operators are strained under the astronomical licensing costs of radio spectrum for third-generation networks, mesh radio provides mobile broadband at speeds greater than 3G for a fraction of the price.
- Whereas conventional wireless networks endure "dead" areas and line of sight issues, mesh radio provides 100% coverage of an area.
- As carriers slowly meet Federal requirements for being able to locate wireless 911 calls (E-911) and ultimately enter the market for location-based services, Mesh Networks has built-in geo-location capabilities to pinpoint positions within 10 m in under 1 s.
- With the controversy over the medical effects of microwaves, mesh radio uses more focused, higher frequencies with towers requiring only 1 W of power, compared with 8 W needed for cell-phone towers.
- Mesh radio is capable of DSL-like speeds of up to 25 MB per second for uploading and downloading.

Mesh architecture overcomes the three major issues facing point-to-multipoint (PMP) systems: firstly, limited customer coverage due to the need for a "line of sight" between base station and customer; secondly, the difficulty of providing high data rates to large numbers of subscribers within available radio spectrum; and third, the large amount of capital investment required for network infrastructure, needed to provide coverage in advance of customer acquisition.

13.8 Seamless migration and multiple wireless systems

Recent development of various wireless technologies such as WLAN and PAN has made a great impact on the future roadmap of wireless technology and service. Former wireless systems of 1G, 2G and 3G have been initiated and developed based on the standardization bodies such as ITU and ETSI or strong leadership of dominant national wireless carriers. AT&T developed the digital TDMA system and ETSI/GSMA developed GSM.

3G standards have long been discussed in ITU where powerful national carriers have made sufficient contributions. In this sense, the history of former wireless technology development can be compared with the *Planned economy* of socialist countries. We can say that the former wireless system roadmap acts as the planned development scenario by several national carriers.

But now the environment is dramatically changing in the wireless world, in just the same way that the wired world has already entered the era of *market economy* by the low-cost ADSL. The wireless world will gradually moving from *planned economy* to *market economy* by WLAN and PAN. Bluetooth and UWB (Ultra Wideband) can be candidates. (Note: we will discuss possible future scenarios in more detail in Chapter 14.) One expert said that "3G will be the final system which tries to cope with all the possible applications. After 4th generation, optimum wireless systems will be combined to cope with applications".

NTT DoCoMo started a commercial WLAN service called *M-Zone* in July 2002, and the concept is to complement FOMA and WLAN and to obtain more value for both services. The target of the combination is not only WLAN, but telematics and the digital broadcasting service. Another service trend is that there is more variety in the applications, such as video, mail, pictures and music. Wireless systems of the future must satisfy the customer's demand for the variety of applications.

So, what is required next? It is the functionality that enables an end-user to select and enjoy the best and optimum wireless system depending on the application he/she wants to use. An end-user has different preferences according to the situation.

For example, when the user wants a voice conversation with a friend, the high-quality voice service of a 3G circuit switch is needed. When the user downloads a video-clip image at a coffee shop, they prefer to use WLAN. When they are on a bus, they may check email on their i-mode phone. When the user wants to save on cost, they may choose a cheap portable VoIP phone. The requirements for these situations are different, and therefore are difficult to integrate with a single system. The important issue for the future is to establish the mechanism to provide seamless function over multiple wireless networks. Here are some ways to achieve this functionality.

(1) SIM / UIM card

The SIM (Subscriber Identity Module)/USIM (User Subscriber Identity Module) card is a mechanism used for security and roaming in GSM and 3G. The card contains information such as terminal number, subscriber telephone number, PIN, PUK, and directory. By using the card, users can enjoy roaming with different operators and enhance security.

(2) Software

Some advanced software technology can achieve seamless capability over different wireless systems. KDDI and Cisco Systems recently announced the development of a seamless handover function between 1xEV-DO (3G) and WLAN networks. With this, a user can use both systems with an automatic switching function. The users can use data communication in a car with 3G and at the office and when parked with WLAN. Fujitsu Labs also announced the development of a seamless handover function between FOMA (3G) and WLAN networks. Mobile-IP is used for location registration of the terminals. HA (Home Agent) server is used to register the terminal IPs. Solutions for WLAN to achieve roaming with different WANs are being developed, for example by Seattle-based venture Netmotion Wireless (www.netmotionwireless.com).

These solutions are effective for a seamless service in office, outdoor and indoor public places using multiple wireless access technologies.

13.9   Multimodal user interfaces

Multimodal access to applications and services is difficult in today's networks as it is not easy to create services with simultaneous voice and display characteristics. There are two trends happening which are encouraging. First, people are building mobile micro-browsers with speech capability and, second, 2.5G and 3G networks promise to offer greater bandwidth, always-on connection, and simultaneous voice and data channels with reduced latency. W3C has instigated work on a multimodal market language, which will help in moving applications to the next evolutionary stage. In the meantime, there will be some proprietary technologies coming to the market in this space.

On the markup language front, two standards seem to be emerging, one is XHTML+VoiceXML based on the work done by the VoiceXML forum, and other is the newly minted SALT (speech applications language tags) formed by Microsoft (see Figure 13.8).

In either case, there are three user-case scenarios that need to be addressed: (a) server-based, (b) distributed voice recognition-based, and (c) client-based. In (a) the speech processing is primarily done on the server after the extraction has taken place on the client. In (b) the extraction and processing can take place on the client itself.

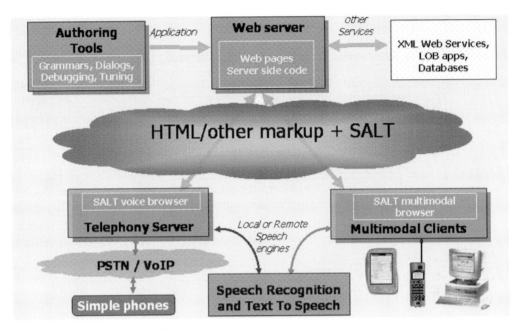

Figure 13.8. SALT architecture.

In (c) the synchronization helps in distributing the tasks of extraction and processing between the server and the client. Figures 13.9 shows these scenarios.

ETSI is working on a standard for distributed speech recognition (DSR) "front-end" of speech recognition. In a DSR architecture the recognizer front-end is located in the terminal and is connected over a data network to a remote back-end recognition server. DSR provides particular benefits for applications for mobile devices such as improved recognition performance compared with using the voice channel and ubiquitous access from different networks with a guaranteed level of recognition performance. Because it uses the data channel, DSR facilitates the creation of an exciting new set of applications combining voice and data. To enable all these benefits in a wide market containing a variety of players including terminal manufacturers, operators, server providers, and recognition vendors, a standard for the front-end is needed to ensure compatibility between the terminal and the remote recognizer. The Speech Transmission Quality (STQ)-Aurora group within ETSI has been actively working on developing this standard.

The main benefits of DSR are: improved recognition performance over wireless channels, ease of integration of combined speech and data applications, and ubiquitous access with guaranteed recognition performance levels.

So far we have primarily talked about speech, keypad and stylus as input mechanisms. There is a lot of research being done to explore "gesture recognition" which essentially processes complex gestural, emotional, and affective input such as face

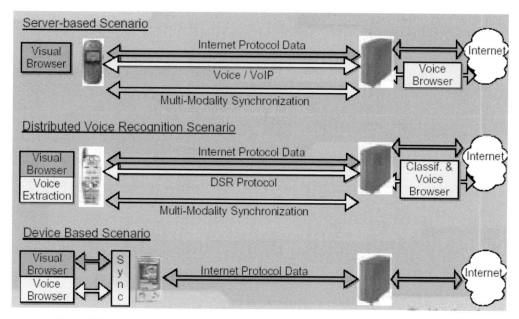

Figure 13.9. Multimodal access scenarios.

and lip movement. Humans often use gestures, change in tones, shifting eye gaze, etc., as cues to direct a conversation. Gesture recognition can help do the same with human–computer interactions. Figure 13.10 shows a multimodal application framework wherein various input signals arrive at the application via an intermediary multimodal integration middleware piece in parallel. During recognition, these modality-specific understanding components produce a set of attribute/value meaning representations for each mode. These values generated for an incoming signal provide alternative meaning hypotheses for that signal, each of which is assigned a probability estimate of correctness. The multimodal integration step combines symbolic and statistical information to enhance system response robustness and synchronize interactions.

13.10 Wearable computing

In the broad spectrum encompassed by speech technology, wearable computers have to this point quietly moved from their own small niche. That niche, however, is beginning to draw its own share of attention by posting significant growth of its own. These computers are designed to be worn and used on the body – from wired jackets to wireless watches to displayable glasses.

So far wearable computers have found their use mostly amongst industrial applications but slowly and steadily we are moving towards consumer applications as well.

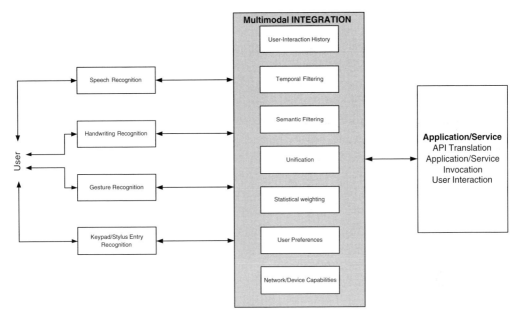

Figure 13.10. Multimodal integration framework.

Industrial applications of wearable computing include systems for warehouse management, automation, performance support, etc. The technology is already being heavily used in industries such as aerospace, law enforcement, aircraft maintenance, field data collection, insurance, emergency services, securities trading, point of sales, etc. The IDC expects the worldwide value of wearable computer shipments to grow from $3.5 million in 2000 to $579 million in 2004.

The obstacles for this industry are miniaturization, cost of specialized parts, user interface, noisy environments, and consumer acceptability. With developments in the areas of wireless, position location, processing, and speech recognition technology, wearable computing is slowly gathering traction. Even for industrial applications, most of the input is via keypad or stylus but gradually speech is being integrated as one of the input mechanisms. There are certain situations where users have to use hands-free wearable computers and speech is the only way they can interact with the computer or backend infrastructure.

13.11 Ultrawideband technology

In a controversial decision in 2002, the Federal Communications Commission approved the commercial use of "ultrawideband" technology. Ultra Wideband (UWB) radio is a revolutionary wireless technology for transmitting digital data over a wide spectrum of frequency bands with very low power. It can transmit data at very high

rates (for wireless local area network applications). Within the power limit allowed
under current FCC regulations, UWB can not only carry huge amounts of data over
a short distance at very low power, but also has the ability to carry signals through
doors and other obstacles that tend to reflect signals at more limited bandwidths and
a higher power. At higher power levels, UWB signals can travel to significantly greater
ranges. Instead of traditional sine waves, UWB radio broadcasts digital pulses that are
timed very precisely on a signal across a very wide spectrum at the same time. Transmitter and receiver must be coordinated to send and receive pulses with an accuracy
of trillionths of a second. UWB can also be used for very high-resolution radars and
precision (sub-centimeter) radio location systems. Using discrete radio-frequency
(RF) components, Intel demonstrated a UWB-enabled system that supported data
rates at speeds of 100 megabits per second. The company aims to push this wireless
technology to 500 Mbps.

UWB technology provides a faster and more secure way of sending wireless transmissions. Automakers could use the technology to build collision-avoidance systems
or improve airbags. Firefighters, police officers and emergency personnel could use
it to detect what is behind walls, buried underground or even inside the human
body. Consumer products, from laptops to personal digital assistants, could use the
equipment to send and receive video or audio. But the US military and mobile phone
carriers say the signals are so powerful they could cause disruptive interference to
their wireless operations.

The FCC decision took the interference complaints into account by setting limits
on what radio frequencies UWB devices can be operated in, including avoiding those
frequencies used by the military and companies that sell global positioning services,
technology that pinpoints a person's location.

With the market already ripe with technologies such as 802.11X, Bluetooth,
HomeRF, and HiperLAN, how will UWB fare? To answer the question, let us take a
quick look at the differences amongst these LAN/PAN technologies.

> *802.11a.* This WLAN technology carries a data transfer rate of 54 Mbps, accord-
> ing to the standard, and can reach roughly twice that speed using proprietary
> "turbo" architectures. It is currently considered to be the only serious con-
> tender to Ultra Wideband, particularly in its throttled-down state. In fact, as
> it stands now, 802.11a has a hypothetically greater range than UWB, although
> once obstacles are figured into the equation that rate/range advantage rapidly
> diminishes. The diminished rate is caused by protocol inefficiencies and system
> overhead, according to Xtreme Spectrum.
>
> Also, according to Xtreme Spectrum and other researchers, the fundamental
> design of the MAC is problematic, because 802.11, as an Ethernet derivative, was
> designed as a packet-based data networking protocol. This makes it inherently
> unsuitable for intensive multimedia applications, which depend on data packets
> arriving in order and in time.

Another fundamental flaw in 802.11a technology is that its power consumption requirements of around 1.5–2 W makes it almost completely unsuitable for battery-dependent devices like PDAs, and even many laptops with short battery lives.

At the present time, the cost of consumer NIC cards for 802.11a is roughly $150–200, which could be too high for broad consumer adoption. In its current manifestation, 802.11a technology appears to be mature enough in the market to be suitable for demanding enterprise or public access markets, where power consumption issues are less important, and data transfer still takes precedence over multimedia. In terms of residential home networking, though, 802.11a is not particularly appealing, given the limited bandwidth and potentially prohibitive costs involved.

802.11b. At 11 Mbps, or even in its 802.11g incarnation at 54 Mbps, 802.11b (Wi-Fi), is perfect for file sharing and email access in a simple business environment unencumbered by too much physical infrastructure that causes interference. One of the most interesting things to note about 802.11b, and 802.11x in general, is that the strength of the industry lobby might also cause it to be perceived as the most viable, robust solution for the enterprise market.

Wi-Fi, however, especially compared with UWB, is not in any way a contender for the residential home networking market as envisioned with multimedia streams racing from stereo to DVD player to flat-panel monitor to interfaces with in-car telematics. The same issues that plague 802.11a, in terms of packet delivery and power consumption, also prohibit 802.11b from becoming a truly sublime home networking technology.

Bluetooth is great cable replacement technology for personal area networks (PANs), and with solutions starting to enter the market, it should continue to play a role in minimizing the number of cables and wires one has to deal with on a daily basis. But other than that, its limited data rate (under 1 Mbps) does not suit it for anything else that UWB can not do exponentially better. Bluetooth's true niche is that its low power consumption requirements and potentially cheap chip sets make it truly excellent in its techno-niche.

HiperLAN2. Primarily being implemented in Europe, HiperLAN2 has similar characteristics to 802.11a. Experts do point out that its MAC layer is superior to 802.11, in that it is network and application independent. HiperLAN2 was designed with support for asynchronous services and is similar to Bluetooth in that manner. Both HiperLAN2 and 802.11a use OFDM (orthogonal frequency division multiplexing). This modulation scheme is considered to be quite robust, but too complex to be broadly disseminated into home networking products.

Figure 13.11 shows the spatial capacity comparison of the 802.11, Bluetooth, and UWB.

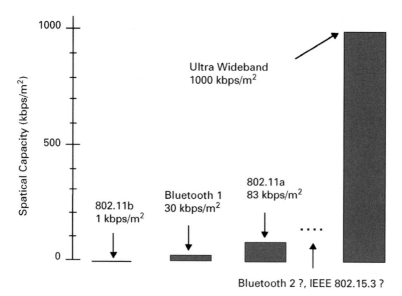

Figure 13.11. Spatial capacity comparison between IEEE 802.11, Bluetooth, and UWB (source: *Ultra-Wideband Technology for Short- or Medium-Range Wireless Communications*, Jeff Foerster, Evan Green, Srinivas Somayazulu, David Leeper, Intel Corp.).

So, UWB is a high-speed transmission technology that has the following advantages over rival technologies:

- virtual immunity to interference
- very low power requirements
- radar-like ability to see through objects
- less likelihood of creating interference for narrow-band transmissions since UWB transmissions are generally perceived as very low-level background noise
- very high capacity
- relative simplicity means low cost for developing receivers.

All this makes UWB attractive for potential new applications and services although it has to clear cross the technical and business challenges in the coming years.

13.12 Applications and services

So, what applications and services are going to be important 5–10 years from now? It is a difficult task to predict such trends as we noted earlier in the chapter. However, we can make some assertions based on what we have seen and experienced over the past decade.

We clearly are moving towards an "always connected" subscriber environment where the user seamlessly has access to the necessary and desired bandwidth. The

Table 13.3. Key to Tables 13.4–13.6

Key	Service category	Segmentation
VS	Rich voice (simple)	Business + Consumer
VE	Rich voice (enhanced)	B + C
LS	Location-based services	B + C
MB	Multimedia messaging service	B
MC	Multimedia messaging service	C
IA	Mobile Internet access	C
IE	Mobile intranet/extranet access	B
CI	Customised infotainment	C

user would be served by the most appropriate network in accordance with the applications' requests as there is no single technology that covers in a cost-optimized manner low/high density, low/high bandwidth applications. In terms of which applications and services will be prominent in the next decade, UMTS Forum did a study and found that mobile intranet/extranet access (by the business sector) and customized infotainment (by the consumer sector) will become the dominant wireless data services. The results of the study are summarized in Table 13.4–13.6 (with the corresponding key in Table 13.3).

Tables 13.4–13.6 list some of the application and service concepts that will become prevalent in the future. As we have seen in the past, the popularity of these applications and services is sometimes geography dependent, for example, conferencing and messaging applications are likely to become popular in Japan, Korea and Europe first before they become a trend in North America and South America. Similarly, regulatory applications will arrive in the USA before they migrate to other parts of the world. In the general enablers' category, pre-paid and number portability will continue to prosper in Europe and Japan while the USA will lag behind (see Table 13.7).

The trend seems to be clear in terms of characteristics of the mobile applications and services:
- requiring more bandwidth
- extremely personalized, adaptive, and context-sensitive content
- interactive multimedia
- always on real-time access (AORTA)[1].

The basic idea is that computing devices will always be "connected" and have "access" to "real-time" personalized information (content) irrespective of its complexity. The devices can be disparate – from laptops to TV to cell phones and everything in between. And this information is not only static but also dynamic. Dynamic information extends to much more than what we are accustomed to today, for example data-mined rich information that can today take hours to obtain and analyze being

[1] AORTA was first coined by Mark Anderson, editor SNS (see Chapter 12 for his interview).

Table 13.4. Top service categories based on revenues

Rank					Worldwide					
1st	VS	VS	VS	VS	VS	VS	VS	VS	VS	VS
2nd		IE	CI	CI	CI	CI	CI	CI	CI	CI
3rd		CI	IE	IE	IE	IE	IE	IE	IE	IE
4th		MC	MC	MC	MC	MC	MC	MB	MB	MB
5th		IA	IA	LS	LS	MB	MB	MC	MC	VE
6th			LS	IA	MB	IA	IA	IA	VE	MC
7th			MB	MB	IA	LS	LS	VE	IA	IA
8th				VE	VE	VE	VE	LS	LS	LS
Year	01	02	03	04	05	06	07	08	09	10

Table 13.5. Revenues (in billions of US$) for services listed in Table 13.4

Rank					Worldwide					
1st	0.1	1	6.4	12	19	36	54	60	72	88
2nd	0	0.9	5.8	11	17	32	48	54	64	86
3rd	0	0.7	3.1	5.9	8.5	15	24	34	47	61
4th	0	0.2	1.6	3.6	5.1	8.8	12	17	22	25
5th	0	0.2	0.8	1.8	2.7	5.7	10	13	15	21
6th	0	0	0.7	1.5	2.2	3.9	6	8.7	14	18
7th	0	0	0.4	0.9	2.2	3.9	5.8	8.1	9.6	14
8th		0	0	0.1	0.7	1.4	4.2	6.8	7.8	9.9
Year	01	02	03	04	05	06	07	08	09	10

Table 13.6. Revenues (in %) for service revenues listed in Table 13.4 (source: UMTS Position Paper Number 2: Ranking of top 3G services)

Rank					Worldwide					
1st	100	33	34	33	33	34	33	29	29	28
2nd	0	30	31	30	30	30	29	27	25	26
3rd	0	23	16	16	15	14	14	17	19	19
4th	0	7	9	10	9	8	7	8	9	8
5th	0	7	5	5	5	6	6	7	6	7
6th	0	0	3	4	4	4	5	5	5	5
7th	0	0	2	2	3	3	4	4	4	4
8th	0	0	0	0	1	1	2	3	3	3
Year	01	02	03	04	05	06	07	08	09	10

Table 13.7. Future services and applications

	Service concepts and applications	Description
Voice/conference	Videophone	Consumer, person-to-person or person-to-mutliparty, real-time video plus voice
	Basic voice (IP-based)	Conversational voice service, comparable to existing mobile voice, which can transmit to and from all types of networks
	Multimedia group call	Voice plus data, images, and/or video broadcast from a single user to multiple users
	Local services	Services provided by the local network, home or roaming
	Mixed media interactive communication	Service where the sender and receiver in a multimedia session can be communicating with different types of media (e.g. voice for sender and text message for receiver)
Web access/me	Instant messaging	Exchange of content, usually text, between a set of participants in real time
	Peer-to-peer gaming service	Service where users play mobile interactive games amongst themselves while still communicating with each other using voice or text without suspending the game
	Server-based gaming	Service where a number of individuals play an interactive mobile game that is managed by a server
	Mobile virtual private network	Business service that provides secure, single sign-on access to the company's information management systems
	Presence	Provides access to availability status to other users or services
	Interactive customer care	Ability for customer service to scan the current user settings and reset terminal settings
	Basic multimedia service	Ability for user terminals to address, access, and present different types of multimedia objects
	Prepaid services	Payment option where users pay a predefined amount of money for mobile service in advance
General enablers	DSR voice portal	Ability to access information and conduct transactions with voice commands
	Downloading multimedia objects	Ability to download and store multimedia objects from mobile websites to the local terminal
	Mobile number portability	Ability for the user to move from one network operator to another without changing the phone number
	Global text and total conversation	Real-time character-by-character text conversion in a multimedia conversational service
	Multimedia based voice response	Ability for users to change which type of media they are using in the middle of a call without terminating the call
	Lawful intercept	Ability for law enforcement agencies to legally intercept any type of media in a call
Regulatory	Emergency call	A "911" or "112" type service that will route calls to emergency facilities and provide location information to the emergency agency
	Priority service	Ability to give priority call completion to designated agencies or individuals in the event of an emergency
	Malicious call trace	The ability for the network to track the events and service involved by a calling party when the user deems the call to be threatening or malicious

Source: 3GPP.

available literally at our fingertips. If you think about it, "information" is the basic building block of any civilization/revolution.

Centuries ago, it was the printing press that spread knowledge and started the revolution. To get information from point A to point B, there are often multiple steps. For example, a few decades ago, if you were looking for a nice Italian restaurant in a new neighborhood, you probably would have just to roam around in the area to find one or find a friend who might be able to recommend something. After that, yellow pages (directories) became popular; they helped in narrowing down the list but were not something you want to carry in your wallet. In came the Internet, and it further shortened the length of time it took to get to the information, a few clicks here and there and you had your information. However, you almost never had access to the information when you needed it the most – lost and wandering in an alien city.

Well, wireless and the Web met, got married at first sight, and thus started an incredible revolution. After the short honeymoon was over, the two industries have been working hard at making the AORTA dream possible. Today, you could potentially get the information you need on your wireless Web-enabled phone in any decent city, BUT coverage needs to be there and you still have to plow through 15 steps to find it. Is it quicker than calling a help-line? In most cases, probably not.

So, what is missing? "Context-sensitivity" – meaning that if Tom is looking for Food, information databases should consider the fact that it is close to midnight in the middle-of-no-where and Tom is speeding at 80 mph to escape those pesky flying bullets, so the system needs to figure that out in a matter of seconds and present Tom with the Top 3 (with the first one being directions to the nearest police station) choices available under the circumstances. If the system gives 50 different options and 20 layers of menu-items, it is not information, it is garbage.

Information is what is fine-tuned for the user based on the user's preferences, context, and usage device. The key is for Tom to get to the information in a minimum number of steps and without delay. You see where we are going with this.

Information needs to be delivered (and of course processed) at the "speed of thought". Above was a simple consumer application, but what if the same demands were placed on enterprise data which is harder to process, is all over the place, and most of the times, we cannot make sense of it anyway.

So, what do we see as key ingredients of an AORTA or pervasive computing ecosystem – data mining of bits and bytes to convert them into information, effective transcoding techniques to deliver information to appropriate situation, agents at client, network, and servers, and the infrastructure technologies to make it all happen. We are not quite there yet, but are steadily getting there. With the building blocks discussed in this chapter in place, we will be empowered with information access that is nothing short of revolutionary in human evolution (see Figure 13.12).

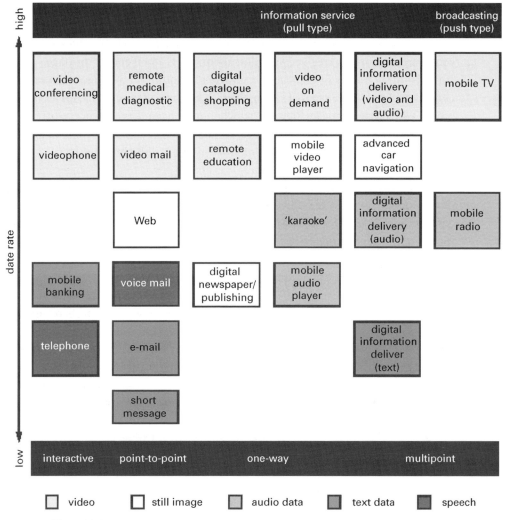

Figure 13.12. Future applications and services (source: *BT Technology Journal*, January 2002).

13.12.1 M-government anyone?

M-government is a complex strategy for the efficient utilization of all wireless devices for the benefit of all involved parties – governments, citizens and businesses. The most visible aspect of m-government is the provision of information through mobile devices in two categories: (1) general information or alerts that are not legally binding for citizens, businesses or government and (2) important or legally binding communication and information, provided by businesses or individuals to government and vice versa.

The first group includes primarily sending SMS/SMS email/email alerts regarding important issues from government to citizens and businesses. This model works on a global scale. An example of this can be found in the USA in California where concerned citizens can receive SMS information alerts about energy shortages. Local administration bodies in Central Europe also send out SMS messages to interested citizens about current issues affecting their area. These services work on an opt-in model, where a user chooses topics on which they wish to receive further information through a website or in paper form. People who sign up for the information service are also asked for the mobile number or email address, where such information is to be sent.

Other enhancements of this service could include the provision of WAP information on any topic or files modified for viewing on other wireless devices (PDAs, handhelds, etc.). GPRS and data-based billing with new WAP specification give further opportunities to this technology at least till 3G comes. A major problem with G2C SMS information alerts is the fact that it is relatively easy to "hack" a mobile phone number, which could cause serious difficulties. The encryption of SMS messages is, however, relatively safe. Once sent, it is nearly impossible to view or modify content of a message. The new generation of wireless messages – EMS and MMS – have many more capabilities; however, their current utilization is for leisure purposes only.

Another important factor is the low penetration of capable devices. These new technologies can, however, frame new opportunities and business models for new sources of information and G2C communication, such as entries from the Land and Immovables Register. The arrival of 3G devices and the commercial operation of wireless broadband (Wi-Fi , UMTS) will give this m-government initiative another boost. Despite the advantages and disadvantages of these information services, m-government has an undisputed chance of becoming a reality in countries with high Internet penetration (the USA, Nordic countries, and Western Europe), not to mention low Internet penetration countries that have considerable levels of mobile penetration such Central European countries and Malta.

13.13 The promise of Web services

Web services are self-contained business functions that operate over the Internet. They are written to strict specifications to work together and with other similar kinds of component. Some of the more established functions at this stage are messaging, directories of business capabilities, and descriptions of technical services. But other functions are in the works as well. Web services are important to business because they enable systems in different companies to interact with each other, more easily than before. With businesses needing closer cooperation between suppliers and customers, engaging in more joint ventures and short-term marketing alliances, pursuing opportunities in new lines of business, and facing the prospect of more mergers and acquisitions, companies need the capability to link up their systems quickly with other companies. Thus Web services give companies the capability to do more business electronically, with more potential business partners, in more and different ways than before, and at reasonable cost.

Because Web services are written according to standards, all parties work from the same basic design. Companies then add value and business advantage to the basic design to meet the needs of their customers. For example, a company can offer its suppliers the capability to view inventory levels of products that the suppliers provide so that they can replenish the stocks without the customer cutting separate purchase orders. Web services provide the basic messaging and service-description functions for this kind of electronic relationship, but the suppliers could build on these basic features to provide better services to the customer. And companies can extend these capabilities to other trading partners, since they are built on standards.

Also, because Web services are built on standards, they make it possible for many systems developers to enter the market, which increases competition and brings down the costs. The competition among vendors also encourages more innovation in the products and services offered to business customers. And basing systems on standards helps to prevent being locked-in to a specific vendor or type of computer or software.

Web services are still a work in progress. Some of the standards are new and not fully tested, and many of the potential business uses are still getting started. But companies should start planning for Web services, and asking vendors for their plans to support Web services.

Web services are based on certain industry standards that are intended to raise the overall level of interoperability and integration in the industry itself. Some of the major standards include the following.

XML describes the transported data.

SOAP (Simple Object Access Protocol) standardizes the way in which the Web service is called. It enables an application running on one operating system to communicate with a program running on another.

WSDL (Web Services Description Language) enables the Web service client dynamically to learn the structure of a Web service to be invoked.

UDDI (Universal Description, Discovery and Integration) serves as the registration and discovery mechanism, allowing companies publicly to register their Web services in a standard way, and enabling other Web services or users to find them.

Figure 13.13 depicts the Web services framework architecture. Some of the components like XML and XML Schemas are already well defined and developed while others like SOAP, UDDI, and WSDL are currently being worked on. With the support

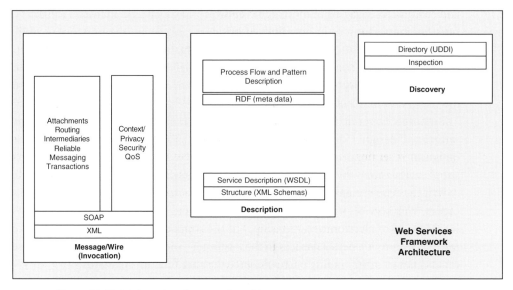

Figure 13.13. Web services framework architecture.

of industry heavyweights behind it (especially Microsoft betting its future .Net strategy with it), Web services are going to be entering the corporate infrastructure in the coming years.

13.14 Intelligent agents

Mobile agents are software programs that can move through a network and autonomously execute tasks on behalf of users. These devices are useful because of the quantity of information content on the Internet, which is exploding at a breathtaking pace. Hundreds of websites are created every minute; countless documents are loaded onto intranets and extranets. Furthermore, seemingly useful information resides in disparate databases. Hence we need intelligent information agents that can act like secretaries and help do the leg work that is often required in information mining. *Information agents* can go even further than mobile agents and can accomplish a sequence of tasks.

We have already begun to see some form of mobile agents being implemented by various online businesses. For example, many online comparison engines utilize agents that go out into cyberspace and bring back pricing and item availability information in real time or on a predetermined frequent basis. Mobile agents are also used for discovering, updating, and indexing content for various search engines. These mobile agents are also referred to as *bots* or derivatives (such as *shopbots*), a term related to *robots*. According to a survey conducted by Neilsen NetRatings, shopbots were the second-fastest-growing e-commerce market segment during the 2001 holiday shopping season. Agents are also used for personalizing Web content. Software agents can mine the plethora of user- and device-specific databases and quickly generate a dynamic page as the user clicks through the site.

These mobile agents are generally based on fuzzy logic or artificial intelligence algorithms that train the agents to behave in a certain way, depending on the request or input they receive from the user's navigation patterns. In some of the more sophisticated set ups, mobile agents of one system can interact with mobile agents of another system (even systems belonging to different companies) to negotiate for the best rates or deals on products or services. In the future, mobile agents will be able to handle much more complex transactions. So far, agent technology has been limited to the desktop environment, but agent technology is gradually moving in its natural progression to wireless devices.

Consider the following scenario. You step off a plane in New York City and turn on your personal communicator. The device is an electronic organizer with an integrated wireless modem and a phone. Here is what you receive via a wireless network, based on your personal profile:

- email that you have designated for wireless transmission from your corporate server or Internet service provider;
- a list of half-price front-row NBA tickets for a New York Knicks and Los Angeles Lakers game and the ability to immediately purchase tickets wirelessly;
- the current value of your stocks and bonds, with the ability to transfer funds, purchase stocks, and pay bills wirelessly;
- a map of midtown Manhattan, with your bank's ATMs pinpointed and, based on your calendar information, the location of your hotel and the meeting place;
- the headlines and first paragraphs of very specific news articles;
- the five-day weather forecast for New York; and
- today-only discount coupons for shopping at Macy's, Border's books, and several favorite computer stores; discounts are personalized based on your past shopping experience and history.

Then you switch to phone mode and speak to your mobile agent to find out whether or not your colleague who was coming in from Europe has landed and where is he staying so that you can contact him to get together for a dinner meeting. Your mobile agent then goes and negotiates with your colleague's mobile agent to set up an appointment at a restaurant of your mutual choice, which is not far from either of your hotels. Your mobile agent automatically books a cab for you. And all this activity happens in a mere few minutes! In this futuristic but realistic scenario, multiple technologies interfaced and exchanged information with each other intelligently.

13.15 R&D

In general, any groundbreaking technology follows a typical path to its commercialization. Researchers in educational or research institutes or researchers in corporate R&D labs make the breakthrough; the technology is patented, and then licensed for commercial engineering and development before it hits the market. The research community is sometimes decades ahead of trends that might hit the market in the future. Also, corporate R&D budgets have been shrinking over the years and it is only the giants like IBM, Motorola, Nokia and Sony who can afford to do any R&D on technologies that might become mainstream 10–15 years in the future. As such, as we have pointed out earlier, there is growing pressure for corporations either to join rivals to share R&D costs or look towards universities for research. Almost all major corporations have ties into academia in the hope of getting to the breakthrough first and also to getting assistance from enormous university talent. In this section, we will look at the research labs and science parks focused on next-generation wireless technologies. The research topics can be broadly divided as follows.

(1) Communications technology,
(2) network technology, and
(3) applications.

Communications technology
(1) Full IP adaptation and high-speed transmission
    (a) full IP adaptation by communications networks
    (b) next-generation high speed satellite communications
    (c) base station design
(2) Telematics
    (a) voice/image recognition
    (b) ITS user interface
    (c) ITS communications architecture
    (d) wireless link mounting under ITS-specific conditions
    (e) traffic modeling and infrastructure design
    (f) biometrics
(3) Enterprise expansion
    (a) technology for value-added services
    (b) convergence of WAN/LAN
    (c) expanding enterprise to WAN data networks

Network technology
  (1) Wireless access network deployment
  (2) next-generation long-distance optical communications at the terabit level
  (3) access network deployment and operation

Applications
  (1) Vertical segments – medicine, retail, field force and sales force automation, etc.
  (2) enterprise applications and services
  (3) multimedia services
  (4) LAN for home/education/entertainment
  (5) multimedia platform technology

  R&D in the wireless industry has also gone through significant changes that are a reflection of the dynamic landscape. R&D is much more a joint effort between different interested entities, even rivals in most instances. The convergence of computing and communications has also had a big impact on design and implementation of R&D projects. Players like Microsoft who five years ago had nothing to do with wireless are one of the biggest spenders of R&D dollars in the industry. Similarly, a lot of government and academic institutions and organizations are joining forces with their commercial brethren to accelerate the commercialization of new concepts and technologies. Major research universities like the University of California not only have partnerships with players such as Ericsson, Nokia, Siemens, HP and others to look at specific problems and innovations, but their Office of Technology Transfer (OTT)[2] also works closely with the investment community to provide funding for their ventures. Fast-paced evolution and shrinking R&D budgets have forced the industry, like many others, to approach innovation in a different light. Corporate groups, academia, partners, and customers all contribute to the cycle of bringing new concepts and technologies to light (Figure 13.14).

13.15.1 Major centers for R&D

This section lists the major R&D institutes around the world (Tables 13.8–13.10). Research groups from large groups and technology providers (Ericsson, Microsoft, Vodafone, etc.) are not listed here.

13.15.2 How to measure R&D investments?

One of the biggest challenges R&D institutes, groups and organizations face is the question of measurement – how do we ensure that we are getting a good ROI on

---

[2] http://www.ucop.edu/ott/

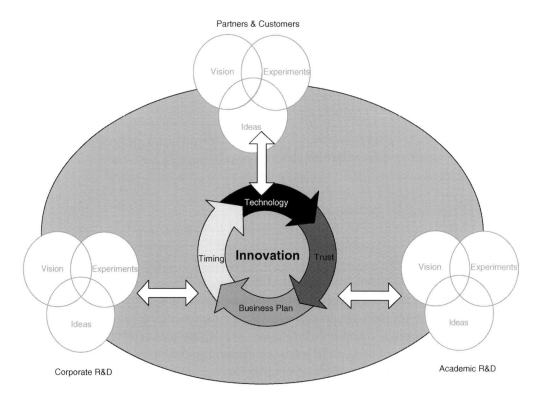

Changing Paradigms of R&D

Figure 13.14. Changing paradigms of R&D.

the investment? There are several good texts dedicated to the discussion of R&D performance measurement and management of global innovation (see References). Here we will describe what R&D investments boil down to – ROI and their contribution to the bottom line over a period of time. Even elite R&D organizations such as Bell Labs face enormous pressures to commercialize their innovations and breakthroughs quickly – the engineering cycles are shrinking by the day. There are certain key factors to R&D effectiveness. Quality of research can be measured from the group's ability to do technology development before engaging in new product development. The number of patents and publications is usually a good indicator of an R&D unit's ability to conduct excellent research. Next, market requirements are best considered by an R&D unit when it sells a high percentage of its projects to business units. Getting an order from a business unit of field operations means going with the market pull. It is a good idea for at least one business unit to fund part of the project, even at the research stage. Concurrent engineering is required during product development to refine and improve the process efficiency to get the product to the market quickly. So, a rough indication of the productivity of an R&D unit is given

Table 13.8. Major R&D institutes in North America

University/organization	Group	URL
UC Los Angeles	AAGROUP	http://www.icsl.ucla.edu/aagroup
Worchester Polytechnic Institute	Center for Wireless Information Network Studies	http://ece.wpi.edu/Research/cwins
Rutgers University	DATAMAN	http://athos.rutgers.edu/dataman/
UC Davis	DWCL (Digital Wireless Communications Lab)	http://www.ece.ucdavis.edu/dwcl/index.html
UC Berkeley	Berkeley Wireless Research Center	http://bwrc.eecs.berkeley.edu/
University of Michigan	CITI	http://www.citi.umich.edu/
Columbia University	Mobile Computing Laboratory	http://www.mcl.cs.columbia.edu/
University of Maryland	Mobile Computing and Multimedia Lab	http://www.cs.umd.edu/projects/mcml
University of Washington	Mobile Computing	http://www.cs.washington.edu/research/mobicomp/mobile.html
Purdue University	Mobile Computing Research	http://www.cs.purdue.edu/research/cse/mobile/
Virginia Polytechnic Institute	Mobile & Portable Radio Research Group	http://www.mprg.ee.vt.edu/home.html
TRLabs		http://www.trlabs.ca/
UC Los Angeles	Wireless Adaptive Mobile Information System	http://www.cs.ucla.edu/csd-gradsgs1/hsiao/intel/wamis.html
Rutgers University	WINLAB	http://winwww.rutgers.edu/
University of Massachusetts	Wireless LAN Group	http://www.ecs.umass.edu/ece/wireless/
Massachusetts Institute of Technology	Oxygen	http://oxygen.lcs.mit.edu/

Table 13.9. Major R&D institutes in Asia

University/organization	URL
Yokosuka Research Park	http://www.yrp.co.jp/en
Kyoto Research Park	http://www.krp.co.jp/HP98Ehtml
Singapore Science Park	http://www.sciencepark.com.sg/

Table 13.10. Major R&D institutes in Europe

University/organization	Group	URL
IMEC	Wireless Research Group	http://www.imec.be/desics/
Technical University of Munich	High Frequency Systems Group	http://www.hfs.e-technik.tu-muenchen.de/ext/d07/hfsites.html
University of Surrey	Mobile Communications Research Group	http://www.ee.surrey.ac.uk/CSER/Mobile/index.html
University of Southampton	Mobile Multimedia Research	http://rice.ecs.soton.ac.uk/index.html
King's College	Mobile Special Interest Group	http://crg.eee.kcl.ac.uk/crg/mmpc_com.htm
University of Cambridge	Mobile Special Interest Group	http://www.cl.cam.ac.uk/Research/Mobile-SIG/
National Technical University of Athens	Telecommunications Laboratory	http://www.telecom.ece.ntua.gr/
Technical University of Berlin	Telecommunications Network Group	http://www-tkn.ee.tu-berlin.de/
Aachen University of Technology	Wireless ATM Project	http://www.comnets.rwth-aachen.de/project/mbs
Oulu University, Finland		http://www.oulu.fi/english/index.html
Sophia Antipolis,		France http://www.sophia-antipolis.net/uk/
NOVI Research Park, Denmark		http://www.novi.dk/novi/eng_web/indexeng.htm

by the number of projects, projects sold to various business units, and the degree of concurrent engineering.

Eventually, it boils down to something represented in the formula below:

$$\text{R \& D Effectiveness Index} = \frac{\text{\% New product revenue} * (\text{Net profit\%} + \text{R \& D\%})}{\text{R \& D\%}}$$

The R&D Effectiveness Index[3] compares the profit from new products with the investment in new product development, using the above formula (where all "%" are a percentage of revenue). As a simple interpretation, the index computes the ratio of profits received from new products divided by the investment in product development. When the index is above 1.0, the return from new products is running at a rate greater than the investment.

---

[3] R&D Effectiveness Index: A Metric for Product Development Performance, Insights, *PRTM*, Volume 5, Number 3.

Whatever the metrics used, it is imperative continuously to evaluate the effectiveness of R&D projects in a dynamic environment.

13.16 Conclusions

In this chapter we have looked at what might become our future. The various technologies that we discussed in this chapter can further enhance our user experience in a tremendous way. Towards the end of this chapter we also discussed how the very nature of R&D is also changing in light of the dynamic and rapid changes. The time a technology or a group of technologies takes to come to market has been incredibly shortened over the past two decades. We can expect the same trend to continue into the next two decades. In the next chapter we will finish up with some final thoughts.

14 Conclusions and recommendations

In Chapter 13 we discussed the future of wireless data technologies, applications, and services. We began our journey with the discussion of the impact of globalization on wireless industry and businesses and ended the conversation by taking a look at the promise of technologies that will enable future wireless data applications and services that will capture our imagination and make our lives simpler. Along the way, we discussed a wide variety of topics to cover different aspects of the wireless data industry. We hope we were able to provide you with discussion on a spectrum of topics and issues that will help you appreciate both the complexity and the promise of wireless data. As was the purpose of the book, we discussed the trends, technologies, business models from a global angle as we have discovered – geography matters, it matters a great deal. The wireless data offerings from the carriers and the developers need to be customized and tailored to individuals. There must be continuous focus on this *mantra* to be successful. Over the past few chapters, we have talked quite a bit about the future of the industry. Let us continue that discussion in this chapter by analyzing various possible scenarios for wireless data for rest of the decade (2003–2010).

As described in previous chapters, the size of the wireless market for voice services has been approaching saturation levels in several developed countries, so the carriers need to create new revenue streams in addition to existing voice services. It is needless to say that wireless data present the most promising services potential in the near future, but how can we help to realize the potential in a timely fashion? Yes, we can find a few good examples such as i-mode in Japan and SMS in Europe, but how we can learn the lessons from their successes and use them to devise a successful business strategy for the future?

What if the wireless operators fail to realize the new revenue streams in wireless data; what kind of world is awaiting us? Would we return to the voice-only world? What parameters do we have for future prediction and scenario planning? How will the rapid changes in the industry environment such as globalization and consumer trends impact our business thinking? It is worth answering these questions by discussing some possible scenarios.

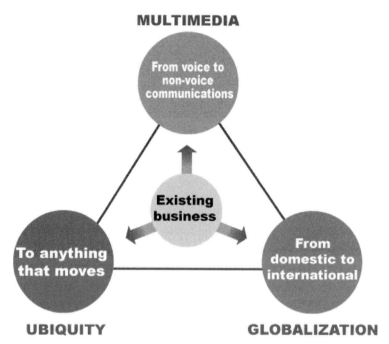

Figure 14.1. Three pillars of progress (source: NTT DoCoMo).

14.1 Background factors for predicting the future of mobile business

Before we get into the discussion of scenarios, let us briefly identify the background factors that will help us in predicting the future of mobile business. We need to keep in mind the recent dramatic movements of end-user behavior, societal shifts, markets, and lifestyles.

14.1.1 Diversification, personalization and ubiquity

For services and applications to be popular they should be available to users irrespective of devices, language, culture, networks, protocols, OS, location, etc. – they should be ubiquitous.

User requirements for wireless services are steadily moving towards diversification and personalization. The big hit of downloading services such as a ringing tone on a cell-phone indicates the strong desire for diversification and personalization of the device. Java-enabled phones make each cell-phone a "unique" device by downloading animation and games based on personal choice (see Figure 14.1).

This is based on a strong desire for personalization. It is a very important marketing factor for young users to select their wireless carriers since the design and functions

Next generation mobile travel solutions

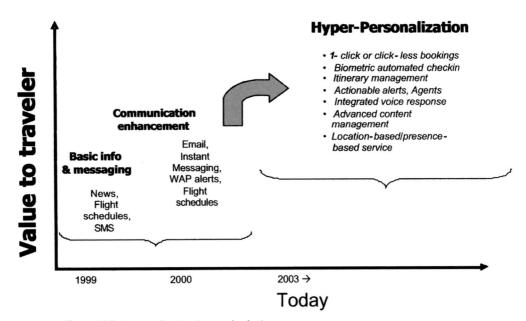

Figure 14.2. Personalization in travel solutions.

of their cell-phone are sometimes a good topic for chatting about in a coffee shop or a bar (Figure 14.2 shows the evolution of personalized mobile travel solutions). We need to remember that the principal motivation to buy is not the technology but the service that the customer is willing to accept and pay for.

14.1.2 Multigeneration and multimedia

Carriers have been upgrading wide-area cellular systems from 1G to 2G and now from 2G to 2.5G/3G. Each transition of the generations took about 10–15 years to mature. It seems that in the periods from 2003 to 2010, multiple-generation systems, devices and services will coexist and coverage areas will overlap. From the user point of view, the service and system will become more complicated owing to this multigeneration situation than 10 years ago when each continent had a single analog system and consequently the calls were automatically transferred to 2G digital systems. Evolving technologies such as WLAN and WPAN will accelerate this complicated situation by convergence with 2G, 2.5G, 3G and B3G systems.

Also, the start of multimedia services on wireless networks will further complicate things. Users will demand services (and hence technology that can support those

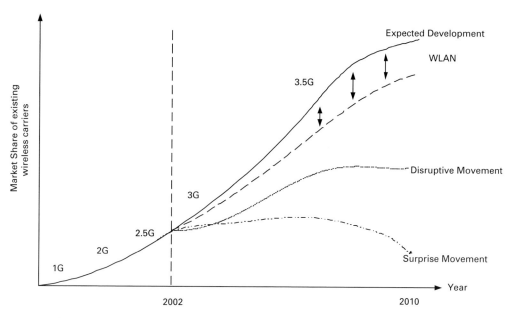

Figure 14.3. Where does wireless go from here?

services) based on their personal user requirements, so systems will need to be adaptive in nature.

14.1.3 Globalization

As already discussed, globalization in economy, consumer culture and technology development will accelerate more in the coming decade than it did in the past decade. This indicates that the winner in some specific service and technology has a great chance to be a global winner much more quickly, but it is also true that the competition is not limited to one nation, but one has to compete globally.

In order to predict the wireless business trend from a global perspective, China has a very important position since the country has huge potential market size.

Next, we will use the above parameters to predict future scenarios.

14.2   Scenarios

Although one can contemplate several future scenarios for the wireless data industry, they can easily be categorized into three main topics (Figure 14.3):
(1) expected development,
(2) disruptive movement, and
(3) surprise movement.

Let us discuss each of these in some detail.

14.2.1   Scenario 1: expected development

This scenario is the most orthodox one. Major wireless carriers such as Vodafone, NTT DoCoMo, Telefonica, Verizon, AT&T Wireless, TIM, China Telecom, etc., keep the majority of their market share as it is now and the total penetration rates of cellular subscribers in each country will increase gradually, especially among the elderly, housewives and children. Wireless data services using 2.5G and 3G systems are widely used for both consumer and corporate users. The most popular wireless data services are the video-mail service and location-based application services such as meeting-arrangement using GPS.

Wireless Internet access services are based on global standards and they mature with time. All the cell-phone terminals have browser functions with wide and color displays and the cost is reduced owing to the large global market scale thanks to a global standard. However, miniaturization of devices does not progress so much because of wide and color displays. The most popular phone design is the "shell" type with a wide and color display.

The content provider business becomes an oligopoly and a few providers dominate most of the contents market. The winners are service providers who are under the control of major wireless carriers or their joint ventures. Independent and PC-based small/middle-sized providers try to "eat the apple", but very few can make it. Portal services by wireless carriers become a large portion of the business and the culture of wireless companies become more like ISPs rather than old-style telephone companies.

Integration and aggregation of cellular and WLAN services progresses gradually and steadily. Hot-spot services become popular, but owing to the furious competition among WLAN service providers on coverage and pricing, few providers can survive by themselves and finally aggregation with existing cellular companies develops.

WLAN service are mostly provided as a value-added service of the 2.5G/3G service and this helps in making the 2.5G/3G service a more economical operation, since 3G operators need a lot of investment to make perfect indoor coverage. The main business structure for the cellular service remains the same as now; for example, most voice and data traffic are transported on the internal network of a cellular company.

The 3G global roaming service starts in major cities like London, Tokyo and New York and it creates new revenue from business users. The SIM card roaming mechanism overcomes the differences of 3G frequency allocation of each continent, especially between the Americas, Europe and Asia.

The telematics service becomes popular and useful in car-centric markets like North America in accordance with 3G network development, and major wireless carriers have alliances with major car companies across the board. A wireless security system for cars become the most popular application of telematics. Cars become the second largest market for cellular companies and a new ecosystem including car-device manufactures is generated.

14.2.2 Scenario 2: disruptive movement

The second scenario is based on disruptive movement of technologies and business models. The major cellular carriers gradually decrease their market share and MVNO operators take off and increase the market share rapidly. The total cellular market itself grows as mentioned in Scenario 1; however, the market share of major carriers is disturbed by MVNOs.

The main reasons for this happening are the delay in 3G commercial service introduction by major cellular carriers and that the brand image of 3G does not attract the customers. As a result, the customer shift from 2G to 3G does not proceed as planned and the 3G service itself can not generate enough income for carriers to continue their investment in 3G network construction.

On the other hand, MVNO operators focus their management resources on creating new markets by making full use of existing 2G infrastructures. Especially, very aggressive marketing strategy focused on young users in big urban areas of developed countries contributes greatly to the increase in market share.

MVNOs are very aggressive in jointly promoting with the top brand companies in other consumer/business areas such as music, car, home-electronics and sports clubs. This strategy appeals greatly to young urban users. MVNO service development strategy focuses on multimedia rather than high-speed data service. Multimedia services provide combination functions of a cellular phone with an MP3 music player and a digital camera. The strategy is learned from the popularity of (5 million users) "Sha"- mail by J-phone in Japan (camera-embedded phone).

The competition in wireless Internet access services becomes more severe and there are no dominant portal services by wireless carriers. PC-based providers become popular again among wireless users, thanks to the wired–wireless integration.

What about the service and tariff plan strategy? Portal services by cellular carriers become unpopular because spam-mail is a big social problem in most countries and it gives a negative image to email using a cellular phone. Terminal functions are more focused on location-based services rather than contents access services. GPS becomes an indispensable function for a "ubiquitous" service and the increased demand for daily security helps the trend. Development of voice recognition and battery technology accelerates the miniaturization of devices and "wrist-watch" type phones become the mainstream design.

Major wireless carriers still keep their high market share; however, "brand switching" from major carriers to MVNOs becomes quite common among the high ARPU users. As a result, profitability of major wireless carriers decreases. Consequently, consolidation of wireless carriers progresses rapidly across the globe.

14.2.3   Scenario 3: surprise movement

The third scenario is the least likely, but it is worth considering. The major impact factor to make it happen is WLAN. The deployment of the 3G service by major cellular carriers does not proceed because of an economical recession in the global markets and financial problems of wireless carriers. The market demand for 3G is only limited for high-end customers such as business executives who require global roaming and high speed/quality data service. MNVOs try to come into the 3G business, but can hardly make enough profit owing to the heavy investment involved. As a result, the financial situation of the cellular business worsens and some major vendors are forced out of the industry.

On the other hand, the WLAN business is tremendously successful and a lot of newcomers try to capitalize on the new business opportunity. Most computer companies rush into the WLAN business and these new players expand the coverage area from just "hot" spots in metropolitan cities to continuous nationwide service coverage. Thanks to the roaming capability among multiple WLAN service providers, WLAN service coverage finally becomes equivalent to that of 2G/3G cellular systems.

Development of Voice over IP (VoIP) technology and the introduction of an all-IP network enabls the WLAN service to reduce the communication tariff dramatically. WLAN chips are cheap enough and WLAN function is included into normal PCs like a CPU chip. VoIP is also developed in the area of wired/fixed communication and basically all communication becomes IP-based for both wired and wireless access.

As a result, VoIP-based WLAN phones are greatly accepted by the market as a substitute for expensive 3G phones and mass/consumer users prefer to use the WLAN service than 3G. This leads to the birth of a totally new ecosystem that is initiated by WLAN technology. This scenario is similar to the beginning of the Internet in the 1990s when voluntary networks connected to each other and the progress by self-reproduction finally helped in the development of global IP standardized networks.

14.3   Lessons

So, what conclusion can we draw from the above scenarios? First, it is not easy to predict the future, especially if we also try to predict the future from social and business viewpoints since the future is decided by the complex combinations and interactions of multiple parameters such as technology development, industry trends, market behavior, user maturity and globalization.

14.3.1   Effort to enlarge the wireless market and to create new services

In the case of Scenario 1, the basic configuration of the wireless market itself does not change dramatically. In this case, players need to make a steady effort to enlarge the market size and also to create a new services menu for higher ARPU. For heavy users (business executives, young users), advancements in communication function (high-performance email, etc.), e-commerce and contents distribution (text, music, video) will be important. For average users (basic service receptors) and non-users (the elderly), social networking, security, safety and health are important keywords.

14.3.2 Listen to the market and to the customer's voice carefully and periodically

It is extremely important for business players to listen to the customer's voice and to use the feedback for service and product development. Sometimes we can predict a symptom of disruption before it actually happens by carefully analyzing messages from users and market.

Consider this example. In Japan, J-phone (Vodafone group) has a strong brand image among young users, especially university and college students. DoCoMo, on the other hand, has the highest brand image among business users. So, when a college student graduates from a school and starts to work for a company, they will often switch from J-phone to DoCoMo. The brand switching phenomenon is actually happening and explains DoCoMo's current high market share. It is important for DoCoMo (and also for J-phone) to check the brand image with college graduates regularly and use the feedback to check their perceptions.

Market leaders and top runners in each business area such as Microsoft, Sony and Toyota, are always doing this kind of periodic fixed-point observation on specific target generation and location. A hybrid-car by Toyota and AIBO by SONY are typical products of such great efforts. Top runners must keep on running at top speed by providing a full package of up-to-date services and functions.

14.3.3 Pursuit for diversification of new business opportunities and revenue sources

In the case of Scenarios 2 and 3, diversification of new business opportunities and revenue sources are essential factors for wireless players. It is difficult for new players like MVNOs and WLAN providers to implement full billing mechanisms owing to lack of time and multi-network service conditions including roaming. In the case of Scenario 3, where WLAN plays a dominant role, there will be multiple new revenue sources; however, flexible and reliable billing functions must be prepared in advance to collect the money.

Traffic and service are still two major revenue sources for existing wireless carriers. Traffic includes man-to-man, man-to-machine, and machine-to-machine communications. Related market revenues such as contents distribution, mCommerce, system integration and telematics will provide new revenues for non-wireless carriers.

Diversification of new business opportunities and revenue sources are an inevitable path for which wireless players (current and future) must start to prepare. Someone who achieves this difficult challenge can be the "Microsoft" of the Wireless Empire in the next decade.

14.4   What is next?

What is next is the question that keeps the thinkers, analysts, and students of the industry engaged all the time. As previously mentioned, it is always difficult to predict the specifics, although educated guesses can be made with regards to the general directions of evolution. To understand the future potential and direction of the wireless industry, it is important to think out of the box and concentrate more on the user rather than the technology – what technology can do for the user rather than what the user can do with the technology, is the question that players in the industry should be answering. Two projects that form the foundation of this thinking are "DoCoMo 2010 Vision" by NTT DoCoMO and "Oxygen" by Massachusetts Institute of Technology. To continue our discussion on the future of the wireless industry, let us discuss these two visions in a bit more detail.

14.4.1   DoCoMo 2010 Vision

Articulated in 1999, "DoCoMo Vision 2010" is a key component of NTT DoCoMo's future strategic plan as they believe that the future of mobile communications holds great promise. This expresses the company's aim to generate growth from the mobile communications field through the development of mobile multimedia, thereby helping to enrich consumer lifestyles and add extra vitality to industry. Within "DoCoMo Vision 2010", the company's five core businesses are embodied in the acronym MAGIC, while the five core action principles for employees are expressed in the acronym DREAM (see Table 14.1). So, when the company's employees realize their collective DREAM, it will create MAGIC. The "E" in DREAM

Table 14.1. DoCoMo Vision 2010

Five Pillars of Business	**M**obile Multimedia
	Anytime, Anywhere, Anyone
	Global Mobility Support
	Integrated Wireless Solution
	Customized Personal Service
Five Action Principles	**D**ynamics
	Relationship
	Ecology
	Action
	Multi-view

stands for Ecology, which represents DoCoMo's contribution to environmental preservation.

DoCoMo's medium- to long-term growth strategy is based on three strategies: (1) "multimedia", or the shift from voice to non-voice communications; (2) "ubiquity", or extending our business to virtually anything that moves; and (3) "globalization", or expansion into overseas markets. "Multimedia" involves the expansion of business to encompass non-voice media such as music and images. "Ubiquity" entails targeting new areas of demand, such as "person-to-machine" and "machine-to-machine" communications. "Globalization" signifies enlarging the scope of operations so that people can take their handsets to other countries and receive the same level of services as they do at home. Alongside these modes of business expansion, DoCoMo is also designing products and services that are aimed specifically at helping to preserve the environment, in the hope that these can help shape a more environmentally-friendly society.

Mobile communications today are mainly for voice, but with the rapid spread of PC usage and the Internet, it is likely that there will be an increasing need for multimedia services such as data and images in the mobile environment. In anticipation of this, DoCoMo has been actively introducing the third-generation mobile communications system (IMT-2000: International Mobile Telecommunications-2000), as well as doing research work on a fourth-generation mobile communications system. They are also working actively to develop terminals capable of receiving a variety of services, improving network and gateway functions, and providing a wide range of applications for utilization of diverse contents. DoCoMo believes that this will in turn create a mobile multimedia market that outgrows the voice market by the year 2010.

In 2002, over 70 million people in Japan used mobile communication services, with the number of people using DoCoMo services exceeding 35 million. Between 2002 and 2010, DoCoMo would like to create an environment where everyone from children to senior citizens will be able to use mobile communications anytime and anywhere. To achieve this goal, DoCoMo will continue to offer attractive tariff plans and terminals suitable for each generation, so that mobile communications can be utilized throughout various stages of life.

14.4.2   Project Oxygen

MIT Project Oxygen is centered on humans. The project's vision is that computation will be pervasive, like batteries, power sockets, and the oxygen in the air we breathe. Configurable generic devices, either handheld or embedded in the environment, will bring computation to us, whenever we need it and wherever we might be. As we interact with these "anonymous" devices, they will adopt our information personalities. They will respect our desires for privacy and security. We will not have to type, click, or learn new computer jargon. Instead, we will communicate naturally, using speech and gestures that describe our intent, and leave it to the computer to carry out our will. Computers will understand and adapt to human needs instead of the other way around.

According to the founders of the project, to support highly dynamic and varied human activities, the Oxygen system must master many technical challenges. It must be:

- *pervasive* – it must be everywhere, with every portal reaching into the same information base;
- *embedded* – it must live in our world, sensing and affecting it;
- *nomadic* – it must allow users and computations to move around freely, according to their needs;
- *adaptable* – it must provide flexibility and spontaneity, in response to changes in user requirements and operating conditions;
- *powerful, yet efficient* – it must free itself from constraints imposed by bounded hardware resources, addressing instead system constraints imposed by user demands and available power or communication bandwidth;
- *intentional* – it must enable people to name services and software objects by intent, for example, "the nearest printer", as opposed to by address;
- *eternal* – it must never shut down or reboot; components may come and go in response to demand, errors, and upgrades, but Oxygen as a whole must be available all the time.

Oxygen's device, network, and software technologies dramatically extend our range by delivering user technologies to us at home, at work or on the move. Computational devices, embedded in our homes, offices, and cars sense and affect our immediate environment. Handheld devices empower us to communicate and compute no matter where we are. Dynamic, self-configuring networks help our machines to locate each other as well as the people, services, and resources we want to reach. Software that adapts to changes in the environment or in user requirements helps us to do what we want, when we want to do it (Figure 14.4).

This discussion has summarized where we as an industry need to be to deliver on the promise of wireless communications. There will be several technologies working together to make this a reality but to the end-user it should not matter, they should not even think about or be made aware of underlying complexities and technologies.

14.5 Epilog

Our journey is approaching its end. We hope we were able to spark some neurons that will help enable the applications and services of tomorrow.

HOW OXYGEN WILL WORK

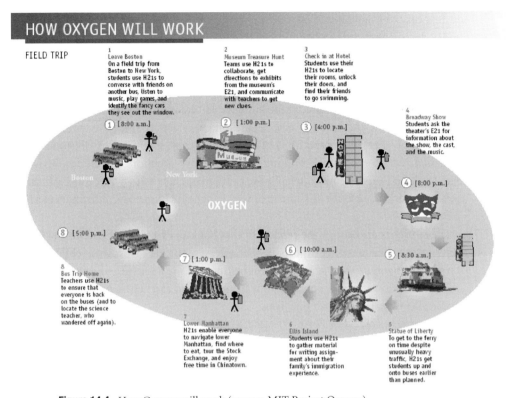

Figure 14.4. How Oxygen will work (source: MIT Project Oxygen).

As we described at the beginning of this book, the key theme of the journey is the insight into mobility which ties wireless and globalization. A wireless system accelerates the mobility and mobility in turns accelerates the globalization. Based on this, we have looked at the wireless and globalization from different perspectives.

We have considered the impact of globalization on businesses, economies, and value chains. We have looked at the latest trends of wireless services by region, wireless business aspects, provided technology updates, and have learned the lessons from successful and failed business models. We also looked at the challenges that this industry faces – from both the technology and business points of view.

To complete the picture, we provided case studies and also got perspectives from wireless business leaders. Lastly, we considered future trends of wireless technology and also made some predictions of future scenarios of wireless businesses.

Wireless engineers have a lot of things to do between now and the future. Wireless communication is now having a great paradigm shift from voice-centric traffic to IP-based data traffic. New wireless technologies such as wireless LAN and PAN are going to be introduced rapidly into the market across all geographies. Digital broadcasting and communications are emerging. All the information is becoming digital and non-humans are also becoming potential telecom users.

It is easy to imagine that people in 2010 might never use the word PHONE since every device, such as the vending machine, will have a phone function. The present authors believe wireless is one of the most powerful tools to achieve a Global Village in the twentyfirst century, since wireless technology gives us a huge chance of enjoying freedom by its basic virtue of mobility.

The most important people in the industry would be those who are challenging the existing mindset and creating new themes in every aspect since the wireless world in 2010 may be very competitive, multi-generational, multimedia, diversified and personalized. Stereotypical thinking does not work anymore. A cellular device in 2010 will no longer be just a telephone, but a magic box which everyone will carry, like a driver's license for ID, and from which they can get any information from anywhere.

The authors encourage the readers of this book to engage in discussion with not only their colleagues and experts in the industry but also with their children and family for it will be they who will have the final say in shaping the future of wireless communications. We do not have to wait till 2010 to start envisioning services and applications that will touch our lives.

The time is now.

References and recommended readings

BOOKS

Andersson, Christoffer, 2001. *GPRS and 3G Wireless Applications*. USA. John Wiley & Sons.

Balston, D. M. and Macario, R.C.V. 1993. *Cellular Radio Systems*. USA. Artech House Publishers. Mobile Communications Series.

Bergman, E. 2000. *Information Appliances and Beyond. Interaction Design for Consumer Products*. USA. Academic Press.

Bigun, J., Chollet, G. and Borgefors, G. 1997. *Audio- and Video-based Biometric Person Authentication. First International Conference, AVBPA '97 Crans-Montana, Switzerland, March 1997 Proceedings*. Germany: Springer.

Boutellier, Roman, Grassman, Oliver and Zedtwitz, Maximilian von, 2000. *Managing Global Innovation, Uncovering the Secrets of Future Competitiveness*, 2nd Edition. Germany. Springer Verlag.

Burgelman, Robert A., Maidique, Modesto A. and Wheelwright, Steven C. 2001. *Strategic Management of technology and Innovation*. USA. McGraw-Hill.

Christensen, Clayton M. 1997. *The Innovator's Dilemma, When New Technologies Cause Great Firms to Fail*. USA. HBS Press.

Cochrane, Peter, 1998. *Tips for Time Travelers*. UK. McGraw-Hill & Companies.

Cook, John L. 2001. *WAP Servlets*. USA. John Wiley & Sons.

DeRose, J. 1995. *The Wireless Data Handbook*, Fourth Edition. USA. John Wiley & Sons.

Dhawan, C. 1997. *Mobile Computing: A Systems Integrator's Handbook*. USA. McGraw-Hill Series on Computer Communications.

Doraswamy, N. and Harkins, D. 1999. *IPSec The New Security Standard for the Internet, Intranets, and Virtual Private Networks*. USA: Prentice Hall Internet Infrastructure Series.

Drucker, Peter. 2000. *The Essential Drucker on Indivisuals: To Perform, To Contribute and To Achieve*. (Japanese). Diamond.

Duato, J., Yalmanchili, S. and Ni, L. 1997. *Interconnection Networks an Engineering Approach*. USA. IEEE Computer Society Press.

Edwards, W.K. 1999. *Core JINI*. USA. Prentice Hall.

Endres, Al, 1997. *Improving R&D Performance The Juran Way*. USA. John Wiley & Sons.

Faigen, George S, Fridman, Boris and Emmett, Arielle, 2002. *Wireless Data for the Enterprise*. USA. McGraw Hill.

Feinleib, D. 1999. *The Inside Story of Interactive TV and Microsoft WebTV for Windows*. USA. Morgan Kaufmann.

Fine, Charles H. 1998. *ClockSpeed, Winning Industry Control in the Age of Temporary Advantage.* USA. Perseus Books.

Fouke, J. 1999. *Engineering Tomorrow Today's Technology Experts Envision the Next Century.* Canada. IEEE Press.

Friedlob, George T. and Plewa, Franklin J. 1996. *Understanding Return on Investment.* USA. John Wiley & Sons.

Gallagher, M. and Snyder, R. 1997. *Mobile Telecommunications Networking with IS-41.* USA. McGraw-Hill Series on Telecommunications.

Garg, V. K., Smolik, K. and Wilkes, J. 1997. *Applications of CDMA in Wireless/Personal Communications.* USA. Prentice Hall. Wireless & Digital Communications Series.

Gates, Bill, 1999. *Business @ THE SPEED OF THOUGHT.* USA. Warner Books.Inc.

Gershenfeld, N. 1999. *When Things Start to Think.* USA. Henry Holt and Company, LLC.

Gibson, Jerry D. 2001. *Multimedia Communications, Directions & Innovations.* USA. Academic Press.

Giguere, E. 1999. *Palm Database Programming – The Complete Developer's Guide.* USA. John Wiley & Sons.

Glisic, S. G. and Leppanen, P. A. 1997. *Wireless Communications TDMA versus CDMA.* Netherlands. Kluwer Academic Publishing.

Hansmann, Uwe, Merk, Lothar, Nicklous, Martin and Stober, Thomas, 2001. *Pervasive Computing Handbook.* Germany. Springer Verlag.

Heiskala, Juha and Terry, John, 2001. *OMFD Wireless LANs; A Theoretical and Practical Guide.* USA. SAMS.

Helal, A., Haskell, B., Carter, J., Brice, R., Woelk, D. and Rusinkiewicz, M. 1999. *Anytime, Anywhere Computing Mobile Computing Concepts and Technology.* USA. Kluwer Academic Publishing.

Heldman, R. 1995. *Telecommunications Information Millennium.* USA. McGraw-Hill.

Hendtry, M. 1997. *Smart Card Security and Applications.* USA. Artech House Publishers.

Hjelm, J. 2000. *Designing Wireless Information Services.* USA. John Wiley & Sons.

Holma, Harri and Toskala, Antti, 2002. *W-CDMA for UMTS.* USA. Wiley.

IBM Systems Journal, 1999. *Human Interaction with Pervasive Computing,* **38** (4), 501–698. USA.

Information & Communications in Japan 2001, 2001. Japan: InfoCom Research, Inc.

Wang, Jiangzhou, 2001. *Broadband Wireless Communications – 3G,4G and Wireless LAN.* USA. Kluwer Academic Publishers.

Kaplan, Robert S. and Norton, David P. 2001. *The Strategy Focused Organization.* USA. HBS Press.

Kelley, Tom and Littman, Jonathan, 2001. *The Art of Innovation.* USA. Currency Doubleday, Random House.

Klusch, M. 1999. *Intelligent Information Agents. Agent-based Information Discovery and Management on the Internet.* Germany. Springer-Verlang.

Kurzweil, R. 1999. *When Computers Exceed Human Intelligence – The Age of Spiritual Machines.* USA. Penguin Books.

Macario, R.C.V. 1993. *Cellular Radio Principles and Design.* USA. McGraw-Hill Communications.

Mann, S. 2000. *Programming Applications with the Wireless Application Protocol – The Complete Developer's Guide.* USA. John Wiley & Sons.

Matsunaga, Mari, 2000. *i-mode Jiken.* Japan. Kadokawa Shoten (in Japanese).

Matsunaga, Mari, 2001. *The birth of i-mode.* Singapore. Chuang Yi Publishing Pte. Ltd.

McGrath, S. 1998. *XML by example Building E-commerce Applications.* USA. Prentice Hall.

Miller, William L. and Morris, Langdon. 1999. *Fourth Generation R&D, Managing Knowledge, Technology, and Innovation.* USA. John Wiley & Sons.

Milojicic, D., Douglis, F. and Wheeler, R. 1999. *Mobility Processes, Computers, and Agents.* USA. ACM Press.

Mitsutoshi Hatori, Takeshi Hattori, Nobuno Nakajima, 2001. *Mobile Global Communications.* Japan. Corona Publishing Co. (in Japanese).

Natsuno, Takeshi, 2000. *i-mode strategy.* Japan. Nikkei-BP (in Japanese).

Nielsen, J. 2000. *Designing Web Usability.* USA. New Riders Publishing.

Oboshi, Koji, 2000. *NTT DoCoMo kyu-seicyo no keiei.* Japan. Diamondo (in Japanese).

Ojanpera, Tero and Prasad, Ramjee, 2001. *WCDMA: Towards IP Mobility and Mobile Internet.* USA. Artech House Publishing.

Oram, Andy, 2001. *Peer-to-Peer, Harnessing the Power of Disruptive Technologies.* USA. O'Reilly.

Porter, Michael E. 1980. *Competitive Strategy, Techniques for Analyzing Industries and Competitors.* USA. Free Press.

PriceWaterHouseCoopers Technology Forecast. 2000. *From Atoms to Systems: A Perspective on Technology.* USA.

Rankl, W. and Effing, W. 1997. *Smartcard Handbook.* Great Britain. John Wiley & Sons.

Rappaport, T. 1996. *Cellular Radio & Personal Communications Advanced Selected Readings Vol 2.* USA. The Institute of Electrical and Electronics Engineers, Inc. (IEEE).

Redl, S., Weber, M.K. and Oliphant, M.W. 1998. *GSM and Personal Communications Handbook.* USA. Artech House Publishers. Mobile Communications Series.

Rischpater, R. 2000. *Wireless Web Development.* USA. Apress.

Sadeh, Norman, 2002. *M-Commerce.* USA. John Wiley & Sons.

Sawhney, Mohan and Zabin, Jeff, 2001. *The Seven Steps to Nirvana, Strategic Insights into eBusiness Transformation.* USA. McGraw-Hill.

Seichi Toshida and Hideo Tamura, 2000. *Network Capitalism.* Japan. Nihon-Keizai Shimbun (in Japanese).

Shafi, Mansoor, Ogose, Shigeaki and Hattori, Takeshi, 2002, *Wireless Communications in the 21st Century.* USA. John Wiley & Sons, IEEE.

Sharma, Chetan and Kunins, Jeff, 2002. *VoiceXML: Strategies and Techniques for Effective Voice Application Development with VoiceXML 2.0.* USA. John Wiley & Sons.

Sharma, Chetan, 2000. *Wireless Internet Enterprise Applications.* USA. John Wiley & Sons.

Spekman, Robert E., Bruner, Robert, Eaker, Mark E. *et al.* 1998. *The Portable MBA.* USA. John Wiley & Sons, Inc.

Tachikawa, Keiji, 2001. *W-CDMA.* Japan. Mruzen (in Japanese).

Tachikawa, Keiji, 2002. *W-CDMA Mobile Communication System.* USA. John Wiley & Sons.

Tanenbaum, A.S. 1996. *Computer Networks, Third Edition.* USA. Prentice Hall.

The Operator Guide to WAP, 2000. UK. Mobile Communications International.

Wireless Application Protocol Forum, Ltd. 1999. *Official Wireless Application Protocol.* USA. John Wiley & Sons.

Wong, P. and Britland, D. 1995. *Mobile Data Communications and Systems.* USA. Artech House Publishers. Mobile Communications Series.

Wyzalek, J. 2000. *Enterprise Systems Integration.* USA. CRC Press. The Auerbach Best Practices Series.

Yacoub, M.D. 1993. *Foundations of Mobile Radio Engineering.* USA. CRC Press, Inc.

MAGAZINE ARTICLES AND WHITEPAPERS

Adrian, M., Zetie, C. and Rasmus, D. Jun 1999. *Connections are Everything. Intelligent Enterprise.*

Andersen, M. 1999–2002. *Strategic News Service.*

Binding, C., Hild, S. and O'Connor, L. 1999. *Research Report. E-Cash Withdrawal using Mobile Telephony.* IBM Research Division.

Biometric Industry Products Buyer's Guide. International Computer Security Association (ICSA).

BiStatix Whitepaper Version 4.1.
  http://www.motorola.com/LMPS/Indala/bistatix.htm.

Blackwell, E. and Fitchard, K. 2000. *M-Commerce The New Face of Fraud.* TelecomClick.

Book of Visions. Wireless World Research Forum.

Bourrie, S.R., June 5, 2000. Fraud is Down, But Not Out. *Wireless Week,* 44–46.

Business Week, Jan. 17, 2000.

Cabri, G., Leonardi, L. and Zambonelli, F. Feb. 2000. Mobile Agent Coordination Models for Internet Applications. *IEEE Spectrum,* 82–89.

Collins, J. June 5, 2000. *Barbed Wireless.* Tele.com.

Constance, Scott and Gower, Jeff, 2001. *A Value Chain Perspective of Economic Drivers of Competition in the Wireless Telecommunications Industry.* MIT

Developer's Guide. *Nokia Mobile Entertainment Service.* Nokia.

Developer's Guide. *Web Clipping.* Palm Computing.

FAQ about MPEG-7. http://www.darmstadt.gmd.de/mobile/MPEG7/FAQ.html.

Fitchard, K. 2000. *Subscription Fraud: The Battle Continues.* TelecomClick.

Harrow, J. F. 2000. Apr. 10, 2000. *The Rapidly Changing Face of Computing (RCFoC) – Reaching Out and Touching.* http://www.compaq.com/rcfoc/20000410.html.

Hendricks, C. 2000. *Website Architectures for a Wireless World: Designing Internet Accessibility for Multiple Non-PC Devices.*

http://public.itrs.net/Files/2001ITRS/Home.htm.

http://www.intel.com/research/silicon/moorespaper.pdf.

http://www.nttdocomo.co.jp.

http://www.qualcomm.com/main/whitepapers/WirelessMobileData.pdf.

http://www.rttonline.com/hottop_frame.htm.

http://www.sega.co.jp.

http://www.w3.org/P3P/.

Foerster, Jeff, Green, Evan, Somayazulu Srinivas, and Leeper, David. *Ultra-Wideband Technology for Short- or Medium-Range Wireless Communications.* Intel Corp.

Joy, B. Apr. 2000. Why the future doesn't need us. *Wired Magazine,* 238–262.

Jurvis, J. Apr. 2000. Choosing A Handheld Is A Matter of Trade-Offs. *Enterprise Development.*

Jurvis, J. and Grehan, R. Aug. 1999. Reach out with handheld apps. *Enterprise Development,* 16–31.

Kaku, M. June 19, 2000. What will replace silicon? *Time Magazine Visions 21 Special Issue – The Future of Technology,* 98–99.

Koshima, H. and Hoshen, J. Feb. 2000. Personal locator services emerge. *IEEE Spectrum,* 41–48.

Lindhe, L. May 29, 2000. Waiting for wireless. *The Industry Standard,* 228–233.

Mann, C. C. May/June 2000. The end of Moore's Law? *MIT's Magazine of Innovation, Technology Review,* 42–48.

Mattis, M. and Daly, J. Dec. 1999. Web visions: shaping tomorrow. *Wired Magazine,* 162–177.

Miller, B.A. Feb. 2000. *Bluetooth Applications in Pervasive Computing – An IBM Pervasive Computing White Paper.* IBM.

Mobile e-business. *Extending SAP systems to pervasive computing devices.* IBM. http://www-3.ibm.com/pvc/mobile/sap.shtml.

Mobile Internet: Content, commerce and applications. Apr. 10, 2000. Mobile Communications.

MPEG-7 Applications Document. July 1999. International Organisation for Standardisation. http://www.darmstadt.gmd.de/mobile/MPEG7/Documents/W2860.htm.

Newmarch, J. 2000. *Jan Newmarch's guide to JINI technologies.* http://pandonia.canberra.edu.au/java/jini/tutorial/Jini.xml.

Pankanti, S., Bolle, R. and Jain, R. Feb. 2000. Biometrics: The future of identification. *Computer*, Volume **33**, No. 2.

Pappo, N. May 2000. Middleware Bridges Internet, Wireless. *Telecommunications Magazine.*

Passani, L. Mar. 2000. *Building WAP Services – XML and ASP will set you free.* WebTechniques.

Pehrson, S. 2000. WAP – The catalsyst of the mobile Internet. *Ericsson Review* No. 1, 2000.

Phillips, M. *Technology Backgrounder Whitepaper.* SpeechWorks.

R&D Effectiveness Index: A Metric for Product Development Performance, Insights. *PRTM*, Volume **5**, Number 3.

Sharma, Chetan. 2000. *Wireless Internet Applications.* Luminant Worldwide Corporation.

Sherman, L. Dec. 1999. The knowledge worker unplugged. New systems for wireless information extend the knowledge chain to mobile devices. *Knowledge Management*, 46–53.

Smith, J.R., Mohan, R. and Li, C.S. May 1998. *Transcoding Internet Content for Heterogeneous Client Devices.* IBM T.J. Watson Research Center. Proceedings IEEE International Conference on Circuits and Systems.

Smith, M.W. and Leung, H.-T. Jan. 2002. Finding the killer application – the role of the Broadband Applications Laboratory. *BT Technology Journal.*

Special Report on Biometrics. Feb. 2000. *IEEE Computer Magazine.*

Special Report on the future of Translation. May 2000. *Wired Magazine.*

Special Report. Mobile Wireless in Europe – The new wireless Internet services in Europe, and what they mean for the United States. April 2000. *Red Herring*, 167–258.

Stanley Kubrick's *2001, A Space Odyssey.* Turner Entertainment Co. 1968. A Time Warner Company.

State of the Union of Wireless Industry. 2002. Yankee Group.

UMTS Position Paper Number 2: Ranking of top 3G services.

Voice eXtensible Markup Language (VoiceXML) Version 1.00. Mar. 2000. VoiceXML Forum.

Welcome to 2010 – Special Report. March 6, 2000. *Business Week.* http://www.businessweek.com/reprints/00-10/design9.htm.

What Distribution Scalability Means for Internet Business. A KPMG and Inktomi White Paper.

White Paper. *2000. Building an Industry-Wide Mobile Data Synchronization Protocol.* SyncML Forum. http://www.syncml.org.

White Paper. 2000. *Enabling the Wireless Internet.* Phone.com.

White Paper. 2000. *Pervasive Management: Expanding the Reach of IT Management to Pervasive Devices.* Tivoli.

White Paper. SpeechSite: *Bringing the Web Model of Self-service to the Telephone.* Speech Works.

Whitepaper. 2002. *Lessons From Metricom and MobileStar: Success Factors for The Portable Internet Access Market.* The Shosteck Group.

Wired Magazine. Jan. 2000. The future gets fun again. Special Anniversary Issue.

Wireless in Cyberspace – Special Report. May 29, 2000. *Business Week – International Edition.* http://www.businessweek.com/datedtoc/2000/0022.htm.

REPORTS

Dalton, J.P. Apr. 2000. *The Web's Speech Impediments.* The Forrester Report.

Digital Economy 2000. USA. The Department of Commerce.

George Foster, Summer 2001. Executive Program for Growing Companies: Stanford University.

Global Equity Research, 2001. UBS Warburg.

Godell, L. Mar. 2000. *Mobile's High-Speed Hurdles.* The Forrester Report.

Kasrel, B. June 2000. *Many Devices, One Consumer.* The Forrester Report.

Lessons from Japan. 2000. Jupiter.

Long Term Potential Remains High For 3G Mobile Data Services. 2001. The UMTS Forum.

May, 2002. *2002 World Wireless Congress.* USA.

Perspectives on Business Innovation. Issue 3. Electronic Commerce. The Ernst & Young Center for Business Innovation.

Proceedings of the WWRF 5. March, 2002. USA.

Rhinelander, T. July 1999. *The Information-Rich Consumer.* The Forrester Report.

Smarter Phones: An Analysis of BREW, J2ME, and Wireless Application. April 2001. Yankee Group.

Stanford Facts 2001. Stanford University.

Technical Issues Involved in Implementing Next-Generation Networks. Nov. 2001. Yankee Group.

Technology and Telecommunications Quarterly. Feb. 2000. Salomon Smith Barney.

The Shifting Wireless Billing Environment. July 2001. Yankee Group.

The Wireless Marketplace in 2000. Feb. 2000. Peter D. Hart Research Associates.

US Wireless Telecom 2001. A Good Wager? 2001. ABN-AMRO.

Wireless Data – Issues & Outlook 2000. 2000. Goldman Sachs Global Investment Research.

Wireless Data. Oct. 2000. JP Morgan, Arthur Andersen.

Index